Peasant Resistance in India 1858–1914

OXFORD IN INDIA READINGS

Themes in Indian History

GENERAL EDITORS
- C. A. Bayly
- Basudev Chatterji
- Neeladri Bhattacharya

PEASANT RESISTANCE IN INDIA 1858–1914

Edited by
DAVID HARDIMAN

DELHI
OXFORD UNIVERSITY PRESS
BOMBAY CALCUTTA MADRAS
1993

Oxford University Press, Walton Street, Oxford OX2 6DP

Oxford New York Toronto
Delhi Bombay Calcutta Madras Karachi
Kuala Lumpur Singapore Hong Kong Tokyo
Nairobi Dar es Salaam Cape Town
Melbourne Auckland Madrid

and associates in
Berlin Ibadan

First published 1992
Oxford India Paperbacks 1993

ISBN 0 19 563390 3

Typeset at Spantech Publishers Pvt. Ltd., New Delhi 110060
Printed at Rekha Printers Pvt. Ltd., New Delhi 110020
and published by Neil O'Brien, Oxford University Press
YMCA Library Building, Jai Singh Road, New Delhi 110001

Contents

Contents

General Editors' Preface

This series focuses on important themes in Indian history, on those which have long been the subject of interest and debate or which have acquired importance more recently.

Each volume in the series consists of, first, a detailed Introduction; second, a careful choice of the essays and book-extracts vital to a proper understanding of the theme; and, finally, an Annotated Bibliography.

Using this consistent format, each volume seeks as a whole to critically assess the state of the art on its theme, chart the historiographical shifts that have occurred since the theme emerged, rethink old problems, open up questions which were considered closed, locate the theme within wider historiographical debates, and pose new issues of inquiry by which further work may be made possible.

David Hardiman's volume looks at peasant resistance in India between the Rebellion of 1857 and the mass movements of the nationalist era. This was a period when peasant movements remained isolated and fragmented, each confined to a specific issue.

In his Introduction, Hardiman seeks to classify these movements in a new way. He looks at the structures of domination to which peasants were subject and against which they rebelled. If the variations are looked at in these terms, the isolated movements seem to fit into a meaningful order. Each movement discussed within the volume then becomes representative of a class—Pabna as a revolt against high landlordism, indigo rebellion as an opposition to *planter* domination, the 1907 movement in Punjab as a protest against high taxes, forest movements as a critique of colonial forestry.

The targets attacked and the issues contested were different in each movement. Yet beneath the variations, in Hardiman's argument, there existed a broad pattern. He shows that they were all struggles based on community solidarity forged through pre-existing bonds of clan, tribe, kinship, caste and religion. This 'community' dimension, according to Hardiman, has been neglected by historians who operate only with concepts like 'class' and 'faction'.

Hardiman's notion of 'community' may seem problematic and his classification of peasant movements may be questioned. The function of the volume is precisely to open up such questions.

Acknowledgements

The publishers thank the following for granting permission to reprint the articles/extracts included in this volume: *The Journal of Peasant Studies,* for Ranajit Guha, 'Neel-Darpan: The Images of a Peasant Revolt in a Liberal Mirror'; *The Indian Economic and Social History Review,* for Kalyan Kumar Sen Gupta, 'The Agrarian League of Pabna, 1873' and I. J. Catanach, 'Agrarian Disturbance in Nineteenth-Century India'; The University of London, for Conrad Wood, 'Peasant Revolt: An Interpretation of Moplah Violence in the Nineteenth and Twentieth Centuries'; *The Journal of Asian Studies,* for Ravinder Kumar 'The Deccan Riots of 1857'; *Modern Asian Studies,* for Neil Charlesworth, 'The Myth of the Deccan Riots of 1875' and N. Gerald Barrier, 'The Punjab Disturbances of 1907: The Response of the British Government in India to Agrarian Unrest'; *Past and Present* for Ramachandra Guha and Madhav Gadgil, 'State Forestry and Social Conflict in British India'.

Introduction

The period covered in this collection of essays begins with the defeat of the revolt of 1857, a revolt in which a major driving force had been peasant resistance to the system of rule imposed on India by the British. Once British power had been destroyed in northern India by the army revolt, many peasant grievances coalesced with explosive power. The defeat of the revolt left most of these grievances unresolved. As the imperial power consolidated its hold and expanded its control over new areas of Indian rural life during the course of the next half century, new areas of conflict emerged. These conflicts were, however, more easily contained by the state. Improvements in communications, the development of the machine gun at the same time as the Indian people were being systematically disarmed, and the expansion of the police and military, all these made it easier to crush peasant insurgency before it could spread beyond a fairly local area. Conflicts therefore tended to remain localized and confined to particular grievances. The history of peasant resistance during the period 1858–1914 is, necessarily, disjointed—a collection of histories of local agrarian relationships and struggles, each of which had its own timetable of revolt. Only with the development of new forms of leadership at the national level after 1918—with Gandhi, and with the Congress championing peasant demands more militantly—did peasant resistance begin to link up once more across the subcontinent to pose a formidable challenge to the colonial state, becoming, once again, something more than a collection of isolated struggles.

Although peasant resistance did not thus pose a direct threat to British rule between 1858 and 1914, it was a force which continually worried colonial officials. Always in the background lurked the spectre of 1857, a reminder of what could happen if the peasants once more came together in revolt. Such resistance also provided a continuing rebuttal to the claims made by colonial officials that India had become a more prosperous and stable society under the British.

British administrators, as well as colonial historians, frequently sought in their reports and writings to deny the rationality of such resistance. Revolts were labelled as being 'backward-looking' and 'unprogressive', the blind hitting out of a people enslaved by a 'primordial' or 'superstitious' consciousness. Colonial officials believed that they knew what was in the best interests of the peasantry, and that peasants had, for their own good, to be forced to accept the system imposed on them by the state. This attitude has continued to be expressed by many historians even after Indian independence, with peasant movements being treated largely in terms of how they related to British policy in India. Little understanding or sympathy was shown for the peasants' motives, nor analysis undertaken of the way in which they mobilized and fought. The derogatory language of the colonial officials was often regurgitated in a wholly uncritical manner in such studies.

From the 1960s onwards an interest began to emerge in the peasantry of nineteenth-century India, not as the mere 'objects' of colonial rule, but as the subjects of their own history. This development was spurred by an awareness that a class of dynamic richer peasants had emerged in post-independence India who owned their land, grew high-yielding crops using modern inputs, and marketed their produce at advantageous rates. Many believed that this class provided a hope for the development of India into a more 'westernized' and prosperous society.[1] Historians began to explore the historical roots of this class.[2] They argued broadly that the British had provided a legal and economic infrastructure which allowed such a class to begin to develop in the nineteenth century. The old communal forms of rights in land were replaced with rights of individual ownership, which allowed a landowner to cultivate and improve his holding as he pleased, if necessary by borrowing capital on the security of this land. In the zamindari areas, such as Bengal, the grant of such ownership rights to

[1] For instance, Kusum Nair, *Blossoms in the Dust: The Human Element in Indian Development* (London: 1961).

[2] See in particular Ravinder Kumar, 'The Rise of the Rich Peasant in Western India', in D.A. Low (ed.), *Soundings in Modern South Asian History* (London: 1968); Eric Stokes, 'The Return of the Peasant to South Asian History', *South Asia*, 6 December 1976; Neil Charlesworth, 'Rich Peasants and Poor Peasants in Late Nineteenth Century Maharashtra', in C. Dewey and A.G. Hopkins (ed.), *The Imperial Impact* (London: 1978).

landlords rather than peasants provided an initial obstacle to the development of a class of rich peasants. However, by struggling for and winning occupancy rights in the 1860s and 1870s, the more substantial peasants gained for themselves an effective right of ownership.

In the ryotwari areas, such as the Bombay and Madras Presidencies, the peasants were given such rights by the British from the start. The development of this class was furthered, it was also argued, by the opening of new markets for agrarian produce, in part through the development of better communications, and in part by the huge growth in demand from metropolitan processing industries. The British also developed canal irrigation, raising the productivity of the land. From all of this emerged a stratum of richer peasants farming primarily for the market. According to Eric Stokes, there were, as a result, two basic types of agrarian protest in late-nineteenth and early-twentieth-century India: (1) by rich peasants in areas of 'high farming' who were seeking to extend their power, and (2) by pauperized peasants who had suffered badly as a result of colonial rule.[3] Seen in this perspective, the resistance against British planters and indigenous landlords in eastern India during the late nineteenth century represented a bid for rural power by an upwardly mobile class of rich peasants, as did the protest against high land-tax rates in Maharashtra and Gujarat in the late nineteenth century, and the protest against the canal colony administration in the Punjab in the early twentieth century.

It was also argued by some historians that, in many cases, the richer peasants used their control over the rest of the peasants to force the latter to take part in agitations which were really in the class interests of the rich. They did this by using ties of patronage, caste or religion. As a result, the normal unit of peasant mobilization was a faction—consisting of a rich peasant and his clients and dependents amongst the poor. Peasant resistance which was not in the interests of the richer peasants, such as revolts by marginalized adivasis, were seen as 'primitive' and 'pre-political'—a desperate gut-reaction of the destitute and hungry, without clear-cut strategy or aim. Even these sorts of outbreaks could at times be manipulated by the rural rich to serve their class interests.[4]

A problem with this general line of argument is that it tends to lump together two very different social classes under the label 'rich peasant'.

[3] E. Stokes, 'The Return of the Peasant', pp.101-2 and 110.
[4] For a review of writings in this vein see D. Hardiman, 'The Indian "Faction": A Political Theory Examined', in R. Guha (ed.), *Subaltern Studies I* (New Delhi: 1982).

One was a class of rural magnates who sought to build up their power essentially as landlords; the other was a class of peasant farmers who sought to free themselves from the domination of superior classes, such as landlords and merchant-moneylenders, so as to be able to maximize their profits from agriculture by farming for the market. Strictly speaking, only the latter should be described as a 'rich peasantry'.

The first is the class which David Washbrook has described as the 'rural-local bosses'. In the area which he has studied—the dry areas of Madras Presidency—he estimates that this group made up about two to three per cent of the rural population. As a rule, they farmed a home estate using hired labourers (who were often bound by ties of debt), and let out the rest of their holdings to tenants on a crop-share basis. They were also the local moneylenders, providing almost all rural loans and holding most mortages. Washbrook shows that the chief ambition of members of this class was not to maximize their capital through commercial agriculture but to use their power to extend their local empires. For them, 'economic considerations always were subordinate to, and an instrument of, political activity.'[5] A similar class was found in Bengal. There, prosperous peasants who had resources to spare did not employ their capital in improving their farms. A successful cultivator who had saved a little money became almost, as a matter of course, a moneylender. By combining the role of creditor, trader and rentier, the richer villager could ensure the perpetual indebtedness of the cultivators and force them to sell produce at low rates. 'In these circumstances peasant agriculture had no chance of developing into capitalist farming.'[6] The chief ambition of such men was to become like a zamindar, living a life of ease on rents, rather than successful commercial producers.[7]

In the case of the second group—those whom we may more accurately describe as a 'rich peasantry'—it is hard to accept the argument that it was asserting itself 'as a class' in the late nineteenth and early twentieth centuries. As yet, it was a class which hardly existed. An assumption is made that because the British had established a peasantry free from landlord domination in areas such as the Deccan, Gujarat or the Punjab, the

[5] D.A. Washbrook, *The Emergence of Provincial Politics: The Madras Presidency 1870-1920* (Cambridge: 1976), p.83.

[6] Rajat Ray, *Social Conflict and Political Unrest in Bengal 1875-1927* (New Delhi: 1984), p.51.

[7] B.B. Chowdhury, 'Agrarian Relations in Bengal (1859-1885)', in N.K. Sinha (ed.), *The History of Bengal (1757-1905)* (Calcutta: 1967), pp.318-20.

richer peasants of these regions were able to market their produce to their own advantage. The cotton boom in the Deccan in the 1850s and 1860s—which enriched many peasants—is cited as an example. In fact, this boom was an exception and short-lived. On the whole, the bulk of agricultural profit went to commercial capitalists such as merchant-moneylenders, local processors of agrarian produce, urban traders and indigenous bankers. From the 1870s onwards there was, in any case, a slump in agricultural prices due to the opening up of huge new areas of virgin farmland in North and South America, Australia and Russia. Britain, unlike most countries in the world, refused to impose tariffs to protect small agriculturists from the influx of cheap grain.[8] The development of communications and market networks therefore prevented rather than encouraged the emergence of a class of peasants able to benefit from commercial agriculture. The British also failed to provide on an adequate scale the inputs essential for such a development. Irrigation projects were limited for the most part to regions such as the Punjab and the north Gangetic plain. Technological improvements were minimal; the chief mechanical power continued to be the bullock—to pull the plough, cart goods to market and draw water from wells. Few advances were made with improved seed varieties, chemical fertilizers or agricultural implements. What is more, farms were generally diminishing in size during the nineteenth century as a result of the equal division of holdings amongst sons over each generation.[9] For all these reasons, it is wrong to try to identify an assertive 'rich peasant' class in areas such as the Deccan, Gujarat or the Punjab during this period.[10] Such assertion has been, on the whole, a post-independence phenomenon, seen in the so-called 'farmers' movements'.

The emergence of the Naxalite movement in the late 1960s forced many historians to revise their opinion about peasant resistance. Through this

[8] E.J. Hobsbawm, *The Age of Empire 1875-1914* (London: 1987), pp.39 and 48.

[9] Sumit Guha, *The Agrarian Economy of the Bombay Deccan 1818-1941* (New Delhi: 1985), p.148. In his later work Neil Charlesworth—one of the original proponents of the 'rise of the rich peasant' theory in the context of western India in the late-nineteenth century—has conceded that there was, in fact, no clear-cut polarization between rich and other peasants in the late-nineteenth and early-twentieth centuries in the region. See his *Peasants and Imperial Rule: Agriculture and Agrarian Society in the Bombay Presidency, 1850-1935* (Cambridge: 1985), pp.174, 202-5, 224-5, 232-6, and 294-5.

[10] I have advanced this argument in more detail for Gujarat in David Hardiman, *Peasant Nationalists of Gujarat: Kheda District 1917-1934* (New Delhi: 1981), pp.246-50.

revolt the poor and landless demonstrated that they could be every bit as assertive and political as the more prosperous peasantry. The Naxalite movement let to a revival of interest in the history of peasant insurgency in India. It was found that this history had been either denied or marginalized in most existing histories. A pioneer in this respect was Kathleen Gough, who wrote an influential article in 1974 on 'Indian Peasant Uprisings'.[11] She argued that the peasant militancy seen in the Naxalite movement had a long history. This had been denied in many studies, which argued that rural resistance had in most cases been led by the upper classes of the countryside, such as landlords and rich peasants, who had used ties of patronage and caste to mobilize the poorer peasants. The great revolt of 1857 was often explained in such terms. In addition, when the poor had rebelled, their resistance was often described as mere outlawry or a form of 'communalism', rather than as a protest against harsh agrarian conditions. In this way, most peasant resistance had been written out of Indian history. Gough argued that there was in fact a strong history of rural revolt against colonial rule and the indigenous beneficiaries of colonial rule. British rule had led to an extreme disruption of rural relationships throughout India. In the late-eighteenth and early-nineteenth centuries the peasantry had been impoverished through extremely high rates of land tax levied by the East India Company. After these taxes were lowered, new forms of expropriating the surplus of the peasants through intermediaries, such as landlords, merchant-moneylenders and British planters, were evolved. These intermediaries often acquired property rights in land of a capitalist kind, which allowed them to exploit the peasants more thoroughly. The British also destroyed much local handicraft industry, forcing the handicraft producers to turn to agriculture and compete for land with the existing peasants. The same was true for pastoralists and forest-dwellers, who had their pasture lands and forests expropriated by the colonial state. There was, as a result, a massive impoverishment and marginalization of the rural population of India during the nineteenth century. This was compounded by famines, which became more frequent and more terrible during the colonial period, as the people lacked the resources to withstand failures of the monsoon rains.

Gough attempted to divide peasant resistance into five types: (1) restorative rebellions to drive out the British and restore earlier rulers and social relations; (2) religious movements for the liberation of a region or an ethnic group so as to establish a new form of government; (3) social

---

[11] *Economic and Political Weekly*, Special Number, August 1974.

banditry; (4) terrorist vengeance, with ideas of meting out collective justice; (5) mass insurrections for the redress of particular grievances. Subsequent scholars, such as D.N. Dhanagare, have compiled similar lists, with modifications as they think fit.[12] A problem with such lists is that they repeat many of the colonial categories, such as 'outlawry' or 'religious fundamentalism', in only slightly modified ways. Also, they isolate certain features of a movement so as to slot it into one or the other category, even though any one movement is likely to have elements to it which cut across the categories. Gough, in fact, admits as much, and does not claim her list to be anything more than a 'rough classification'. Such lists are at best useful for ordering a mass of evidence; they cannot explain—except in a tautological manner—why peasant resistance took the form it did. Also, they do not explain how such resistance develops and evolves over time.

Another scholar who was moved by the Naxalite upsurge to re-examine the history of peasant resistance in nineteenth-century India was Ranajit Guha. In his book, *Elementary Aspects of Peasant Insurgency in Colonial India*,[13] he has tackled the problem of classification of peasant movements in a more sophisticated manner. In this book he advanced two main arguments. The first was that there were certain underlying structures to peasant uprisings in nineteenth-century India. These he examined under a number of heads, such as 'negation', 'ambiguity', 'solidarity', 'transmission' and 'territoriality'. Structural similarities—often of a striking kind—were found between movements which on the surface might have appeared to have been very different. The other argument was that peasant consciousness was revealed most clearly in revolt. This was in part because the historical record was much richer during periods of revolt, and in part because the peasantry for a brief period spoke and acted in an unrestricted manner. Guha argued that by reading their actions one could start to understand their minds. Although Guha's approach allowed for a far richer understanding of peasant resistance in India, it did not attempt to provide any theory of the evolution of such resistance over time. Guha did not try to address the sort of problems raised by Ravinder Kumar and Eric Stokes.

One element of peasant resistance which was often acknowledged but subsequently marginalized in much analysis of the 1960s and 1970s was

---

[12] D.N. Dhanagare, *Peasant Movements in India 1920-1950* (New Delhi: 1983), pp.213-14.

[13] (New Delhi: 1983).

the religious and community content of this resistance. In part, this was a reaction to colonial historiography, which depicted Indian history as a story of never-ending conflict between primordial groups, such as castes and religious communities, which failed to develop and change with the times. Historians have rightly challenged this view, showing that changing socio-economic realities have provided much of the underlying dynamic to such conflict and solidarities. In the process, however, they denied too much. Community became a residual category which explained why— given the same socio-economic conditions—mobilization was successful in some instances but not in others. Thus, it was said, ties of caste or religion could help the richer peasants mobilize the poorer in what were essentially the class interests of the former; poorer peasants might use ties of caste, tribe or religion to achieve greater unity in the struggle for what was essentially their class interest. As the reality of the struggle was held to be its class content, community sentiments were not accordingly studied in their own right.

The attempt to explain all peasant resistance in terms of 'essential' class interests has been challenged by Partha Chatterjee. He has argued that in India peasants have generally conceptualized relationships of power in terms of the idea of the community. They conceive of the community as a collective which exists in a close relationship to the land which it controls. By right, the individual is allowed access to the land only as a member of the community. Chatterjee states that: 'the place of the individual in the social ordering of rights is determined by his membership in the community; the collective is prior to the individual parts and its authority larger than the mere sum of the parts.'[14] He argues that political power in such a community is organized as the authority of the entire collectivity. It is not a 'democratic' authority, in the bourgeois sense of the term, for it is usually vested in patriarchal elders or chiefs. They cannot, however, go counter to the general will of the collectivity to any degree.

Chatterjee goes on to say that both feudal and bourgeois states have sought to destroy peasant communities and the bases of peasant-community ideology, as they have provided an alternative focus of power and loyalty. The absolutist monarchs of medieval India waged a constant struggle with these communities—described as 'tribes' or 'clans' in the literature. They were never entirely successful, always having to reach some sort of compromise with them. The colonial state, representing the

[14] Partha Chatterjee, 'Agrarian Relations and Communalism in Bengal, 1926-1935', in R. Guha (ed.), *Subaltern Studies I* (New Delhi: 1982), p.12.

class interests of the bourgeoisie, likewise attacked the solidarity of these groups. The process continues to this day.

According to Chatterjee, it is wrong to try to analyse collective actions by peasants in terms of narrow class interests. He argues:

> The language of organized politics often characterizes such forms of [communal] mobilization as an alliance of various classes or strata within the peasantry, but ideologically the notion of the alliance is hardly ever relevant in collective action of this kind; it is always the concept of the community as a collective whole, a form of authority incapable of being broken down into constituent parts, which shapes and directs peasant politics *vis-a-vis* the state.[15]

Community consciousness is very different from bourgeois consciousness. The latter operates from the premise of the individual and a notion of his or her interests, with solidarities being based on alliances of common interest. In peasant consciousness people are required to act together because of existing bonds of solidarity. 'Collective action does not flow from a contract among individuals; rather, individual identities themselves are derived from membership in a community.'[16] Chatterjee argues that this understanding poses a challenge to the methodology of bourgeois economists and sociologists (including those of the Chayanovian and 'moral economy' varieties), who search for the 'rational peasant', as well as to many Marxist scholars who write on the agrarian question.

In any discussion of community-based political action, it is important to make a clear distinction between communities as actual social groups (such as members of a village, an endogamous caste or followers of a particular religion) and 'the community' as a form of consciousness.[17] Partha Chatterjee is talking, essentially, about the latter. 'The community' in this respect exists in a relationship of opposition to those who are not of the community. Its boundaries will shift and change according to context and circumstance. Thus, in a conflict between moneylenders and peasants, the 'peasant community' would include all those who are exploited by the moneylenders. Subsequently, in the same region, there might be a conflict between a dominant landed peasantry (a category which often coincides with a caste or congeries of castes) and agricultural labourers (who tend to

[15] *Ibid.*, pp.34-5.

[16] Partha Chatterjee, 'For an Indian History of Peasant Struggles', *Social Scientist*, 16:11, November 1988, p.11.

[17] Raymond Williams has pointed out the importance of making such a distinction in his discussion of the term 'community' in *Keywords: A Vocabulary of Culture and Society* (London: 1983), p.75.

be from a congeries of subordinate castes and outcastes). The two groups will see themselves as communities, rather than classes, in conflict. Because of this, an agricultural labourer who happens to be a member of the dominant caste is likely to side with the dominant group; a better-off peasant who is of low caste is likely to side with the subordinate group. Once again, the peasantry coalesces around a particular consciousness of what makes up 'their' community, even though the 'community' of the first conflict—against moneylenders—is very different from the two 'communities' of the second conflict. It is important to note that such communities rarely, if ever, coincide in any exact sense with economic classes.

Because of this, it is often argued that community-based solidarities tend to favour the interests of wealthy and powerful members who control the group. Although this might sometimes be true, it is by no means always the case. This is because there is a tension between the idea of 'the community'—which assumes an equality amongst members—and the fact that there will be in any one such community a variety of divisions, which may be along lines of caste, class, religion, age or gender. This tension provides the basis for continuing struggles within peasant communities. For instance, within a village community there may be a struggle in which a low-caste group tries to assert itself through a combination of economic action—withholding labour or goods—with aggressive claims to a higher ritual status through the renunciation of 'low' habits such as meat-eating and liquor-drinking. Likewise, women may fight against patriarchal domination through a silent refusal to conform to male demands.[18] Within a caste or religious group a section of younger people may fight to change the practices and rules enforced by the elders.[19] In all these cases a struggle is going on to redefine the nature of the community. Often, these struggles become most intense during periods of upheaval and rapid change, as during a peasant movement against an enemy who is not of the community. Community-based resistance does not therefore preclude the self-assertion of subordinate groups within the community.

The present book focuses on peasant resistance to those seen as being outside the peasant community. It does not attempt to examine struggles

[18] See Ranajit Guha, 'Chandra's Death', in R. Guha (ed.), *Subaltern Studies V* (New Delhi: 1987).

[19] For an example of such a struggle within a peasant caste see Anil Bhat, 'Caste and Political Mobilization in a Gujarat District', in Rajni Kothari (ed.), *Caste in Indian Politics* (New Delhi: 1977).

within peasant communities along lines of class, caste, religion, age or gender. These are all subjects of great importance which need to be concentrated on in their own right. Other books in this series will take up some of these issues.[20]

As pointed out in the first paragraph, the history of peasant resistance during the period 1858–1914 must necessarily be the history of many disparate movements. Resistance never linked up with that cumulative explosive power which transformed local struggles into wide-ranging rebellion. It had happened in 1857; it occurred less dramatically in 1920–22; it happened on a wide scale once more in 1942. The approach which I shall therefore adopt in this introduction is to try to reveal wider unities to peasant resistance during the 1858–1914 period by demarcating particular *areas of resistance*. These relate to specific relationships of domination and subordination. Each of these relationships had, in its own region, its distinct chronology of revolt.

The chief areas of resistance to be looked at here are (1) peasants against European planters; (2) peasants against indigenous landlords; (3) peasants against professional moneylenders (*sahukars*); (4) peasants against the land-tax bureaucracy; and (5) peasants against forest officials. These five areas are not exhaustive. There was considerable resistance by peasants to colonial liquor laws, which sought to stop people from drinking home-made liquor and toddy in favour of centrally-manufactured and taxed liquor.[21] Also worth mentioning was the struggle against draconian measures taken by the British to prevent epidemic diseases—most notably the virulent plague which appeared in India during the 1890s.[22] I have omitted several excellent studies of adivasi revolts which occurred during this period, such as the movement of Birsa Munda,[23] or the *fituris* of Rampa and

[20] See in particular the volumes to be edited by Rosalind O'Hanlon, Gyanendra Pandey, Gyan Prakash, Sumit Sarkar and Tanika Sarkar.

[21] I have myself written an essay on this theme: 'From Custom to Crime: The Politics of Drinking in Colonial South Gujarat', in R. Guha (ed.), *Subaltern Studies IV* (New Delhi: 1985).

[22] For an essay on this theme see I.J. Catanach, 'Plague and the Indian Village, 1896-1914', in Peter Robb (ed.), *Rural India: Land, Power and Society under British Rule* (London: 1983). David Arnold also refers to such resistance in 'Touching the Body: Perspectives on the Indian Plague, 1896-1900', in R. Guha (ed.), *Subaltern Studies V* (New Delhi: 1987).

[23] Suresh Singh, *The Dust-storm and the Hanging Mist: A Study of Birsa Munda and his Movement in Chhotanagpur (1874-1901)* (Calcutta: 1966).

Gudem[24], because a separate volume is planned in this series to cover adivasis. The important writings on this topic are, however, included in the annotated bibliography at the end of the book. Moreover, resistance by adivasis is dealt with to some extent in the essay by Madhav Gadgil and Ramachandra Guha.

In this introduction I shall look at each of the five main areas of resistance in turn. In each section I shall examine the relevant essays reprinted in this collection in a critical manner, putting them in their wider context and seeing the extent to which such resistance was replicated in other parts of India. I shall also look at the attempts made by early nationalists to link their movement with the peasantry. Although this enterprise was marginal to the peasant struggles of this period, nationalist leadership was to provide a new unity to such resistance after 1918, and it is worth asking why no such leadership emerged to any important extent in the preceding period. I shall conclude by examining some of the long-term implications of this history of peasant resistance.

II

The first essay concerns the clash between peasants and planters in Bengal in the period 1859-62, and the reaction of the middle classes of Calcutta to this dispute. The plantation system which had come into being under British rule and against which the peasants protested in 1859-62 was confined almost entirely to eastern India, that is Bengal, Bihar and Assam. It was not a genuine plantation system in the normally accepted sense, with European capitalist farmers cultivating large estates with hired labour, as in the tea gardens of Assam or Sri Lanka, or the sugar plantations of the West Indies. In eastern India, European capital financed the establishment of warehouses and processing factories at strategic points, and the raw or semi-processed produce was acquired from small peasant cultivators. These cultivators were normally tenants of indigenous landlords (*zamindars*). Through such a system British capital sought to expand cash crop cultivation while keeping intact the unequal relationship between zamindar and peasant. The planters profited by processing and selling the peasants' crops, while rural relationships remained largely unchanged.

The planters acquired two main crops under this system—indigo (used for a blue dye) and opium. Attempts were made to produce sugar in this

---

[24] David Arnold, 'Rebellious Hillmen: The Gudem-Rampa Risings, 1839-1924', in R. Guha (ed.), *Subaltern Studies I* (New Delhi: 1982).

manner in the 1830s and 1840s, but it was unable to compete with West Indian sugar on the world market.[25] Neither indigo nor opium had to compete with major rival international centres of production, so that they were able to yield a good profit despite their relatively inefficient system of peasant production. Jute, the other major non-food cash crop of colonial Bengal, became important only after 1860 when the plantation system was in decline there, and jute production never came under the plantation system.

Opium was a government monopoly, and it could not be acquired by private entrepreneurs. The government dealt with leading villagers, who took the opium from the actual cultivators. The peasants were relatively free to decide whether or not they should grow opium, and when the government reduced the price paid to them many switched to more remunerative crops. As a result, there was no overt struggle against this system, as there was with indigo.[26]

In the case of indigo, local factories were operated by European entrepreneurs or managers for larger companies. These men were known as 'planters'. As a rule, they advanced money to the peasants so that they could pay their rent to the zamindars and provide for their subsistence needs, on the understanding that they would cultivate indigo on a certain proportion of their land and hand over the harvested crop to the factory. The peasants were often bound by contracts which ranged in time between one and five years. When they were unwilling to enter into such a contract, the peasants were often forced to do so either by the planter or his hired toughs, or by leading villagers who were in the pockets of the planters, or by their zamindars. The zamindars favoured the system because it ensured for them a regular payment of their rents in cash. Over time many zamindars allowed planters to become managers of their estates. Many zamindars were also in debt to the planters, and in some cases they gave them long, even permanent, mortgages of parts of their estates. The peasants themselves had to pay usurious rates of interest on the sums advanced to them by the planters, and, as they could never hope to repay, were tied to them also by bonds of debt.

The planters were hated throughout eastern India because of their racial arrogance and their contempt for the law. They maintained small private armies of strongmen, whom they would use to coerce the peasantry,

[25] B.B. Chowdhury, 'Agrarian Relations in Bengal (1859-1885)', p.252.

[26] B.B. Chowdhury, *Growth of Commercial Agriculture in Bengal (1757-1900)*, Vol.1 (Calcutta: 1964), pp.17-57.

forcing them to grow indigo.[27] The zamindars had an ambiguous attitude towards the planters. On the one hand they needed the planters' money, on the other hand they resented the way in which the planters operated as a rival centre of power in the rural areas. Likewise, the leading peasants profited by acting as intermediaries between the planters and the ordinary peasants, but they were also tyrannized by the planters. Leading peasants who tried to assert any degree of independence were beaten up or illegally imprisoned in the factories. Over time, as the planters became more rapacious (in part because of growing economic difficulties after the collapse of the Union Bank, their chief financier, in 1847), the zamindars and leading peasants began to turn against them. The actual revolt came in 1859, when the Bengal government began to refuse to support the planters in their depredations. The government had for long been embarrassed by the excesses of the planters but, so long as indigo continued to be important for the economy as a whole, it looked the other way. By the 1850s indigo was losing its importance: by then only about ten per cent of exports from Bengal were of indigo.[28] The coercive methods of the planters violated the principles of *laissez faire* and freedom of contract between entrepreneur and labourer, and liberals in both Britain and India were demanding that the peasants be free to sell their produce to the highest bidder. In 1859 John Grant, who supported this liberal position, became Lieutenant-Governor of Bengal. He demanded that the planters conform to the law. Even though the attitude of the Bengal government thus changed, the attitude of most Europeans in Bengal—including the judiciary—remained firmly pro-planter. Out of this developed the conflict of 1859–62.

There was already a history of peasant resistance to the planter system. This was associated in particular with the Fara'idi sect of East Bengal. This Muslim sect was founded around 1820 by Haji Shariatullah and carried on after his death in 1839 by his son Dudhu Miya. The Fara'idis sought to Islamicize the customs and practices of the Muslim peasantry. The confident and assertive manner in which they carried on their activities frightened both the Hindu zamindars and the planters. They adopted a

[27] These strongmen—*lathials* (men wielding sticks), *durwans* (guards), *chaukidars* (watchman)—were recruited normally from untouchable castes, such as the Chandals and Bhuinmalis, or from poor Muslim tenants. See Sumit Sarkar, 'The Kalki-Avatar of Bikrampur: A Village Scandal in Early Twentieth Century Bengal', in R. Guha (ed.), *Sublatern Studies VI* (New Delhi: 1989), p.12.

[28] Blair B. Kling, *The Blue Mutiny: The Indigo Disturbances in Bengal 1859-1862* (Philadelphia: 1966), p.25.

militant line against such exploiters, holding that 'God made the earth common to all men, the payment of rent is contrary to his law.'[29] When the zamindars and planters tried to suppress them by force, they organized their own bands of strongmen to fight back. There were clashes between the Fara'idis and the planters, zamindars and government officials. These conflicts continued throughout the 1840s and 1850s. The leader, Dudhu Miya, ran a tight organization with a network of local leaders. A tax was levied on all followers. This helped to pay for the legal defence of Fara'idis who had been taken to court by the zamindars and planters. Many of the peasants who took part in the anti-planter movement of 1859-62 were Fará'idis.[30]

The revolt began in Nadia district when peasants, who had heard that the new Lieutenant Governor was sympathetic to their woes, began to refuse to accept their customary advances from the planters. The protest soon spread to other districts. Attempts at coercion were repulsed by the peasants' use of force. By early 1860 the movement had become strong throughout the entire delta region of Bengal. The zamindars were often happy to see the planters being thus humbled, though they did not, as a rule, actively support the peasants. Leadership was provided by the more substantial peasants—the class which had, in the past, acted as intermediaries between the planters and the smaller peasants. The peasants took the offensive, marching to the factories of particularly notorious planters and beating them up. When the planters retaliated by sending their toughs to attack the peasants and loot and destroy their houses, the peasants, men and women alike, fought back with *lathis*, spears, bows and arrows and brickbats. In many cases the hired toughs were beaten back.[31]

In March 1860 the Bengal government, frightened by the level of violence, passed a temporary law which made it a punishable offence to breach contracts to cultivate indigo. The planters immediately filed hundreds of cases of contract violation against the peasants in court. Magistrates took the side of the planters, accepting their testimony against the peasants, and passed heavy sentences on those found guilty. Often the leading peasants of a village were singled out for such punishment, in a attempt to break the back of the resistance. Many peasants were forced, as a result, to agree to cultivate indigo during the 1860 season. Protests

---

[29] Narahari Kaviraj, *Wahabi and Farazi Rebels of Bengal* (New Delhi: 1982), p.90.

[30] B. Kling, *The Blue Mutiny*, p.61.

[31] *Ibid.*, pp.78-102.

continued; when Grant toured Bengal by boat in August 1860, the banks of the rivers were lined with men, women and children calling for justice. When the temporary law to enforce contracts expired in October of that year, Grant refused to renew it. Orders were sent to the courts forbidding them from trying to force the peasants to grow indigo. The peasants soon found that they could now use the courts to their advantage. Once more, the planters were forced on the defensive.

After this the movement took a more radical turn, developing into a no-rent campaign. This was directed against both European and Bengali landlords. The latter became frightened and went over to the side of the planters. The planters who were landlords began to evict their tenants on a large scale and demand higher rents. The peasants fought back, in early 1861, refusing in huge numbers to pay their rents. Threats of eviction were challenged by the better-off peasants in the courts. They claimed that they were occupancy tenants who could not, under the terms of the Tenancy Act of 1859, be evicted under the law. The courts had to work overtime to settle so many disputes. In the process the rights of the peasants under the law were defined more clearly, in a way that left the planters with less room to manoeuvre.

By mid-1863 the struggle was largely over. The planters had to raise the rates paid to the peasants for their indigo. Many planters gave up the work entirely, as they knew that without government support for their extra-legal methods, the business could not be sustained. The revolt of 1859–62 effectively destroyed the indigo plantation system in lower Bengal.

The best book-length history of this movement is the study by Blair Kling. It is, however, centred too much on the leading officials of the Bengal government; it is not rooted in the peasantry and their resistance. For this book I have chosen Ranajit Guha's powerful and provocative essay on *Neel-Darpan*. This provides an analysis of the movement as well as a critical examination of the role of the Bengali intelligentsia in these events. Guha shows how the grievances of the peasants were used by various superordinate classes to press their own demands. The richer peasants wanted to free themselves from the oppression of the planters so that they could operate their own *mahajani*—i.e. moneylending and usury—freely. The zamindars were pleased to see the power of the planters undermined. The intelligentsia sought to establish themselves as the true friends of the peasants and thus their legitimate political representatives. In all of this the peasants' own voice was largely ignored, and in the end they gained very little from the struggle. The essay is particularly convincing in its critique of the intelligentsia, whose attitude to the peasants was 'a curious concoction of an inherited, Indian-style paternalism and an acquired western-style humanism.'[32] Guha shows how bogus many of their sentiments were for the peasants. His argument that the richer peasantry did not give sustained support to the poorer peasantry, and that they tended to back down when the going became rough,[33] can however be questioned. There was an understandable tendency for all peasants, regardless of class, to back down in the face of harsh repression, and, from what evidence I have seen, it is hard to single out the richer peasantry as being particularly weak in this respect. During the movement there was a remarkable solidarity which cut across class divisions amongst the peasants. The peasant community fought the planters and zamindars as a collective. After the movement died down, it became apparent that the chief beneficiaries were the richer peasantry, but this is to use hindsight and it is wrong to say that they abandoned the struggle when their particular class demands had been fulfilled, leaving the rest of the peasants in the lurch.

The revolt of 1859-62 sounded the death-knell of the indigo plantation system in Bengal, but not in Bihar. Indigo cultivation was expanding rapidly in the latter region, and the process was not checked at all by the hiatus caused to the system in Bengal in the early 1860s. In fact, a lot of capital which would have been invested in Bengal indigo was, after 1862, invested in Bihar instead. The expansion continued in Bihar till checked by the invention, in Germany in 1898, of an artificial dye which replaced indigo, after which the industry went into a sharp decline. Initially the system which operated in Bihar was identical to the one which existed in Bengal up until 1860. After 1860 the Bihar planters (increasingly) acquired landlord rights usually by leasing holdings from the indigenous landlords. They were then able to force the peasants either to grow indigo or face ejection.[34]

We may wonder why the 'liberal' government of India, which was supposed to have been so upset by the ways of the Bengali planters that it turned against them in 1859-62, continued to tolerate such a system in an adjoining region for another half century. Neither Blair Kling nor Ranajit Guha addresses this question, and I can only suggest some answers. It may

---

[32] See below in this volume.

[33] *Ibid.*

[34] Colin Fisher, 'Planters and Peasants: The Ecological Context of Agrarian Unrest on the Indigo Plantations of North Bihar, 1820-1920', in Dewey and Hopkins, *The Imperial Impact*, pp.114-25.

have been that it was particularly embarrassing to have such a rapacious system operating in the countryside near the capital of India, plain for all to see and protest against. Bihar was a relatively remote and backward region in which such oppression was less easily observed by what passed for 'public opinion'. But also it must have been the particular conjuncture of events in Bengal in 1859-62 which made the struggle against the planters irresistible. This was not the case in Bihar, where a similarly strong movement against the planters by the peasants of Darbhanga and Champaran was suppressed with no changes of substance being made to the system. In 1874 protest erupted once again in Champaran, after peasants who refused to grow indigo at a time of grain shortage were attacked by the planter's toughs, their houses looted and some even murdered. The Lieutenant-Governor of Bengal at the time, Richard Temple, refused to appoint a commission of enquiry, arguing that it 'would create a considerable disturbance, it would excite feelings which would not readily subside; it would shatter vested interests, place capital in jeopardy and bring proprietary status and occupancy rights into uncertainty for a time.'[35]

In Bihar the struggle against the planters continued into the twentieth century. By then, the industry was in any case in decline—though it revived briefly during the First World War. Again, resistance was led by the richer peasants. In 1907-8 there was a powerful movement in which the indigo factories were boycotted and factory employees intimidated. Meetings were held and planters threatened with violence. The government intervened and forced the planters to give substantial concessions to the peasants. The system, however, remained and the struggle continued. It attracted support from local lawyers, teachers and political activists, who managed to win increasing support for the cause at an all-India level after 1912. This led to Gandhi's intervention in 1917 and the Champaran satyagraha, which brought an effective end to the system in Bihar.[36]

III

The second broad area of resistance was that of peasants against landlords.

[35] B.B. Misra, 'Introduction' to B.B. Misra (ed.), *Select Documents on Mahatma Gandhi's Movement in Champaran 1917-18* (Patna: 1963), p.11.
[36] See *Ibid.*, and Jacques Pouchepadass, *Planteurs et Paysans dans l'Inde Coloniale: L'Indigo du Bihar et le Mouvement Gandhien du Champaran (1917-18)* (Paris: 1986).

I have included two essays on this theme, covering movements in two regions of India geographically far apart—Bengal and Kerala. Despite the many differences between the two movements, there was an important similarity: both received their support from a Muslim peasantry whose religion helped to forge bonds of solidarity against largely Hindu land-lords. In both cases, the tenants were chiefly from groups of low-status peasants who had converted to Islam over the centuries, and who had suffered a growing degree of exploitation by landlords who had been given new powers by the British. Let us look first at Bengal.

During the 1870s there was a series of agitations in Bengal by peasants against the zamindars, in which demands were made for moderate and fixed rents, firmer rights of occupation for tenants and an end to various forms of feudal oppression. The movement was strongest in central Bengal, in the districts of Pabna and Bogra, but peasants in the districts of Dacca, Mymensingh, Tippera, Faridpur, Bakarganj, Rajshahi, Hooghly and Midnapur were also involved. The agitations were the result of two main processes: first, attempts by the zamindars to extract an ever-increasing rent from the peasants, and second, the self-assertion of a new class of 'occupancy ryots'.

By the mid-nineteenth century the large-sized zamindari estate was on the decline in Bengal. The great estates created as a result of the Permanent Settlement in the late eighteenth century had by the mid-nineteenth century been split up, in most cases through inheritance. Zamindars had also sold off portions of their estates to raise money, either selling the zamindari rights or selling secure tenancy rights. Investment in such holdings was popular for those with cash to spare.[37] Although their incomes were declining, the zamindars sought to maintain their accustomed lifestyle by taking more from the peasantry. This involved pushing up rents, demanding extra charges from the peasants on a whole number of pretexts (often to pay for specified items of social expenditure of the zamindars), and reclassifying rights in land to the disadvantage of the peasants. These demands were enforced through peasant intermediaries—who were let off certain charges if they collected the demands from the rest of the peasantry—through the filing of cases against peasants in the British courts of law, or by direct coercion by the gangs of toughs which the zamindars invariably employed. As British officials and the courts normally sided with the zamindars against the peasants, they were able to use

---

[37] Partha Chatterjee, *Bengal 1920-1947*, Vol.1, *The Land Question* (Calcutta: 1984), pp.10-12.

both the law and extra-legal methods of coercion with impunity.

Until 1859 the British followed an official policy of non-interference in landlord-tenant relations. This changed in 1859 with the Rent Act, which was a measure designed to stop some of the more glaring acts of oppression by the zamindars. Under the existing law, zamindars could confiscate the crops and property of the peasants for arrears of rent in a manner which caused grave injustice to the peasants. Under the 1859 Act, zamindars were not allowed to confiscate crops of property for arrears of over one year's standing, or in cases where the peasants had promised to pay up within a reasonable length of time. The tenants were classed into two main groups: those who had rented a plot for at least twelve years, and who were now considered to have a right of occupancy, and those who had rented a plot for a shorter time, who were not considered to have an occupancy right. Legal checks were placed on rent increases by zamindars. The twelve-year limit was not one which existed in local custom, and large numbers of peasants were deprived of their customary rights of occupancy at a stroke, merely because they had not occupied the same plot of land for twelve years or could not prove that they had. The new law created two peasant classes in rural Bengal—the occupancy tenant and the non-occupancy tenant.[38]

Initially the 1859 Act did not benefit any of the peasants very much. Although legal checks had been placed on rent-enhancement and arbitrary eviction of occupancy tenants, in practice the landlords enhanced rents, either at will or by using the compliant local courts. They also forged documents which denied occupancy status to tenants. In 1869, because of these abuses, rent cases were transferred from the courts of the local officials to the ordinary civil courts. As the latter followed the letter of the law more strictly, rent increases became harder to obtain through the courts. The zamindars retaliated by widespread falsification of documents and the use of bribed witnesses in the courts, as well as through direct coercion. This created a situation of extreme tension between the zamindars and peasants in the years leading up to the anti-zamindar agitations of the 1870s.[39]

Whereas the zamindars saw the 1859 Act as destroying what they believed to be the previously harmonious relations between them and the peasants, many peasants took the Act as a charter of rights. They realized that their rights had to be fought for in the courts and that they needed to

[38] B.B. Chowdhury, 'Agrarian Relations in Bengal', pp.295-7.
[39] *Ibid.*, pp.274-6.

unite to assert these rights.[40] This feeling was particularly strong amonst the Muslim peasantry of East Bengal who were members of the Fara'idi sect. Fara'idis were heard to assert that they had got rid of the planters through the revolt of 1859-62 and that they would now get rid of the zamindars.[41] The first agrarian league to resist zamindar oppression was started in Dacca district in 1870-1. This was followed by the Agrarian League of Pabna of 1873. Pabna was a Muslim majority area of central Bengal which had played a leading part in the indigo revolt. Over half of the tenants there had occupancy rights under the 1859 Act, and it was the attack by the zamindars on these rights which was the direct cause of the movement of 1873.[42] The protest was supported most strongly by the occupancy tenants.[43] It involved either the withholding of rents or the payment of what peasants judged to be a 'fair rent', and a demand for the abolition of illegal charges and other malpractices and acts of coercion by the landlords. The peasants collected money from amongst themselves to fight rent cases in the courts, in the process demonstrating the inadequacies of the existing rent law. The movement was largely non-violent, though cases were reported of retaliation against the violence of the police or the landlords' toughs. The sentiments of the peasants were anti-zamindar; no anti-British feeling was reported. On the contrary, peasants demanded that they be made 'ryots of Queen of England',[44] which in effect meant that they be granted the sort of fixed land-tax settlements enjoyed by the peasants of British India. The movement spread from Pabna to Bogra, and then, in a weakened form, to many other districts of southern and eastern Bengal.

The article by K.K. Sengupta reproduced in this book is a shortened and somewhat altered version of the third chapter of his book on the Pabna movement of 1873. Sengupta argues that the movement was not concerned merely with high rents, but attacked all forms of zamindari oppression. He shows that the peasants gained the confidence to act because they believed the British to be sympathetic to their cause. This was because rent cases had to a large extent been judged on their merits by the courts since 1869, rather than given a pro-landlord slant. Some British officials had also taken action against oppressive landlords. For instance, Peter Nolan had pun-

[40] Kalyan Kumar Sen Gupta, *Pabna Disturbances and the Politics of Rent 1873-1885* (New Delhi: 1974), p.37.
[41] B.B. Chowdhury, 'Agrarian Relations in Bengal', p.287.
[42] K.K. Sen Gupta, *Pabna Disturbances*, pp.10-15.
[43] *Ibid.*, p.149.
[44] *Ibid.*, p.42.

ished a zamindar who had kidnapped a peasant.[45] Sengupta goes on to show how the peasants organized their resistance by forming agrarian leagues. The leaders were petty landlords, village headmen and substantial peasants. Some of the leaders were Hindu. Sengupta argues that this reveals the non-communal nature of the movement. Another indicator of the lack of communal sentiment was, according to Sengupta, the fact that the peasants were mostly converts from lower Hindu castes, 'most of whom were completely ignorant of even the elementary doctrines of the Koran.' He quotes a contemporary report which held that this was nothing like a 'Fara'idi rising'.

Although Sengupta is justified in arguing that this was not a battle between Hindus and Muslims as such, his treatment of the issue is not satisfactory. As Ranajit Guha has pointed out in a critique of Sengupta, converts often rise against their former co-religionists, especially if the conversion was as a result of an oppressive higher class group using religion to further their exploitation (as when low caste groups are considered 'polluting').[46] One might add that ignorance of scriptures is hardly an adequate reason for lack of religious fervour. Although Sengupta might have stated his case better, his argument still has the problem of amounting to no more than a statement, in essence, that conflicts between Hindus and Muslims of the sort seen in the twentieth century did not occur in the 1870s in Bengal. In this, Sengupta's approach is typical of much recent historical writing which, in claiming to be progressive and 'secular', tries to impose such values onto the subject of their study in an anachronistic manner. The peasants of nineteenth-century Bengal were not 'secular'; religion always played an important role in shaping their consciousness. However, as argued in section I, the boundaries of the perceived community were never closed: peasants could at one moment be inspired by their religion as well as fight for the peasant community as a whole against those seen as their exploiters in a way that cut across religious boundaries. So there was no necessary contradiction between Fara'idi militancy and a struggle by the peasant community, as a whole, led in part by Hindus. Sengupta could have paid more attention to this question by examining the role of the Fara'idis in more detail. This aspect of the movement needs further study. The struggle continued during the 1870s. The zamindars found it increasingly hard to coerce the peasants, as is

---

[45] *Ibid.*, p.35.
[46] R. Guha, *Elementary Aspects of Peasant Insurgency*, pp.172-3.

reflected in a lament by a zamindar published in the *Amrita Bazar Patrika* in 1877:

> For some years past, the rebellious tenants, having stopped payment of the rent due to the owners of the land, have caused untold losses and humiliation to them. Far from getting the proper rent in due time from the tenants, the collecting agent during his tours for realizing the rent is turned back with various abuses... The tenants are now refusing to pay more than one fourth of the rates at which they formerly paid rent.[47]

In 1879 the British decided that landlord-tenant relations in Bengal needed to be placed on a more satisfactory footing. The revolt by the Irish peasants at the time showed the dangers of allowing continued rackrenting of an impoverished peasantry. The Indian Famine Commission Report of 1880 blamed the rent laws for much of the destitution of the peasants in Bengal and Bihar. The report concluded that the state had a duty to regulate the relations between tenants and landlords in India. The government was impressed by this argument and it was decided to formulate a new tenancy law.[48] The zamindars strongly opposed the proposed legislation. As a result, the law was enacted only in 1885.[49] Under the Act a peasant was granted occupancy status if he had farmed land in the same village for twelve years. The aim was to circumvent the zamindar's practice of reallocating plots of land within a village frequently so as to prevent occupancy rights being granted. Checks were placed on rent enhancements and it was made harder to eject peasants for arrears of rent. The government also took in hand the official recording of occupancy rights.[50] The Act strengthened the position of the occupancy tenants, but left the non-occupancy tenants without any rights. The power of the zamindars became increasingly confined to a relatively small home estate—often just their village—with most of their land being held by increasingly powerful occupancy tenants.

The resistance of the Mappilas (or Moplahs) of Malabar to their landlords was more dramatic but less successful during this period than that of the Bengal peasants. In this region Hindu landlords had been given unprecedented powers by the British when they annexed the area in 1792. The peasants were deprived of their rights of occupancy and turned into a

[47] Quoted in Rajat Ray, *Social Conflict and Political Unrest in Bengal 1875-1927*, pp.66-7.
[48] B.B. Chowdhury, 'Agrarian Relations in Bengal', pp.303-4.
[49] Sen Gupta, *Pabna Disturbances*, pp. 138-47.
[50] B.B. Chowdhury, 'Agrarian Relations in Bengal', pp.304-5.

mere tenantry-at-will. From the 1830s the Hindu landlords became more and more oppressive, provoking a series of uprisings which continued from 1836 to 1921. In the essay reprinted in this book, Conrad Wood argues that Islam provided a means of solidarity in a region of poor communications and isolated settlements. The mosque provided a rallying point. The Hindu tenants lacked such forms of solidarity, and because they shared the religion of their landlords they had a certain sympathy with them. Also, they were liable to be put out of caste by the landlords for even the slightest show of dissent. Because of this, opposition to the landlords came almost entirely from Mappila Muslims. This led to certain weaknesses in the movement, such as that Hindus were sometimes attacked because of their religion and not because they were oppressors. Also, Wood argues, the attack was mainly against Hindu landlords, rather than the ultimate oppressor, the colonial state. Because these revolts were ostensibly 'religious', the British were able to play down the question of tenant rights. A peasant league fighting for such rights in a legalistic manner may well have achieved better results.

Conrad Wood's argument has been attacked in a book by Stephen Dale called *Islamic Society on the South Asian Frontier: The Mappilas of Malabar 1498-1922.*[51] Dale believes that the chief motivation of the Mappilas was not economic grievance, but the desire to gain paradise by dying in defence of the faith. The Mappilas' fervour could be traced back to the fifteenth century, when persecution by the Portuguese forced them to become a militant community. This established a tradition which continued into the nineteenth and twentieth centuries. Their revolt was merely 'archaic' and offered 'no viable alternative for the future'. Dale denies that these were peasant revolts in which peasants joined specifically to resolve their agrarian problems. He argues that the Mappila rebels were by no means all hungry and landless.

In a review of this book, David Arnold has argued that Dale has failed to examine in any depth the socio-economic conditions in the area and the way in which they related to the consciousness of the Mappila peasantry. Dale never describes the social composition of the community, nor does he note the fact that it grew rapidly in the nineteenth century largely through conversion from low-caste Hindus and former slave castes. In this way a new solidarity of the lower classes was forged. Most of the Mappilas were found in south Malabar, an area in which the high-caste Hindu landlords enjoyed their greatest power. Most of the outbreaks occurred in

[51] (Oxford: 1980).

this area. In these revolts, the Hindu landlords were singled out for attack and their accounts and papers destroyed. Arnold argues that these basically anti-landlord revolts revealed the importance to the peasantry of a legitimizing ideology with a strong sense of belief in a better future. In peasant society in India, this almost invariably took a religious form, and this was the case in Malabar.[52]

Although Arnold on the whole favours Wood as against Dale, his points raise doubts about the tone of Wood's conclusion. In disparaging the Mappilas for what he labels their 'unsophisticated' political techniques, their 'rudimentary' organization, their lack of adequate leadership or well-formulated demands, Wood is, in essence, criticizing the Mappila rebels for not being modern socialists. In nineteenth-century Malabar there was no realistic possibility that they could be this. They rebelled in the manner available, forging a remarkable solidarity through their religion. Their misfortune lay not in this, but in the fact that the British were able to get away with labelling their revolt as an act of 'religious fanaticism' and merely a law-and-order problem. The peasant movement in Bengal—often equally 'religious'—could not be compartmentalized in such a manner. It was far more widespread, it occurred in the vicinity of the capital of India, and it affected a highly vocal class of landlords, many of whom were also members of the intelligentsia and thus with a certain ambivalence of attitude towards the peasantry. South Malabar, by contrast, was a small backwater with weak links with Madras, the provincial capital. As a result, the Madras government was able to get away with this convenient misreading of the Mappila outbreaks. It was only after the greatest of such risings, in 1921, that the British were forced to take the matter more seriously. In 1929 occupancy rights were granted to the tenants of Malabar by law. There were no more Mappila revolts after this.

Nowhere else in India during this period did anti-landlord agitations achieve such levels of militancy. There was, however, continuing conflict of a less dramatic kind in other regions of strong landlordism such as Awadh and Rajasthan. In Sitapur district of Awadh, peasants protested against rent increases and other taxes imposed by the landlords during the 1860s.[53] In Rajasthan, a movement began in the Bijolia estate of Mewar in 1897 when the peasants refused to pay a tax levied on marriages; they

[52] David Arnold, 'Islam, the Mappilas and the Peasant Revolt in Malabar', *The Journal of Peasant Studies*, 9:4, July, pp.258-64.

[53] M.H. Siddiqi, *Agrarian Unrest in North India: The United Provinces (1918-22)* (Delhi: 1978), p.53.

migrated to a different part of the state in protest. The tax was lifted as a result. Encouraged by their success, the peasants pressed for further concessions and rent reductions, some of which were conceded in 1905. The movement continued to gather strength under the leadership of a sadhu called Sitaramdas. He led a strong agitation between 1913 and 1915 in which the peasants refused to cultivate their land for nearly two years. As a result, rents were reduced and certain taxes abolished in 1915. These concessions were by no means adequate, and the peasants continued to press their demands, launching a new movement, even more powerful, in 1917 and continuing to 1922.[54]

All of these anti-landlord movements continued after our period, becoming in most cases more powerful with the passing years. The movement in East Bengal developed most strongly in the years after 1930, with attacks by the Muslim peasantry on the largely Hindu zamindar class.[55] In Malabar the Mappilas rebelled once more in 1921, for a time winning control over large parts of southern Malabar. The British brutally suppressed the revolt in late 1921. In Awadh a massive anti-landlord movement, led by Baba Ram Chandra, began in 1919. This linked up with the Congress campaign of non-co-operation in 1920.[56] In Rajasthan resistance to the jagirdars developed in almost all princely states in the years after 1920.[57] These struggles continued for the most part until after Indian independence.

IV

The third set of essays in this collection relates to the peasant revolt against moneylenders in Maharashtra in 1875. Although grain dealers-cum-moneylenders (*sahukars*) had long been a feature of the Indian country-side, their position had been considerably strengthened under British rule. This was in part because the British had worked towards creating a largely uniform system of landed property, in which an area of land was held to belong to an individual and to be freely saleable on the market. In the past there had been a complex system of interlinking rights in land, some of

[54] Pushpendra Surana, *Social Movements and Social Structures: A Study in the Princely State of Mewar* (New Delhi: 1983), pp.63-103.

[55] For this history see Sugata Bose, *Agrarian Bengal: Economy, Social Structure and Politics 1919-1947* (Cambridge: 1986), pp.181-232.

[56] Siddiqi, *Agrarian Unrest in North India*, pp.104-95.

[57] See Pema Ram, *Agrarian Movements in Rajasthan 1913-1947* (Jaipur: 1986).

which were granted or acquired by charter, but many of which were customary. Although the buying and selling of such rights was possible, and was in the eighteenth century becoming more common, it was not easy for someone not acceptable to the village elite to acquire such rights over peasant land. Under the new system introduced by the British, it became possible for moneylenders to take individual peasants to court and force them to sell their land to pay off their debts. The civil courts established by the British worked in favour of the moneylenders as they favoured the written evidence of the sahukars to the oral testimony of the peasants. The sahukars won their cases by producing carefully prepared and often doctored debt-records in court. The position of the moneylenders was strengthened also by the growth in population in rural areas, as this created a demand for land and made land a valuable security for debt. In many cases moneylenders who had peasants already in their clutches refused to make further advances unless the land of the debtor was made over to them in mortgage. In time, after the peasant had failed to repay by a stipulated date, the moneylender became owner of the land, turning the peasant into a tenant on the land he had once owned. As tenants, peasants had to hand over far larger amounts of their produce to the sahukar than had formerly been the case. It was this growing exploitation by moneylenders which caused extreme resentment amongst peasant communities which in the past had prided themselves on their independence. During the second half of the nineteenth century there were, as a result, a number of clashes between peasants and sahukars in different parts of British India. The uprising which has received most attention from scholars occurred in the Deccan in 1875.

The revolt began at Supe in eastern Pune district on 12 May 1875. Supe was a large village—a local market centre—in which there were many Gujarati sahukars. Peasants gathered from the surrounding villages and attacked the shops of the sahukars, looting them and in one case burning down a house. The sahukars were not assaulted physically. Other villages in eastern Pune district, then Ahmednagar district, began to follow suit, attacking mainly Marwari and Gujarati sahukars. Within a fortnight there were similar attacks on sahukars in an area about sixty-five kilometres from north to south and about a hundred kilometers from east to west.

The object of the rioters was in every case to obtain and destroy the bonds, decrees, etc. in the possession of their creditors: when these were peaceably given up to the assembled mob, there was usually nothing further done. When the moneylender refused or shut himself up, violence was used to frighten him into a surrender or to get possession of the papers.[58]

Many sahukars fled their villages rather than face the wrath of the peasants; others were ordered by the peasants to leave or face social boycott. Peasants who continued to have dealings with sahukars were also threatened with boycott.

The British sent in troops and began arresting those they suspected of having attacked the moneylenders. This was not an easy task, as most peasants refused to inform on their fellows. On the whole, only those who were found with stolen property could be arrested and tried. By the end of May, many of these had been given severe sentences as a deterrent to others.[59] Police posts were established in the most militant villages to overawe the peasants. By early June the attacks had stopped, though the peasants continued to show hostility to the moneylenders. In Supe, for instance, despite the repression carried out by the government, the peasants continued to boycott the sahukars. The sahukars were warned that what had been done to them up to then was nothing and that they were to hand over all their remaining documents to avoid further trouble.[60] The state of extreme tension continued for many months. Peasants refused to have their normal dealings with sahukars, and they insisted that they hand over the remaining debt-bonds. One moneylender of Supe who went to try to collect the money due to him from a nearby village had his nose cut off by his debtors in late November.[61] The bitterness between the two groups continued for many years.

The three essays reprinted in this book on the revolt—often described as the 'Deccan riots'—deal with the event in very different ways. None of them, taken individually, provides a rounded view. Ravinder Kumar's essay is concerned mainly with the socio-economic background. He argues that, before the coming of British rule, village moneylenders had very little power in villages. They had to deal with the village community as a corporate body. Under the ryotwari system introduced by the British, the government began to collect tax from individual cultivators rather than

---

[58] *Report of the Committee on the Riots in Poona and Ahmednagar 1875* (Bombay: 1876), p.5.

[59] Report by G. Norman, Collector of Pune, 20 and 25 May 1875, Maharashtra State Archieves, Bombay: Judicial Department 1875, 82/730.

[60] Petition by 13 sahukars of Supe to Govt. of Bombay, 23 September 1875, Maharashtra State Archives, Bombay, Judicial Department 1875, 82/730.

[61] Mamlatdar of Bhimthari Taluka, 22 November 1875, Maharashtra State Archives, Bombay: Judicial Department 1875, 82/730.

from the village community as a whole, which meant that peasants now had to borrow as individuals to pay their tax demands. Their position was thus weakened; the power of the sahukar was correspondingly strengthened. From the 1840s the moneylenders began to acquire the land of peasants in settlement of their debts. British law and the civil courts helped them in this respect. Peasants who had previously held large areas of land became mere tenants of the moneylenders. In this way the 'harmonious relationship' (as Kumar puts it) between peasant and sahukar was transformed into one of 'acute antagonism'. The situation became worse in the late 1860s and early 1870s because of a disastrous fall in agricultural prices at a time when the British were increasing land-tax rates. This led to an agitation against the tax increases. It was in attempting to resolve this issue that the British took measures which inadvertently set off the attacks on the moneylenders in May 1875.

Kumar's essay is valuable in providing us with the background to the revolt. A few problems with the essay should, however, be pointed out. It is difficult to accept his picture of the village community of the pre-colonial Deccan. Village communities did not manage themselves free from outside interference in the manner described by Kumar; there was a range of outside forces which controlled what happened in villages.[62] Recent research has shown that the power of the moneylenders was already on the increase in villages before the British conquest.[63] Kumar's equation of 'moneylender' with 'vani', i.e. a member of the Vaniya caste (such as the Marwari and Gujarati Vaniyas), obscures the fact that the sahukars in the Deccan were of three main classes: the alien Marwari and Gujarati Vaniyas, indigenous Vaniyas (the actual 'vani'), and Maharashtran Brahmans. These divisions were of great importance, both in precipitating the revolt (Maharashtran Brahman moneylenders sometimes supported the peasants against rival Marwari or Gujarati moneylenders), and in determining the object of attack (Brahman moneylenders were almost always left alone during the uprising).

A further problem with Kumar's essay is its assumption that once material conditions of exploitation reached a certain point 'a violent clash

[62] See Frank Perlin, 'Of White Whale and Countrymen in the Eighteenth Century Maratha Deccan', *Journal of Peasant Studies*, 5:2, January 1978, and other writings by Perlin.

[63] See Sumit Guha, *The Agrarian Economy of the Bombay: Deccan 1818-1941* pp.14-16; Sumit Guha, 'Commodity and Credit in Upland Maharashtra 1800-1950', *Economic and Political Weekly*, 26 December 1987, Review of Agriculture, p. A-128.

between the *kunbis* and the *vanis* became inevitable'.[64] Kumar, as a result, hardly brothers to describe the actual uprising, the course it took, the mentalities underlying particular forms of action, the exact targets of attack, the reasons why it ended. For him these events were merely 'a reflection of the tensions generated within rural society through the legal and administrative reforms carried out by the British Government'.[65] If this were so, we may wonder why uprisings did not occur all over Maharashtra in the 1870s, for everywhere the peasants were suffering from the exactions of the sahukars.

Such issues, which concern the realm of politics rather than economics, are addressed more centrally by Ian Catanach in the second of the papers reproduced here on the Deccan revolt. Catanach's paper, in fact, goes to the opposite extreme, describing the economic background in one brief paragraph. He looks for patterns in the attacks, such as the tendency for them to have occurred on market days and the 'discriminating purposeful-ness' of the peasants who carefully singled out the Marwari and Gujarati sahukars for attack and then destroyed only the debt-bonds and not the moneylenders. He argues, convincingly, that the revolt failed to spread beyond Pune and Ahmednagar districts because of repression by the police and army. He examines the question of why the revolt occurred in May 1875 and not at any other time. His explanation for the time of year—that, in May, peasants are idle and their tempers become 'frayed' by the heat—is not satisfactory; the reason is more likely that the peasants depended on advances from the moneylenders during the lean months of summer, and, in that particular year, the moneylenders had refused their usual advances for reasons that he spells out. This had infuriated them. Catanach searches for immediate economic causes, such as harvest failure, which may have precipitated the revolt in 1875 rather than in any other year, but fails to find any—rightly in my view. He feels that agitational work by middle-class activists in the two preceding years may have had something to do with the timing. He does not discuss this aspect very adequately, however, and the question remains in the air.

Catanach then continues with an interesting and suggestive discussion of popular beliefs during the revolt. One was that the peasants believed that top government officials were on their side against the Marwaris and that they would not take action if the peasants attacked the moneylenders. He tries to trace a connection between the revolt and the cholera epidemic

---

[64] See R. Kumar, below.
[65] *Loc. cit.*

which was raging in the area at the same time. He fails to make any convincing connections, and, from what I know of the event, it seems unlikely to have existed in any close relationship of cause and effect. Epidemics were seen more as a harbinger of future events of great importance—an omen—rather than as providing an immediate reason for revolt. The example he quotes, of villagers refusing to allow a money-lender's cart to be moved from a village because they were afraid of cholera, is baffling in its meaning. The problem is resolved when we look at the original record, in which it is stated that the villagers refused to allow a moneylender to bring a cart *into* the village to remove grain because they feared that cholera would thereby be brought into the village.[66] The fear was understandable and can hardly be viewed as some form of mysterious sub-text to the revolt.

Catanach's treatment of these issues is tentative and suggestive, no-where reaching the level of sophistication of Ranajit Guha's more recent monograph on peasant insurgency (which has, incidently, many scattered references of great insight on this revolt). Catanach raises important issues, however, and he would be the first to admit the need for more systematic research and writing on the subject. Strangely, although the essay is now a quarter of a century old, nobody has yet provided a more satisfactory history along such lines.

Neil Charlesworth's article on 'The Myth of the Deccan Riots' has been included because it attempts a controversial reinterpretation of the subject. In these comments I shall take into account some refinements which he made to his argument in a book on the agrarian history of this region, published in 1985.[67] Chapter Four of this book is on the revolt of 1875. The argument advanced in the 1972 article and 1985 book is substantially the same, though in the latter it is presented less polemically, and with a more careful use of evidence.

Charlesworth disputes Kumar's argument that British rule had trans-formed Deccan society by the 1870s. In his view, the only important changes were that the ryotwari system had greatly reduced the power of village headmen and that the relationship between peasants and money-lenders was now mediated through courts of law. The moneylenders used the courts to harass the peasants. This was strongly resented, and it was

---

[66] Collector of Pune, 13 July 1875, Maharashtra State Archives, Bombay:, Judicial Department 1875, 82/730.

[67] Neil Charlesworth, *Peasants and Imperial Rule: Agriculture and Agrarian Society in the Bombay Presidency 1850-1935*(Cambridge: 1985).

significant that the bonds used in courts should form a focus for the revolt in 1875.

In this sense, the legal documents formed a new formalization of each peasant debtor's economic dependence, which might previously have been unwritten and undefined. There were now, then, actual physical objects to serve as the focus for tensions within the credit system. In this way, the legal system introduced by British rule went to make the Deccan Riots a more coherent and effective movement, but it had not created an agrarian revolution to cause the disturbances.[68]

Charlesworth argues that a social revolution would have occurred if (1) the peasants had lost a major part of their land to moneylenders, and (2) there had been a commercial revolution in agriculture. In his estimation, only about five per cent of the cultivated land of the Deccan had passed to Marwari and Gujarati moneylenders at the time of the revolt. These sahukars were not interested in acquiring land; they merely threatened to dispossess the peasants so as to frighten them into repaying their debts. The indigenous high-caste moneylenders of the Deccan were far more interested in getting land, yet they were spared in the uprising. On the second issue, commercialization, Charlesworth denies that British rule had by the 1870s brought any significant change. The chief crops continued to be the staples of *bajra* and *juvar*. Cotton had enjoyed a brief boom during the 1860s, but this was an extraordinary event, and much land reverted to subsistence agriculture afterwards. There was little improvement in irrigation or a raising of the productivity of land; peasants lacked the resources to be able to do this. Peasant agriculture remained tied to the vagaries of the monsoon.

Charlesworth argues that the sahukari system had continued unchanged in its essentials for centuries. Always, harvests had been erratic and the peasants had lacked resources. 'Borrowing offered the only easily available facility to mitigate these circumstances, particularly in years of total crop failure.'[69] Availability of loans from sahukars allowed the peasants to survive shortages during the year and tide over bad years. High interest rates did not mean usury; they reflected the risks taken by moneylenders in advancing loans to peasants. Without the moneylenders, crops could not have been marketed on a large scale. Therefore, peasants had relationships with sahukars because it was to their 'essential advantage' to do so.[70] The

[68] *Ibid.*, p.104.
[69] *Ibid.*, p.85.
[70] *Ibid.*, p.87.

problem, according to Charlesworth, was not so much the supposed iniquities of the system as the fact that sahukars did not provide adequate credit to peasants to allow them to improve their land, dig tanks and wells and so on. He thus advances what he calls 'a strictly functionalist view of credit and debt in India.'[71]

He goes on to argue that there were times when the moneylenders themselves faced a crisis—for instance, when prices were falling, so that the crops they took from the peasants were falling in value. This happened in the early 1870s. The sahukars were forced to withdraw credit to remain solvent. This could cause discontent and possible 'grain riots' (as he puts it) against moneylenders. This happened periodically and was not a feature new to British rule. Such 'rioting' was like a 'safety-valve', allowing the peasants to work off their anger, and it was therefore 'a source of stability for the system'.[72] He argues that this was the case in 1875. The uprising occurred in an area of notorious poverty in which prices had fallen and moneylenders were consequently refusing to give loans. This distressed and angered the peasants. As it was, the violence was limited to a small area of only 33 villages. These were therefore 'no more than minor grain-riots, a simple outburst born of deprivation, rather than an articulate protest against a social revolution.'[73]

Despite some good points, Charlesworth's overall argument is not, in my view, sound. First, there is his argument that there had been no funtamental changes in the sahukari system. This is contradicted by much recent research which has shown that commercial capitalists were increasing their economic power at all levels in the eighteenth century. At the village level this meant greater control over the agrarian economy by sahukars.[74] British rule both consolidated and strengthened this process, with the establishment of bourgeois rights of property and civil courts to enforce debt-repayment. Charlesworth fails to examine this historical process and see what it entailed, namely a growing hatred of sahukars.

Second, there is the argument that the peasants lost little land to Marwari and Gujarati sahukars. Charlesworth's figures certainly show

[71] *Ibid.*, p.85.

[72] *Ibid.*, p.99.

[73] N. Charlesworth, below.

[74] See C.A. Bayly, *Rulers, Townsmen and Bazaars: North Indian Society in the Age of British Expansion, 1770-1870* (Cambridge: 1983), pp.170-6 and 462-3; and David Hardiman, 'Penetration of Merchant Capital in Pre-Colonial Gujarat', in Ghanshyam Shah (ed.), *Capitalist Development: Critical Essays in Honour of A.R. Desai* (Bombay: 1990).

that Kumar's assertion that there was wholesale transfer of land needs revision. However, it should be noted that the Marwari and Gujarati sahukars often took the most fertile and valuable land, so that their acquisition of what Charlesworth considers to be a small amount—five per cent of the cultivated land—would have been particularly resented, both for the loss it caused there and then and also as a very visible harbinger for the future. He also fails to see that the process could have continued very rapidly after 1875 *if* the revolt had not taken place and a law subsequently passed to give the peasants greater protection in court. The revolt made the sahukars afraid of the consequences of grabbing land in the future, and the Deccan Agriculturists Relief Act of 1879 made it legally harder for them to do so.

Thirdly, Charlesworth argues that there was a connection between the loss of power by the village elites in the early nineteenth century and the revolt. The old elites resented what had happened and provided leadership for the peasants in the uprising. He contrasts the situation in the Deccan with that in the Konkan and Gujarat where, he holds, the village elites had retained their powers, continuing to act as village officers, moneylenders and traders. He cites the Gujarat talukdars as one such elite. The argument is odd, seeing that the talukdars were confined largely to one district of Gujarat—Ahmedabad—and that they were so notoriously indebted that the government had to introduce legislation in 1862 to save them from the clutches of sahukars. This is not, therefore, an adequate explanation for why there was a revolt in the Deccan but not in Gujarat.

The chief implication of this line of argument is that the revolt had a strong ethnic dimension, being initiated and led by a decaying Maratha elite against a class of *nouveaux riche* usurers from Marwar and Gujarat. In this, it can be argued the elites enjoyed the support of the mass of Maratha peasants. Solidarity was, therefore, along ethnic rather than class lines, for many of the old elites were also moneylenders. The revolt accordingly presented 'a picture, familiar to many students of peasant societies, of a tight knit village community reacting against outsiders who are, *simply*, disliked personally.'[75] Charlesworth concludes that the 'foreignness' of the sahukars was as much a cause of the revolt as their alleged rapacity.

In his discussion Charlesworth fails to distinguish adequately between two main groups amongst the old elites. Firstly, there were the Maratha patils, who had lost their powers in the early nineteenth century and who

[75] N. Charlesworth, *Peasants and Imperial Rule*, p. 107. (My emphasis.)

by the 1870s were deeply in debt to sahukars and were losing land. They undoubtedly provided leadership in the revolt—but as they were suffering from the inroads of the sahukars, this was to be expected. Secondly, there were the Brahman Deshpandes and Kulkarnis. Although the British abolished the office of Deshpande, many ex-Deshpandes became big landlords. They were also big moneylenders. The Kulkarni (village accountant) was an important functionary under the British. The position was dominated by Brahmans who worked hand-in-glove with sahukars of all sorts. Brahmans were powerful all over Maharashtra, both in villages and in the towns and cities, where they were in government service and the professions. The non-Brahman movement, launched by Jotirao Phule at this time, was directed against this power.[76] The role of Brahman land-lords, Kulkarnis and usurers in the revolt appears to have been ambiguous. In some cases, they were happy to see the Marwaris and Gujaratis attacked. More often, however, they were frightened of the implications of such peasant action, and they sided with the forces of law and order. One reason why fewer Brahmans than Marwaris and Gujaratis were attacked was that the former tended to live in towns rather than villages, and were thus better protected by the police. There were, however, a few cases of attacks on Brahman moneylenders, as at Supe where the nose of one of them was cut off.[77] In the years after the revolt the non-Brahman movement gathered in strength, and peasant resistance to Brahman moneylenders became as strong as against Marwaris and Gujaratis.[78] The revolt was not therefore the expression of a 'simple' dislike for sahukars of a different ethnic background, as Charlesworth tries to argue.

Fourthly and lastly, there is the problem of Charlesworth's depiction of these as no more than 'minor grain riots'.[79] He holds that what was no more than a recurring type of grain riot was blown up out of all proportion by paranoic British officials who feared that the discontent against money-lenders could fuel a huge rising against British rule.

For a start, it is wrong to describe the uprising as 'minor'. It occurred in villages in an area sixty-five kilometres north to south and a hundred

[76] See Rosalind O'Hanlon, *Caste, Conflict, and Ideology: Mahatma Jotirao Phule and Low Caste Protest in Nineteenth-Century Western India* (Cambridge: 1985).

[77] Mamlatdar of Bhimthari Taluka, 22 November 1875, Maharashtra State Archives, Bombay: Judicial Department 1875, 82/730; *Report of the Committee on the Riots in Poona and Ahmednagar*, p.5.

[78] R.O'Hanlon, *Caste, Conflict, and Ideology*, pp.278-81.

[79] N. Charlesworth, below.

kilometres east to west. This is by no means a small area. The official report on the revolt noted attacks in only 33 villages. This appears to be a considerable under-statement.

In one file in the archives in Bombay the names of about 90 villages in which there were disturbances of some sort (not always full-fledged risings) are mentioned.[80] Moneylenders tended to be concentrated in small towns and large villages, extending their operations into the surrounding villages. As a result, the attacks tended to occur in these centres, with peasants coming from the nearby villages. The number of villages represented in the revolt was therefore far greater than the number of villages in which attacks occurred. It would be no exaggeration to say that in this whole area there was massive solidarity in almost all villages against the moneylenders. It is likely that if the British had not acted promptly in suppressing the revolt, it would have spread all over Maharashtra. Activists of Pune reported in late June 1875 that disturbances had occurred in Pune and Ahmednagar districts, but were threatened also in Sholapur, Satara and Nasik districts. In some cases isolated outbreaks actually occurred in these latter districts, as at Igatpuri in Nasik and Kokrad in Satara.[81] The fact that only debt-bonds were destroyed and no moneylenders were killed is not an indicator of any feebleness, as Charlesworth believes; rather, it is a reflection of the fact that the attacks were directed very precisely against a clear target, namely the documents which the sahukars used to give them power over the peasantry. For this reason it is quite wrong to describe these as 'grain riots'. Charlesworth's discussion of the attitude of British officials is also unsatisfactory. They were not merely afraid of possible peasant insurrection, but seriously concerned about the effects of unchecked parasitic usury on the health of the society which they were governing.

The Deccan revolt was an important turning point in the history of Maharashtra. The Marwari and Gujarati usurers never again had things so easy. Their exploitation was in future always tempered by the memory of what had happened in 1875. In future years the Maratha peasantry would go on to battle with the Brahman elites as well, in the twentieth century gradually wresting power for themselves in rural Maharashtra. This

[80] Maharashtra State Archives, Bombay:, Judicial Department 1875, 82/730.
[81] Report by Pune Sarvajanik Sabha, 28 June 1875, Maharashtra State Archives, Bombay:, Judicial Department 1875, 82/730. For Kokrad riot see District Superintendent of Police, Satara District, 21 September 1875, Maharashtra State Archives, Bombay: Judicial Department 1875 81/1320.

history, in all of its social, political and economic dimensions, has yet to be written.

There were similar attacks on moneylenders elsewhere in India during the second half of the nineteenth century. During the revolt of 1857, when British power was destroyed for the moment in western U.P., large numbers of peasants attacked their sahukars, plundering their property and destroying their debt-bonds. Eric Stokes has described how the peasants of Saharanpur district rose up and robbed the sahukars of their property or made them pay fines. They forced them to surrender their account books and debt-bonds. The revolt was described by the British as directed against both them and *bania ka raj*. An official reported that the smaller landowners, many of whom had lost land to sahukars through the operation of the British courts, took the lead.[82] It was extremely uncommon for a Bania to be hurt.[83] In 1868 in Baglan taluka of Nasik district, there was a series of attacks on moneylenders by adivasis who had been oppressed by them. The attacks were made to destroy debt-bonds, rather than for plunder. In only one case was a sahukar killed.[84] In 1874 a band of Koli outlaws of the ghat region between Bombay and Pune began attacking sahukars, cutting off their noses and destroying their books and papers.[85] Attacks similar to these continued in this region during the rest of the nineteenth century.[86]

In 1891 the northern part of Ajmer district in Rajasthan was shaken by a series of attacks on moneylenders. No history of this uprising has been written; our knowledge of it comes from official reports compiled at the time.[87] In Rajasthan there were only two areas under direct British rule, the adjoining districts of Ajmer and Merwara. These two districts came under the operation of British law, and it had been possible for the local moneylenders—Banias by caste—to appropriate many of the best fields of

[82] Eric Stokes, *The Peasant Armed: The Indian Rebellion of 1857* (Oxford: 1986), pp.206-8.

[83] Gautam Bhadra, 'Four Rebels of Eighteen-Fifty-Seven', in R. Guha (ed.), *Subaltern Studies IV* (New Delhi: 1985), pp.245-25.

[84] *Deccan Riots Commission, Appendix B* (Bombay: 1876), p.171.

[85] *Report of the Committee on the Riots in Poona and Ahmednagar*, p.105; *Deccan Riots Commission, Appendix A* (Bombay: 1876), p.36.

[86] There is an important history of these Kolis during the nineteenth century yet to be written. The documentation in the Bombay archives on the subject is extremely rich. Files in the Judicial Department contain reports of a long series of attacks on moneylenders.

[87] These reports are contained in a file in the Rajasthan State Archives, Bikaner. Ajmer Commissioner's Office records, English Office Z(12)30, basta 132.

the peasants, turning them into their tenants. They also used the courts to harass the peasants. Nowhere else in Rajasthan were sahukars able to use the courts in this way and, significantly, nowhere else were there attacks on sahukars during this period. In Ajmer district itself the revolt was confined to villages under direct British control; in villages under jagirdars (known locally as *istimrardars*), where sahukars were not permitted to acquire land, no attack occurred. The immediate cause was, as in the Deccan in 1875, a refusal by the sahukars to give the usual advances to the peasants. In this case the sahukars refused because the rains had failed that year and they knew that the peasants would have no crop with which to repay loans. The uprising occurred in September, at the end of the monsoon period. A further precipitating cause was a rumour that the district commissioner had ordered that no peasant would be punished for taking grain from the moneylenders. The origin of this rumour was that the commissioner was aware that the sahukars had been harassing the peasants by lodging false cases of grain robbery against them, and he had ordered that the police should report all supposed grain robberies to a magistrate before taking any direct action. The revolt occurred in a strip of land about seventy kilometres long and forty kilometres broad in the north of Ajmer district, being concentrated in 22 larger villages in which moneylenders were based. In every case there was looting of grain stores and destruction of the sahukar's books and papers. Although the level of violence was somewhat greater than it had been in the Deccan, not a single moneylenders was killed. The revolt was spreading to Merwara district, and one rising did occur there, at Beawar, but by then the police counter-attack was being felt and the resistance died away. Strong resentment against the moneylenders was reported from all over the region, a resentment which did not subside even after order had been restored.

This pattern of resistance to moneylenders continued in the twentieth century. There were attacks on sahukars in the Punjab in 1914, sparked off by a belief that the peasants were about to be liberated by the 'German Emperor,'[88] and in East Bengal in 1930. A British official who was on the spot wrote about this latter upsurge:

> The usual procedure was for a mob of anything from 100 to 1000 men to demand back from the moneylender all the documents in his possession. If he said that the documents were not with him, he was told to have them ready by

---

[88] Malcolm Darling, *The Punjab Peasant in Prosperity and Debt* (London: 1932), p.204.

a certain time, and his house was looted, and in some cases burnt, if he did not produce the document at the time fixed.[89]

Chatterjee says that many peasants believed that the government supported their actions. Some even stated that a government order to that effect had been promulgated. The structural similarity of all of these risings against sahukars is striking. They were a feature of the colonial period; we do not know of any uprisings of this nature before the impact of the British land system and law was felt on Indian rural society. Since Indian independence this form of action has become largely a thing of the past, with the development of a whole range of new sources of rural credit and the decline of the sahukars as a class, often to be replaced as moneylenders by rich peasants of the same community as the rest of the peasants.

V

In contrast to peasant resistance to sahukars, conflicts between peasants and tax-collectors were not new to the colonial period. There was a long history of refusal to pay land tax, an action which had strong political implications. By paying tax to a particular ruler, peasants in effect acknowledged the legitimacy of the ruler. Tax-refusal represented a withdrawal of such recognition. In the pre-colonial period peasants had resisted the tax demands of oppressive rulers either by flight or through armed resistance. Irfan Habib has described how peasants often fled their villages during the Mughal period to escape high tax demands.[90] He also argues that the mass support for the Maratha, Sikh and Jat risings against the Mughals in the late seventeenth century was provided by peasants who resented exorbitant tax demands.[91] Such resistance continued during the colonial period. During the 1830s Maharashtran peasants fled to the Nizam's territories in protest against the heavy tax demands of the British.[92] During the 1860s and early 1870s many peasants of Baroda state migrated to the adjoining British territory in protest against heavy tax

---

[89] Quoted in Partha Chatterjee, 'Agrarian Relations and Communalism in Bengal', pp.29-34.

[90] Irfan Habib, *The Agrarian System of Mughal India 1556-1707* (Bombay: 1963), pp.328-9.

[91] *Ibid.*, pp.330-51.

[92] Ravinder Kumar, *Western India in the Nineteenth Century: A Study in the Social History of Maharashtra* (London: 1968), pp.102-3.

demands.[93] Likewise, the peasants resorted to force. The last major revolt of this kind occurred in 1857. By our period, however, these two main forms of resistance were becoming more difficult. Migration was becoming harder because of the growing pressure of population on the land. After the revolt of 1857 the peasants were disarmed by the British, taxes were lowered in many regions and the local gentry (which had often provided leadership in the revolt) bought off with favoured treatment. Thereafter, peasant resistance to land-tax demands was forced into new channels, and with a new form of leadership.

The area with the strongest no-tax campaigns during our period was Bombay Presidency. This area was only marginally involved in the revolt of 1857, and the authorities had not felt any need to modify the tax system in response to discontent. The Bombay government had lowered land-tax rates in most areas after the protest migrations of the 1830s, but it raised them subsequently in the tax settlements of the 1860s and 1870s. It justified the increases by arguing that prices had risen and that cash crops were being grown over a greater area than before. Unfortunately, these enhancements came into effect just at a time when produce prices had crashed, making in harder for the deeply indebted peasants to pay their taxes. This situation led to the agitation in 1873-4 described by Ravinder Kumar in his article on the Deccan Riots reprinted in this book.[94]

According to Kumar, mass migration was no longer possible because of the increase in population. The peasants looked for support to their traditional leaders at such times of crisis—the Deshmukhs. Kumar argues that these men were no longer able to provide effective leadership. Their appeals to the British were couched in an emotional and moralistic language more suitable to the times of the Peshwas. The British listened only to reasoned argument and hard facts and figures. Such argument was provided in part by sympathetic British officials who were alive to the problems faced by the peasants, and in part by a new middle-class leadership. Their reasoned appeals forced the government on the defensive, so that it had to appoint a commission of enquiry. The middle-class leaders also used agitational tactics which they had observed being used in Britain, such as propaganda tours to encourage the peasants to refuse to pay the increased tax demand. The peasants responded well. In taluka after

[93] David Hardiman, 'Baroda: The Structure of a 'Progressive' State', in Robin Jeffrey (ed.), *People, Princes and Paramount Power: Society and Politics in the Indian Princely States* (New Delhi: 1978), p.113.

[94] See Ravinder Kumar, below.

taluka, tax officials were unable to collect the revised tax demands. Tax refusal was limited not only to those whose crops had failed and were suffering hardship; peasants who could afford to pay the tax also refused in solidarity with those who could not. Rather than have a head-on confrontation with the peasants, the government decided to step down and give certain concessions to the peasants. A clash was thus averted, for the time being at least.

The agitation of 1873-4 failed to change the Bombay land-tax system in any essential respect. The revised rates of tax continued to be collected for the most part, and tax continued to be increased up until the period of the great famines (1896-7 and 1899-1900) which brought the century to a close. Worse, in fact, than high tax demands was the great inflexibility shown by the Bombay government in tax collection. In most other parts of British India, land tax was promptly suspended throughout a region when the harvest failed. In Bombay Presidency the onus rested on the individual peasant who had to prove to a tax official that he could not afford to pay his tax that year. Because it was almost impossible to carry out detailed enquiries in years of widespread harvest failure, most tax officials preferred to ignore the distress and demand payment. There was a strong ethos in the revenue department against tax remission. Officials who failed to collect taxes in full were considered to have acted slackly and inefficiently. Generosity to the peasants during a famine was thus damaging to an official's career. Because of these pressures it was common for lower officials to use illegal methods, even torture, to extract the land tax during bad years.[95] This caused extreme resentment amongst the peasants. During such years it became common for the peasants of Bombay Presidency to refuse to pay their land-tax in protest. The British reacted harshly on all occasions, encouraging local officials to take the law into their own hands to break the resistance, as well as using the legal means at their command, such as threatening to confiscate and sell a peasant's land for failure to pay tax. For example, when the rains failed in Dahod taluka of the Panchmahals district of Gujarat in 1877-8, the Deputy Collector reported:

> It was discovered that there had been a combination among a large number of the well-to-do cultivators of the taluka encouraged by some of the Bania landholders not to pay the revenue, but to demand remissions according to the

[95] These abuses were exposed in an enquiry into the methods used to collect land tax in Broach District of Gujarat during the famine of 1899-1900. See E. Maconochie, *Life in the Indian Civil Service* (London: 1926), pp.124-5.

custom prevailing in the times of the former Government:[96] a little firmness convinced them however that opposition to Government in that manner was useless.[97]

One is left to guess what this 'firmness' may have entailed. Many such no-tax movements were crushed during these years before they could become major political issues.

This was not the case in 1896-7, when the monsoon failed throughout the Deccan, leading to a severe famine and a widespread campaign of tax refusal. Again, the Bombay government was extremely reluctant to remit any land-tax. The agitation was strongest in the coastal districts of Thane and Kolaba where the effects of the famine were, in fact, least. The protest had strong links with Bombay, for many political activists of the city had estates in these two districts. To a large extent it was a movement by speculators who had bought up land from destitute peasants in the preceding years, rather than of the peasants themselves. However, it linked up with a strong protest movement by the local peasants against the new forest and liquor laws, which had caused them extreme hardship. Five thousand adivasi peasants met at Vasai in Thane district in November 1896 to protest against the forest laws, and in the following month there was a riot at Kelve-Mahim in the same district agains the liquor laws.[98]

The other areas of strong protest were Khandesh and Dharwar districts. In Khandesh the movement was in a prosperous cotton-growing region in the heart of the district. The local sahukars—mostly Marwaris and Gujarati Banias—refused to pay the land-tax after the failure of the harvest in 1896. It was common in this area for much of the land-tax to be paid directly to the authorities by the sahukars. They took back the money from the peasants with added interest at harvest time. As in 1896, when there was no prospect of a proper harvest, the sahukars knew that they would not get back their money and therefore refused to pay the tax. In Dharwar district the Lingayat peasantry—many peasants combined farming with trade and moneylending—refused their tax. There was a weaker movement in western and central Pune district, where the Maratha peasantry held out against payment. There was no protest in the more impoverished and

[96] This area was under the rule of the Sindhias of Gwalior until 1860. This was the 'former Government' referred to here.

[97] District Deputy Collector's report for Dahod and Jhalod Talukas, 1877-8, Maharashtra State Archives, Bombay:, Revenue Department 1878, 15/1024.

[98] Richard Cashman, *The Myth of the Lokamanya: Tilak and Mass Politics in Maharashtra* (Berkeley: 1975), pp.133-6.

worst-hit eastern part of the district. The general pattern in 1896-7 was—
in contrast to the 1873-4 movement—for protest to be strongest in more
prosperous rather than poorer areas. This was in effect an agitation by
landlords and rich peasants against the land-tax policies of the Bombay
government.

The authorities were extremely worried by these tax-refusals, largely
because of the support they received from middle-class activists of
Bombay and Pune (we shall deal with this aspect in section VII of this
Introduction). They therefore took firm measures to suppress the agitations. In Khandesh they confiscated the land of the richest sahukars who
had not paid tax on the land they owned. They also raided villages, seizing
property in lieu of tax. The sahukars capitulated immediately, which
demoralized the peasants and caused a quick collapse of the movement. In
Thane district some leading tax refusers were arrested, which frightened
the rest into paying up. In both areas the movements had been broken by
the end of January 1897.[99]

In 1899-1900 Gujarat suffered a severe famine, and again the government demanded that land tax be paid with very little remission. Peasants
throughout the region protested by refusing to pay their land tax. They
were led by the more prosperous peasantry and solidarity was maintained
through caste boycotts of those who paid their tax. Seventy per cent of the
landowners of Kheda district refused to pay their tax during the course of
the year.[100] In Surat district meetings were held, peasant unions were
organised, funds were collected, petitions circulated, and a deputation sent
to Ahmedabad to protest to the commissioner.[101] There was similar tax
refusal in Broach district. In all cases the leaders of the protest were local
people, mostly more substantial peasants; no middle-class urban agitators
played any active role. Government officials broke the movement, in part
through physical coercion—as was revealed in the enquiry by E. Maconochie in Broach district[102]—and in part by threatening to confiscate and sell
the land of the leaders of the movement. Once it became clear that the

[99] *Ibid.*, p.142.
[100] David Hardiman, *Peasant Nationalists of Gujarat: Kheda District 1917-
1934* (New Delhi: 1981), p.57.
[101] Collector's report for Surat District and Deputy District Collector's report
for Olpad, Mandvi and Bardoli Talukas 1899-1900, Maharashtra State Archives,
Bombay, Revenue Department 1901, 55/137.
[102] *Report on Revenue Collection in Broach District* by E. Maconochie, 1901,
pp.25-6 and 64-5. (Published report in Maharashtra State Archives, Bombay; title
page and publication details missing.)

government was serious in its threats, most of these leaders paid up. This demoralized the rest of the peasants, who then paid up themselves.

The moral victory was not, however, to be with the Bombay government. The two commissions which examined the famines of 1896-7 and 1899-1900 criticized it for its harsh and unfeeling attitude towards the peasants during these famines. They advanced three main criticisms: first, that land tax in Bombay was too high and too rigidly applied; second, that remissions were not given in an adequate and proper manner during famines; and third, that the Bombay officials were out of touch with the peasantry. These criticisms were accepted by the Viceroy, Lord Curzon, and the Secretary of State, Lord Hamilton. Although the Bombay government tried hard to defend itself, the evidence against it was strong, and it agreed eventually to change its rules to allow remission of land tax to be made more generously during famines.[103] The mentality of the officials did not however change. Almost exactly the same problems were seen in Kheda district in 1917, when heavy rain destroyed the crops and the local officials refused to remit land tax in an adequate manner. The result then was the Kheda Satyagraha, the first major Gandhian agitation amongst the peasants of Gujarat.[104]

There are, unfortunately, no good essays concentrating on this series of no-tax movements in Bombay Presidency to include in this collection. There is, however, an essay on another important no-tax movement during this period, which occurred in the Punjab in 1907. This is reprinted here. Ironically, those who had attacked the Bombay government for its insensitivity towards the peasants were in most cases admirers of the Punjab government. The so-called 'Punjab tradition' stressed the importance of being in close touch with the needs of the peasants,[105] keeping taxation at reasonable levels, and providing positive protection to peasants from moneylenders (the latter concern was given a strong legislative form with the Punjab Alienation of Land Act of 1900). The Punjab authorities exercised their authority according to local needs and customs, rather than being bound by tight regulations. This gave the local officials great power, which they claimed they exercised in a benevolent and paternalistic manner. By the first decade of the twentieth century a certain complacency had however developed, as was to be revealed in the no-tax movement of

[103] Cashman, *Myth of the Lokamanya*, pp.138-41.
[104] D. Hardiman, *Peasant Nationalists of Gujarat*, Chapter 5.
[105] P.H.M. van den Dungen, *The Punjab Tradition: Influence and Authority in Nineteenth Century India* (London: 1972).

1907.

The agitation was by peasants in the Chenab Canal Colony against a proposed law to allow those who broke canal water rules to the fined, and also against a steep enhancement of the tax on water. The peasants held that the authorities were already abusing their powers and that legislation would make matters worse. Also, they complained that their right of ownership to the land they cultivated in the canal colonies had never been clearly defined. What they wanted, therefore, were clearer laws and regulations to check the arbitrary misuse of authority by officials. The movement was started by the peasants, with richer peasants providing leadership. Middle-class nationalists became involved after the agitation had started. At first, the Punjabi authorities refused to believe that 'their' peasants could be so disloyal, and tried to put all the blame onto these middle-class agitators. The Governor of the Punjab even believed that the movement was a part of a plot to overthrow British rule in India. He therefore deported the nationalist leaders. Contrary to expectations the movement did not then collapse. At this point the Government of India intervened, carrying out an investigation which revealed the Punjab authorities in a bad light. The Government of India then vetoed the legislation which provided for a system of fines for water abuse. It also forced the Punjab government to give full proprietary rights to the peasants in the canal colonies. In this case, in contrast to Bombay, the peasants scored a quick victory. The chief reason for this was that the British relied on the region for the recruitment of a high proportion of soldiers for their army in India, and the Government of India did not want a continuing agitation which could have shaken the loyalty of Punjabi toops. No such consideration applied in Bombay Presidency.

The essay by N. Gerald Barrier reprinted in this collection gives an excellent description of the government reaction to the movement, both at the Punjab and Government of India levels. The description of the peasants' involvement is, however, sketchy. That there was more to the peasants' defiance than canal grievances alone is shown in an article by Ian Catanach on the great plague epidemic which swept India in the 1890s and 1900s. In 1906-7 three per cent of the population of the Punjab died of plague. In May 1907 an itinerant sadhu was prosecuted for spreading a rumour that the government was spreading plague by having special pills thrown into wells and streams. As a result of many such rumours, villagers were at the time guarding their wells. The Secretary of State for India, John Morley, was amongst those who believed that the plague epidemic had added to the anti-government feeling in the Punjab at the time. Catanach also mentions other catastrophes and unusual events suffered by the Punjab peasants at that time. In early 1907 there was a plague of locusts, hailstorms, and rust in the wheat crop. There was a fearfully hot wind before the coming of the monsoon. This series of misfortunes was read as an omen that times were out of joint. This created a fertile ground for the agitation of that year.[106] Another factor which must have been important in the protest was the growing assertiveness of the Sikh Jat peasantry of the Punjab, an assertiveness which was to find a voice in the Akali movement of the 1920s.[107] None of these aspects of the movement are mentioned in Barrier's essay.

The no-tax movements in the Punjab and Bombay Presidency were not the only ones in India during this period. Movements were reported from Awadh in 1879, Cambay State in Gujarat in 1890,[108] Tanjore district of Madras Presidency in 1892-3,[109] and Assam in 1893-4.[110] As elsewhere, solidarity was maintained through the threat of social boycott. In several of these cases the movements turned violent, with tax officials being beaten up in Awadh and peasants rioting in Cambay and Assam. Unlike the movements in the Punjab and Bombay, these other movements did not receive strong support from urban nationalists.

VI

Our fifth area of conflict, which forms the subject of the last essay in this volume, is that between peasants and forest officials. When the British gained control over India in the late-eighteenth and early-nineteenth centuries, large areas were under forest. These forests provided a liveli-hood for a wide range of people. Some lived by hunting and gathering, others by shifting cultivation. Communities of artisans lived in the forests, such as basket weavers—who used bamboo as their raw material—and

[106] I. Catanach, 'Plague and the Indian Village', pp.225-6.

[107] Richard Fox, *Lions of the Punjab: Culture in the Making* (Berkeley: 1985).

[108] Ian Copland, 'The Cambay "Disturbances" of 1890: Popular Protest in Princely India', *The Indian Economic and Social History Review*, 20:4, Oct.-Dec. 1983.

[109] David Washbrook, 'Political Change in a Stable Society: Tanjore District 1880 to 1920', in C.J. Baker and D.A. Washbrook, *South India: Political Institutions and Political Change 1880-1940* (Delhi: 1975), p.41. This appears to have been more of a landlord than a peasant movement. In Tanjore most of the land was under the control of *mirasdars*, a rentier class.

[110] Sumit Sarkar, *Modern India 1885-1947* (New Delhi: 1983), p.53.

iron smelters. Many pastoralists grazed their cattle, sheep and goats there. The forests also provided raw materials, particularly wood, for peasants and artisans who lived outside their boundaries. In years of dearth they could gather food and graze their cattle in the forests. The forests were thus not only a home for large numbers of people, but also a valuable resource for society as a whole.

During the course of the nineteenth century the British government gradually restricted access to the forests. Although environmental considerations were not entirely absent,[111] their chief motive for doing this was to assure a continuing supply of wood for imperial needs. The British needed timber, first for constructing their great colonial cities and building ships, and then for building and running the railways. By the 1860s railway construction required about a million sleepers a year, and large amounts of wood were needed to fire the boilers of the railway engines.[112] The first measures were taken in 1849, when the Governor-General, Dalhousie, ordered certain wastelands to be formed into government estates. In 1852 he issued instructions for the creation of fodder, fuel and forest reserves.[113] The Imperial Forest Department was created in 1864 and the first Indian Forest Act was passed in 1865. This allowed large areas of forest to be demarcated as reserves in which the government would have a monopoly right to all produce. The government was considered to have full right of ownership over these reserves.[114] Over the next decades the forests were surveyed, the reserves mapped out and marked, and the rights of the existing forest dwellers set down on paper. In the process, many of their customary rights were taken away. The second Indian Forest Act of 1878 gave even greater powers to the government. The Act allowed the authorities to take unoccupied or waste lands belonging to villages into the reserved forests, effectively depriving villagers throughout India of their common lands. Forest dwellers and pastoralists were allowed use of the

[111] Richard Grove, 'Conservation and Colonial Expansion: A Study of the Development of Environmental Attitudes and Conservation Policies on St. Helena, Mauritius and in Western India, 1660-1860', Ph.D. thesis, Cambridge: University, 1988.

[112] Ramachandra Guha, 'Forestry in British and Post-British India: A Historical Analysis', *Economic and Political Weekly*, 29 October 1983, pp.1883-4.

[113] Neeladri Bhattacharya, 'Colonial State and Agrarian Society', in R. Thapar and S. Bhattacharya, *Situating Indian History* (New Delhi: 1986), pp.117-18.

[114] Richard Tucker, 'Forest Management and Imperial Politics: Thana District, Bombay: 1823-1887', *The Indian Economic and Social History Review*, 16:3, July-Sept. 1979, p.281.

forest as a 'concession' granted by the grace of government, rather than as a right. Generally, the people were allowed only limited access to the forest, and they were denied all power of management over it.

The forest reserves—which by 1900 covered over twenty per cent of India's land area—were thus created contrary to the needs of peasant farmers, pastoralists and forest dwellers. In the course of decades the old symbiosis between agriculture, cattle husbandry and forests was destroyed, being replaced by an antagonistic relationship between agriculture-cum-husbandry and commercial forestry. Trees which had value to the peasants for their fruit, fodder and other produce were replaced by trees which had a high commercial value.

Peasant resistance to this vast appropriation took many forms. Many forest dwellers merely voted with their feet, migrating to areas in which the forests were not yet reserved (often princely states—though these too gradually implemented British-style forestry measures). Those who remained often refused to recognize the validity of the government's claims, continuing to cultivate at will, cutting trees and burning the forest floor (an old practice which helped the growth of fresh grass each year). In this manner was created a whole new legal category of 'forest crimes'. In some cases there was wanton destruction of the forest, due to a feeling that if large trees were allowed to grow the forest department would be the chief beneficiary.[115] In this manner the erstwhile protectors of the forest became its destroyers, alienated from the environment in which they lived and losing their sense of responsibility to the forest.[116] There was also much resentment against the lower-level forest officials, who acted as a new kind of police force. They demanded gifts and services during their tours and threatened to lodge cases of violation of forest laws unless the people paid them money.[117] Such officials were often assaulted; in some cases even killed.[118] During the late nineteenth century this resentment fed into several major risings against the British by forest dwellers, such as the Rampa and Gudem revolts, the peasant movement in Thane district in

[115] Ramachandra Guha, 'Forestry in British and Post-British India', p.1885.
[116] Ramachandra Guha, *The Unquiet Woods: Ecological Change and Peasant Resistance in the Himalaya* (New Delhi: 1989), pp.56-8.
[117] R. Tucker, 'Forest Management', p.290.
[118] N. Bhattacharya, 'Colonial State and Agrarian Society', p.126.

1896,[119] and the Bastar revolt of 1910.[120] In Uttarakhand in the Himalayas, the imposition of forest laws provoked a series of non-violent agitations from 1906 onwards.[121] It was not, however, until the Civil Disobedience Movement of 1930 that this resistance linked up with the nationalist movement in any very significant manner.

Because of this resistance at many levels, the British were not able to create forest reserves free of people, as they would have wished. They had to allow a limited access to the forest, and in certain areas set aside reserves in which shifting cultivation was allowed so long as the peasants allowed 'valuable' timber trees to be grown alongside their crops. The British were also unable to expand the area of reserved forest far beyond the boundaries designated in the late nineteenth century. In many cases they were forced to grant grazing rights to pastoralists. Thus, by the 1890s, about eighty per cent of the forest area in the Punjab hills was allowed to be kept often for grazing.[122] In the forests of Kangra (now in Himachal Pradesh) villagers were allowed a share in the earnings from timber so as to win their support for the government's forest policy.[123] In Kumaon division of U.P., the forest department changed its policy in 1916 to allow planned and controlled firing of the forest floor each year by villagers.[124] These gains were, however, small compared to the massive loss suffered by peasants, pastoralists and artisans throughout India as a result of the loss of their customary rights over the forests.

This momentous historical development was neglected by historians until recently. As a result, the range of studies available at present is not very extensive. The person who has more than anyone else opened up this subject has been Ramachandra Guha, who has carried out important research on resistance to the forest laws in the U.P. hills during the twentieth century. This has recently been published as a book, *The Unquiet Woods* (Delhi: Oxford University Press, 1989). For this collection I have chosen an overview of the subject written by him and Madhav Gadgil, his senior and a major Indian environmental scientist. Although generally an

---

[119] *Source Material for a History of the Freedom Movement in India*, Vol.II (Bombay: 1958), p.637.

[120] E. Clement Smith, 'The Bastar Rebellion, 1910', *Man in India*, Rebellion Number, December 1945.

[121] Ramachandra Guha, *The Unquiet Woods*, p.71.

[122] N. Bhattacharya, 'Colonial State and Agrarian Society', p.129.

[123] *Ibid.*, p.130.

[124] Ramachandra Guha, *The Unquiet Woods*, p.53.

excellent presentation, two criticisms can be made. The first is that the paper underestimates the extent to which colonial forest policies were dictated by a desire to preserve the forests of India. The growth in population and development of railways, cities and industries were placing an unprecedented pressure on the forests, and protective measures were certainly needed. The measures taken were however bad, showing as they did an arrogant contempt for the experience of communities which had lived and worked in the forests for centuries. The other criticism is that the two authors refer to James Scott's concept of 'everyday resistance' with approval, even though the evidence in the paper contradicts it. James Scott has argued that undramatic day-to-day resistance by peasants against oppressors wins far more in the long term than outright revolt.[125] Gadgil and Guha's article reveals the importance of both covert and overt resistance. If anything, they tend to show that peasants who campaigned openly against forest restrictions won fuller concessions from the authorities than those peasants who relied merely on 'everyday' forms of resistance which minimized the element of confrontation with the state. In fact, those who failed to resist in a positive manner ended up in many cases as impoverished and marginalized groups in areas of settled peasant agriculture.[126]

VII

The attitude of the newly-emerging Indian middle classes towards these peasant movements was, as a rule, ambivalent. They were torn in two directions. On the one hand, they were creatures of British rule—English educated and serving in the bureaucracy and liberal professions. Some also owned landed estates or worked for landlords or lent money to peasants. All of these occupations depended on the protection of the colonial state. They normally came from a high-caste background and had little empathy with, or respect for, the peasants. On the other hand, they were in the process of forging a new ideology of nationalism, within which they wanted to be seen as the true representatives of the people. During these years we see this class groping its way towards such a position. On the whole, this enterprise was carried on in a spirit of compromise and timidity. There were, however, some exceptions—men such as Phadke,

[125] James Scott, *Weapons of the Weak: Everyday Forms of Resistance* (New Haven: 1985).

[126] See Gadgil and Guha, p.175.

Tilak, Lajpat Rai—whose militancy and courage carried the movement forward in a powerful manner.

The middle classes had an often undue faith in the intrinsic good of British rule in India. Abuses in the system were commonly attributed to the maladministration of particular 'bad' officials. They believed that the peasants were best served by resorting to the courts of law established by the British.[127] Both of these elements of middle-class thought were demonstrated during the Pabna movement of 1873, when the intelligentsia of Calcutta condemned the peasants for their 'violence' and advised them to seek redress in the courts rather than take direct action against the landlords.[128]

During the indigo revolt of 1869-72 the general sentiment of the Calcutta middle class was against the planters, who were notorious for their racial bigotry and supposedly 'un-English' ways. However, only a few lawyers actively sided with the peasants, and the Christian missionary James Long, who championed the peasants' cause, complained that the Calcutta intelligentsia was extremely selfish and indifferent, if not hostile, to the welfare of the common people.[129] During the Pabna movement of 1873 the Calcutta middle classes were passive, and often disapproving, towards the peasants in their struggle. Bankim Chandra Chattopadhyay went so far as to condemn the movement in his paper *Bangadarsan*, and he demanded that the popular pro-peasant play *Zamindar Darpan*, written by a Muslim village poet in peasant dialect, be banned.[130] Only with the establishment of the Indian Association in 1876 did a stronger middle-class voice begin to be heard in Calcutta in favour of the peasants. In the 1880s this body organized meetings of peasants to press their demands against the landlords.[131]

The middle-class intelligentsia of Pune was, in contrast with its counterpart in Bengal, more radical. The Pune Sarvajanik Sabha, which was founded in 1870, took up the question of the growing impoverishment of the peasants under British rule from the start. One of the leading members, M.G. Ranade, developed a strong economic critique of British rule in

---

[127] See Ranajit Guha, below.

[128] K.K. Sen Gupta, *Pabna Disturbances*, p.123.

[129] B. Kling, *The Blue Mutiny*, p.197.

[130] Sunil Sen, 'Peasant Struggle in Pabna 1872-73', *New Age*, 3:2, Feb. 1954, pp.73-4.

[131] Partha Chatterjee, *Bengal 1920-1947*, pp.12-13; K.K. Sen Gupta, *Pabna Disturbances*, p.141.

India.[132] Ravinder Kumar has described, in the essay reprinted in this volume, how this association championed the peasants of Maharashtra in their struggle against enhancements in land tax in 1873-4. Activists went to the villages to encourage the peasants in their stand and explain to them the underlying cause of their distress. This was, however, an essentially moderate and legalistic body, as was revealed when its leaders condemned the peasants for taking direct action against their sahukars in the riots of 1875. Many of these leaders were from a sahukar background, and they were clearly frightened by the peasants' militancy.

One of the most remarkable attempts by a middle-class activist to place himself at the head of a peasant revolt occurred in 1879. Vasudeo Phadke was a Chitpavan Brahman of a village of Kolaba District.[133] In 1862 he found employment as a clerk in the Military Finance Office in Pune. During the 1870s he started an organization in which secret military training was carried out for a future revolt against British rule. Phadke developed contacts with low-caste peasants in the area to the south-west of Pune, whom he believed would provide mass support for his rebellion. In February 1879 he formed a peasant army about two hundred strong and began attacking rich villagers to secure weapons and money. In many cases, wealthy moneylenders were the object of attack. Because these moneylenders were extremely unpopular, Phadke and his peasant army received support and shelter from the villagers in the mountainous region to the south-west of Pune.

Phadke was inspired in part by the example of the rebels of 1857, but in particular by Shivaji, who had begun his revolt against the Mughals in these same hills, supported by the local peasantry. The stories and songs of these peasants were full of tales of outlaws who had championed the interests of the peasants and, in time, become great princes. In 1879 rumours spread that an avatar of Shivaji had appeared in the form of Phadke. Phadke issued a proclamation from his hiding place demanding lower taxes for the peasants, more adequate provision for them in time of famine, encouragement for Indian trade and commerce, and lower slaries for Europeans. He sent emissaries to different parts of India to encourage local activists to rise and destroy railways and telegraph lines and liberate prisoners. He failed to get any response in this respect. In May, Phadke threatened to begin looting Britishers, as well as rich Indians, and even

---

[132] V.S. Joshi, *Vasudeo Balvant Phadke: First Indian Rebel Against British Rule* (Bombay: 1959), pp.21-2.

[133] The next two paragraphs are based on *ibid.*, *passim*.

threatened to massacre them so as to precipitate a revolt on the scale of 1857. Scared by this threat, Europeans living in small towns in the region sent their wives and children to Bombay and Pune for safety. By May, however, police counter-measures were beginning to tell, forcing Phadke to flee from the area in which he enjoyed peasant support. He was captured in July, tried and transported to Aden, where he died in 1883, aged 37. In sustaining his revolt for so many months in the face of great odds, Phadke's was a remarkable achievement, and he became a legend in the region. He failed, however, to shake British power there in any significant way.

The Indian National Congress, which held its first meeting in 1885, took an extremely cautious line on the peasant question. During its first thirty years it never tried to act as an agitational body; its aim was to win concessions from the government through reasoned appeals. Although the delegates to the Congress sessions generally accepted that the peasants were suffering increasing impoverishment, the chief blame for this state of affairs was placed on high land-tax demands and the 'drain of wealth' from India to Britain. Even rural indebtedness was attributed to these causes, rather than to the operation of British law and the introduction of individual property rights. The moneylender and landlord classes were largely absolved of blame. Neither Ranade, Gokhale nor Tilak favoured legislation to protect peasants against moneylenders, and the latter on several occasions even made strong pro-moneylender statements.[134] In the Congress sessions of 1893 and 1894 measures proposed by the Bengal government to improve tenants' rights were strongly attacked.[135] At the 1900 session of the Congress many delegates from the Punjab opposed the measures being introduced by the Punjab authorities to prevent the alienation of land to moneylenders. The majority of Punjabi Congressmen were members of the moneylending classes, and their paper *The Tribune* vehemently defended moneylending interests. However, because some Muslim delegates from the Punjab opposed it, no resolution condemning the legislation was put forward.[136] Generally, up until the First World War, the Congress shied away from championing peasant grievances in any positive manner. No attempts were made to discuss radical solutions to rural problems, such as land redistribution, a ceiling on holdings or land

---

[134] See G.P. Pradhan and A.K. Bhagat, *Lokamanya Tilak* (Bombay: 1959), p.134; Cashman, *Myth of the Lokamanya*, pp.147-9.

[135] John McLane, *Indian Nationalism and the Early Congress* (Princeton: 1977), p.228.

[136] *Ibid.*, pp.249-54.

to the tillers. The Indian social structure was never blamed for causing rural poverty.[137]

The timidity of the Congress in this respect was revealed very sharply in 1887, when the leading organizer of the Congress at that time, Allan Octavian Hume, tried to take the movement to the peasants. He printed thousands of pamphlets in twelve Indian languages, which were distributed in rural areas. Although the pamphlets adopted a harsh tone, their demands were moderate. Over a thousand meetings were organized in rural areas to propagate the message of the Congress. The government reacted sharply, accusing the Congress of trying to arouse hatred of the British and attempting to unleash forces which they could not control. Hume argued that unless a safety-valve was found for peasant discontent there would be a flood of future violence. Hume had not received advance approval for his campaign from other Congress leaders, and they had a resolution passed at the Allahabad Congress of 1888 which disclaimed all responsibility for anything their officers and members did or said outside the annual Congress session. They did not, however, pass any resolution formally repudiating Hume's pamphlets. In 1892 Hume again tried to win the Congress leaders over to his demand for a peasant campaign. He warned them in a letter that unless action was taken soon there would be a violent and bloody peasant revolt in India. He prophesied that well-to-do Congressmen would be among the first victims of this revolt. The letter caused a sensation in both India and Britain. Other Congress leaders took strong exception to it, openly disassociating themselves from Hume's views. The rupture which occurred as a result between Hume and most other Congress leaders effectively ended his career as a Congress leader in India.[138]

One of the few Congressmen who was sympathetic to Hume's analysis in 1892 was B.G. Tilak. He and a few other nationalists, such as Lala Lajpat Rai, believed that the nationalist movement had become controlled by self-serving people who could never provide for the masses a worthy example of patriotism and service to the nation. They believed that the leaders should devote their full life to the struggle for the political and cultural regeneration of India. Tilak argued that 'we will not achieve any success in our labours if we croak once a year like a frog.'[139] Besides organizing popular cultural events like the Ganpati and Shivaji festivals, Tilak sought

[137] *Ibid.*, p.242.
[138] This paragraph is based on *ibid.*, pp.118-21.
[139] *Ibid.*, p.158.

to take the movement to the peasants during the famine of 1896. Soon after the failure of the monsoon in Maharashtra, he dispatched activists to the villages to make enquiries as to the extent of the disaster and to encourage the peasants to demand their rights. His lieutenants combed the villages of the Deccan and Konkan between October 1896 and April 1897, visiting every district except Kanara. Detailed enquiries were carried out about the state of the crops and to see whether the government had remitted taxes and provided relief. Demands were made for tax remission in the case of crop failure, and petitions making this demand were printed in Marathi and circulated. Although the activists were careful to keep within the law, the British accused them of sedition and arrested and jailed some of them in December 1896. As we have seen in section V, the Bombay authorities took harsh measures to crush the agitation, breaking it effectively in early 1897.[140]

The only other area in which middle-class nationalists gave strong support to a peasant movement during this period was in the Punjab in 1907. The movement was not started nor encouraged initially by the nationalists; they joined it only when it had got well under way. Leaders such as Lala Lajpat Rai and Ajit Singh toured the affected area, making fiery speeches. Ajit Singh called for the peasants to refuse their land tax. He held meetings at which peasants who paid the new water rates were threatened with social boycott. Soon after, he and Lajpat Rai were arrested and deported. They were released after the government capitulated to the peasants' demands in May 1907.

Although both the Pune and Punjab activists were considered seditious by the British and persecuted, and even though neither received firm support in their actions from the Congress, both were vindicated in the stands they took. The Punjab activist's case was accepted by the authorities within the year, and Tilak's case was accepted implicitly by the Government of India in its Famine Commission Reports which criticized the Bombay government. The lesson was clear: middle-class nationalists could champion peasant grievances with considerable success and the provincial governments could be made to eat humble pie. Before 1917, however, no other leading nationalists attempted to follow the path charted by Tilak and Lajpat Rai. It was only when Gandhi took up the grievances of the peasants of Champaran in 1917 and Kheda in 1918 that this strategy became of central importance to the nationalist movement.

[140] This paragraph is based on Cashman, *Myth of the Lokamanya*, pp.126-7.

VIII

In only one case were the peasants completely successful in their resistance in the period 1857-1914. This was in their struggle against indigo planters, who were forced out of eastern India, never to return. In other cases, the systems of exploitation against which they were fighting persisted, though in an often modified state.

In Bengal, the struggle against the zamindars led to the consolidation of a new class of occupancy tenants, who themselves became a new landlord class, renting out land to sharecroppers. This led to new forms of resistance, seen most notably in the Tebhaga movement of 1946-7, in which the peasants demanded that they be allowed to retain two-thirds of the crop.[141] In the Deccan the resistance to sahukars led to the Deccan Agriculturists Relief Act of 1879, which made it harder for moneylenders to oppress the peasants who used law courts. However, sahukars continued to dominate local credit and produce markets in the region until the mid-twentieth century.[142] It was only after the winning of power by the peasant-backed Congress Party after the Second World War, with a resulting enactment of much stronger anti-moneylender legislation in 1947,[143] and the provision of new sources of co-operative credit and co-operative marketing and processing organizations (most notably the co-operative sugar factories), that the peasantry began to free itself from sahukar domination on a significant scale. A class of capitalist farmers subsequently began to emerge in Maharashtra in a strong manner.

The resistance by forest dwellers won some minor concessions—such as wider rights of grazing and permission to burn the forest floor during the summer months—but it failed to change the overall system, which continues to operate in a perverted form to this day. The forest officials who were meant to act as guardians of the forests have been unable to combat the money-power of timber contractors, which has led to over-exploitation and massive destruction of the forests. This has had grave climatic and ecological consequences, not only for the forest regions, but for the subcontinent as a whole. In this tragic process the only ray of hope has been the continuing resistance by some forest dwellers against the

---

[141] For an account of this movement see Adrienne Cooper, *Sharecropping and Sharecroppers' Struggles in Bengal 1930-1950* (Calcutta: 1988).

[142] For the continuing indebtedness of the Maharashtran peasants in the early twentieth century see Harnold Mann and N.U. Kanithar, *Land and Labour in a Deccan Village*, study no.1 (Bombay: 1917) and study no.2 (Bombay: 1921).

[143] This was the Bombay Agricultural Debtors Relief Act of 1947.

destruction of their forests. Increasingly, it is realized that the problem can be solved only in co-operation with these people, with them being given a role in the protection of the forests.

The resistance of landowning peasants to tax-officials had only limited success before 1914. Soon after our period, however, there was a series of major battles, most notably in Gujarat, where the peasants fought under the leadership of Gandhi and Vallabhbhai Patel. The turning point in this struggle was the Bardoli Satyagraha of 1928, which forced the government of Bombay to, in effect, cancel a steep enhancement of the land tax. Thereafter, the government dared not raise land tax significantly, and today it is, for most peasants, a merely nominal charge. Because the rural rich now have so much political power, no attempt has been made to make them pay income-tax on their earnings, even though many enjoy incomes far higher than most income tax payers in India. They, in turn, exploit low-caste or adivasi agricultural labourers, who are often migrants, in an extremely ruthless manner.[144] From this has developed a new form of resistance—that of agricultural labourers against capitalist farmers.

These were, in all cases, struggles on a broad front against various systems of exploitation imposed on peasant communities by the colonial state and its local allies, the planters, landlords and sahukars. Although each class amongst the peasantry had its own particular experience of such exploitation and its own particular grievances, solidarity in these struggles was achieved not by appealing to sectional interests, but by infusing a spirit of community solidarity against outside oppression. All sorts of bonds of clan, caste, tribe, kinship, religion and locality were invoked in building such solidarities. These community solidarities left their legacy for the future. The movements by the peasants of East Bengal and Malabar helped to consolidate new forms of Islamic identity, which fed the pan-Islamic movement which emerged in the twentieth century. The revolt of the Deccan peasants fed the movement against Brahmanical hegemony which developed in the following decades into a vehicle for the assertion of the Maratha-Kunbi community at an all-Maharashtra level. The agitation by the Punjabi peasants in 1907 can likewise be seen as a precursor of the movement of the Sikh Akalis, which developed rapidly in the rural areas after 1920. An important legacy of these movements was, therefore, the assertion of rural solidarities based on ties of community and religion.

[144] See Jan Breman, *Of Peasants, Migrants and Paupers: Rural Labour Circulation and Capitalist Production in West India* (New Delhi: 1985).

Unfortunately, few of the authors of the essays reprinted here discuss these movements in these terms.

Because these struggles were against specific forms of oppression of the peasants by the ruling class and its allies, these were not solidarities of the sort which are today known as 'communal', i.e. a solidarity based on a perception of group interests in narrowly religious terms. It would be incorrect, also, to argue that this sort of 'communalism' has been an inevitable outcome of peasant mobilization and resistance along lines of community in the past. Present-day communalism has, more than anything else, evolved in South Asia out of a particular history of nationalism. It has emerged from the call by populist politicians, working in the interests of the elites, for a religious identity to the nation. This involved in the first place a call for a 'Hindu nation', which was countered by the call for 'Pakistan', and, more recently, by the demand for the Sikh state of 'Khalistan'. The elites encouraged such appeals in the knowledge that this, of all forms of popular solidarity, was least threatening to their class interests. Modern religious communalism has, in other words, arisen chiefly from the needs of the political systems of modern South Asia, not from the legacy of peasant-community solidarity.

Against this, we may set the history of many twentieth-century agrarian struggles in which community solidarity has not been diverted along such lines. Such struggles continue to this day amongst the hill peasantry of U.P. who seek to save their trees from the axe of commercial foresters; amongst the peasants of Baliapal in Orissa who refuse to be ejected from their fertile lands for the sake of a missile test range; amongst the adivasis and peasants of the Narmada valley whose lands are to be swallowed by massive artificial reservoirs. These and many other struggles have kept alive the tradition of peasant resistance to a system of social organization—that of capitalism—which was brought into being in India by the colonial state and extended by its successors, the independent nation states of South Asia. In this resistance, communities which have, over centuries, evolved a way of life appropriate to a particular local environment, continue to fight a system which has sought to centralize power, destroy local communities, judge all values in terms of market forces, and apply abstract 'scientific' principles in a universal and often inappropriate and highly destructive way.

The peasant movements examined in this collection fit, in all cases, into this broad pattern of community resistance. However, because the forms of oppression—by planters, zamindars, sahukars, government officials—evolved and changed over time, the targets and methods of resistance also changed. For instance, the characteristic action taken against sahukars— that of the forcible appropriation and destruction of account books and papers—was specific to a period in which such documents could be used to control and dominate the peasants. Today, with rural usurers getting little protection from the law and courts, such a form of resistance is no longer appropriate. Likewise, European planters no longer exist, old-style feudal landlordism has either disappeared or is on the wane, and land tax has ceased to be an imposition of any weight on the mass of the peasants. Some of the old grievances remain, such as those against forest laws and forest officials, and resistance continues in a manner similar to that in the past. There are many new grievances—most notably against new forms of commercial farming and state development projects—which generate new forms of resistance.

The strength of such community-based resistance comes in part from the solidarity which such a form of mobilization and organization provides. Over and above this, however, is the assertion implicit in this resistance, the assertion that peasant communities have a right to exist in and for themselves. By their actions, the peasantry questions the claim of the ruling classes to have a right to override local interests for supposed 'wider' needs, such as 'national development' or 'progress'. They thus contest one of the chief legitimizing devices of the ruling classes. It is this assertion of the right to an alternative way of being which has given and continues to give such abiding power, resilience and importance to community-based resistance in India.

Chapter One

Neel-Darpan: The Image of a Peasant Revolt in a Liberal Mirror[1]

RANAJIT GUHA

The indigo plant, the original source of the dye used for bluing cotton textiles, formed the basis of a flourishing sector of commercial agriculture in Bengal by the beginning of the nineteenth century. From the very outset, however, the ryots, that is the tenant cultivators, were made to grow indigo under much coercion, for the surplus appropriated by the planters, mostly Europeans, and the methods they used, made this crop most uneconomic for the producers. A slump in the London prices of indigo between 1839 and 1847, the fall of the Union Bank of Calcutta, a consequent credit squeeze and the takeover of smaller concerns by larger 'indigo seignories'

[1] We have used the text of *Neel-darpan* (which means, literally, 'a mirror of indigo') as given in *Dinabandhu-Rachanavali* (Sahitya Samsad Edition, Calcutta 1967). Translations are our own. We have not used Michael Madhusudan Datta's rendering into English, because it is slipshod, inaccurate and altogether unrepresentative of the brilliant colloquialism of the better parts of the original.

Bibliographical references are indicated by sets of numerals enclosed within brackets. The arabic numerals preceding and following a colon stand respectively for the serial number of a publication listed up at the end of the essay and its page or pages. A roman numeral indicates, when it occurs, the volume number of the given publication.

The abbreviation, *RIC*, stands for *Report of the Indigo Commission Appointed under Act XI of 1860 with the Minutes of Evidence taken before them*; and *Appendix* (Calcutta, 1860). Figures followed by 'e' refer to answers by witnesses to questions put to them by the members of the Commission; those followed by 'r' indicate the relevant parts of the *Report* itself; and those followed by 'a' refer to an appendix in Part One of the *Report*.

A much abridged and rather different version of this study was published in the Calcutta weekly, *Frontier*, in December 1972.

increased the pressure on the ryot and his misery still further. By 1860 the regional grievances and localized acts of resistance among the peasantry snowballed into a general uprising in nine Bengal districts. Later on that year Dinabandhu Mitra published a play, Neel-darpan, with the planters' atrocities as its theme. Translated into English, the play soon became the focus of a legal and political contest between the Calcutta liberals and European planters. For over a century this play has enjoyed the reputation of a radical text. How radical was it? In an exegesis undertaken in answer to this question an attempt has been made in this essay to document the response of Indian liberalism to one of the mightiest peasant revolts in the sub-continent. A glossary of Indian terms has been inserted at the end of the essay.

> There appeared among the ryots a general sense of approaching freedom. They behaved as if about to be released from something very oppressive, and 'as if impatient of the slowness of the process.—W.J. Herschel, Collector and Magistrate of Nadia, in his evidence to the Indigo Commission (*RIC* 2819e).

I

Sivnath Shastri has recorded the impact Dinabandhu Mitra's play about the indigo ryots had on his generation: '*Neel-darpan* involved us all' [21:224]. One wonders why. The author was far from well-known yet. He was one of those young men, patronized by Ishwarchandra Gupta, who managed to have some bad prose and worse verse published in the *Sambad Prabhakar* between 1853 and 1856. Then he seems to have disappeared from the Calcutta literary scene altogether for a number of years until he surfaces again, in 1860, with a play published, far from the metropolis, in Dacca. All of which makes of this a pretty obscure story so far. Was it then the literary and dramatic merits of the work that made such an immediate impression on its contemporaries? The answer, alas, must be in the negative, if Bankimchandra Chattopadhyay's somewhat left-handed tribute[2] and Dwarakanath Vidyabhushan's unenthusiastic response [*11 IV: 689*] to the play are any measure of prevailing critical opinion. Was the impact,

---

[2] In his famous essay on Dinabandhu Mitra's life and works Bankimchandra Chattopadhyay appears to be torn between his critical judgement and his loyalty towards a dear, departed friend. The verdict he reached after a certain amount of beating about the bush, is perhaps best summed up as follows: 'Grantha bhalo hauk ar manda hauk, manushta bado bhalobashibar manush', meaning 'Whether he wrote well or not, he was a most loveable person' [*4: 833*].

then, due to the author's documentation of despotism and distress? Not likely, because there was nothing in *Neel-darpan* in terms of information that could not have been known already to the Calcutta intelligentsia. Many of them came from the indigo districts and continued to maintain close links with the villages there. Some of the leaders of the social and intellectual life of the city—the Tagores of Jorasanko and Pathuriaghata, for instance—were planters themselves. Moreover, an influential section of the press, owned and edited by the Bengalis, had been reporting the plight of the indigo ryots and the tyranny of the European planters at regular intervals throughout the eighteen-fifties. These reports could leave no one in doubt about what was going on in these parts of the province, particularly in Jessore and Nadia districts. Harish Mukherji's *Hindoo Patriot* came out very strongly indeed on this subject, and at least nine of those historic despatches from its brave Jessore correspondent, the young Sisirkumar Ghose, had already appeared in its pages between March and August 1860, that is, during the six months that preceded *Neel-darpan* [*2: 7-36*].

What, then, was it that made the publication of not-so-bright a play in the mufassil by not-too-well-known a writer on not-too-unfamiliar a theme appear like 'a comet on Bengal's social horizon' [*21: 224*]? The answer is simply that it was the European planters' reaction to the play that triggered the baboos' response to it. For, it is indeed curious that although *Neel-darpan* was published in September 1860, it was not until May 1861 that the Calcutta intelligentsia began to take any serious notice of it. During these eight months a number of things happened due to an apparently accidental lapse in communication between the Lieutenant-Governor of Bengal and the Secretary to the Government of Bengal on what was indeed a routine administrative matter [*3: 197-205; 14: 201-02*] and culminated in the planters' decision to make the play a cause for libel. It was at this point that the literati of Bengal came to realize that the defence of *Neel-darpan* could be made to look like the defence of the peasantry without anyone risking his head at the hands of the planters' lathials. So when James Long, inspired equally by his concern for the ryots and his eagerness to shield the Lieutenant-Governor from embarrassment [*14: 202*], stepped out to receive the day's crown of thorns, the leading lights of Calcutta rallied behind the good Englishmen (missionaries and officials) as against the bad Englishmen (planters). And thus *Neel-darpan* became the instrument—one could almost say, pretext—for the fabrication of a nice little middle-class myth about a liberal Government, a kind-hearted Christian priest, a great but impoverished poet and a rich intellectual who was also a pillar of society—a veritable league of Power and Piety and Poetry—standing up in defence of the poor ryot. Coming when it did, this myth did more than all else to comfort a bhadralok conscience unable to reconcile a borrowed ideal of liberty with a sense of its own helplessness and cowardice in the face of a peasant revolt.

II

The response to *Neel-darpan* during its first fifty years could be best characterized as liberal-humanitarian. It was liberal in the politics that inspired it and humanitarian in the idiom that expressed it. Bankimchandra Chattopadhyay, who knew the author very well, defined this idiom thus:

> Dinabandhu used to feel greatly moved at other people's distress and *Neel-darpan* was an outcome of this particular quality. It is only because he sympathized so completely with the misery of the ryots (*projas*) of Bengal that *Neel-darpan* was written and published. [*4: 825*]

It is this empathy that makes *Neel-darpan* the most forceful of Dinabandhu's plays, as Bankimchandra rightly observes [*4: 835*]. How forceful, can be judged from the anecdote about an outraged Ishwarchandra Vidyasagar hurling his sandal at an actor playing the role of the senior planter with a particularly vicious realism. This, as we know, never actually happened [*13: 594*]. Yet like all myths it refers to reality a few removes away. It is, in this case, an idealized description of the manner in which the middle-class audience would itself respond to the play with a mixture of pity and indignation: for Vidyasagar was known to be a most compassionate and fearless man.[3] Sivnath Shastri sums up the emotions evoked by the play when he says:

> *Neel-darpan* engrossed us all; Torap ran away with our affection; Kshetramoni's distress made our blood boil with anger; and we thought that if the planter Rogue fell into our hands, we would, in the absence of any other weapon, tear him to pieces with our sheer teeth.[*21: 224*]

Brave words. In reality, however, this show of righteous anger turns far

---

[3] One wonders if this should not be regarded as a classic example of myth resulting from a complete reversal of reality. For, we have it on the authority of Binodini Dasi that *Neel-darpan*, staged in 1875 by the Great National Theatre Company in Lucknow, was broken up by the European members of the audience when Torap, in the rape scene, forced his way into the planter's apartment and beat him up [*8: 17, 98-100*].

too soon into a tearful supplication. The liberal feels genuinely outraged by the tyranny of the planters. Their predatory ways violate his own attitude to the ryot, which is a curious concoction of an inherited, Indian-style paternalism and an acquired, western-style humanism. To the extent that with all his growing association with the big city he has, still in the eighteen-fifties, vital links with the village, he feels both his economic position (as one who lives off the peasantry) and his socio-cultural authority (as a member of the rural elite) threatened by the planters. He wants to stand up for the peasantry and in doing so defend himself. Yet he is unable to act on his own. He is ready to connive at the ryot arming himself against the planter. But he would not take up arms. That would be as illegal as the outrages committed by the planters themselves. The only way to end oppression is for the law to assert itself. It is the Government, the true custodians of the law, who alone can restore the rule of law. Hence, in a land of superstitions, the new theology of liberalism introduces yet another superstition to fit the politics, the morality and the sensibility of a colonial middle-class: corresponding to the illiterate peasant supplicating the gods against blight and drought we now have the highly literate baboo supplicating the local magistrate, the Lieutenant-Governor, the Governor-General or the Queen—the status of the member of the pantheon addressed depending on the degree of deprivation—for relief from the 'blue monkey' overrunning the countryside.

The stand taken even by the most advanced liberals like Sisirkumar Ghose illustrates this. His involvement with the jacqueries of 1860-61 in Jessore, the hottest district, went further than perhaps that of most other intellectuals of the time. Long before it became customary for eye-witness accounts of popular struggles to be fabricated in cool interiors around Chowringhee, Sisirkumar, at great personal risk, insisted on being where the action was in the deep countryside and reporting it to Haris Mukherjee for publication in the *Hindoo Patriot*. Scion of a rich talukdar family he had been involved in disputes over land with the planters long enough to know their ways. With his intimate knowledge of the Jessore district and his grassroot links with the peasantry he had no difficulty in reading the signs that were there, in the spring of 1860, of a rapid build-up towards armed conflict. The planters were driving the ryots up the wall by terror; the Government had moved in troops who joined the planters in hunting down the ryots; the latter were firm in their resolve not to grow indigo. Yet, hopefully, Sisirkumar described the mood of the ryots on the eve of the clashes as 'passive'. He was obviously indulging in wishful thinking when he wrote that the ryots 'will neither attack nor defend, but are ready to suffer [2:8]—a statement so Gandhian that it seems to have inspired Jogesh Chandra Bagal, the distinguished authority on nineteenth-century Bengal, to characterize the indigo peasants' revolt as 'the first non-violent mass movement' [2: 5]. In a letter narrating in detail how the ryots were getting ready to deal with repression measure for measure—villagers being arrested *en masse*, peasants in north-western Jessore pledging themselves never to sow indigo, nine hundred of them marching from Kalopole to Calcutta—Sisirkumar could still exhort the rest of the population in words like these:

> Rise, rise, ye countrymen with supplicating hands, fall prostrate before the Governor, catch his feet, and do not let him go unless he has granted your requests. [2: 8-9]

His faith in the raj remains unshaken even when he is himself made an object of official persecution. Incensed by his exposure of the collusion between planters and the custodians of the law, a vindictive magistrate starts a district-wide hunt for him forcing him to go to the ground. 'Yet', he writes, 'I hope Mr. Belli, one of the best of Judges... will not leave me unprotected'. When the good judge does not seem to be responsive, and the magistrate, failing to lay his hands on Sisirkumar, starts harassing the rest of his family, the prayer is simply addressed to the next higher authority, the Lieutenant-Governor 'who, I hope, will protect them [his family] from such unjust attack of the Magistrate'[2: 23, 28, 32].

In the event the Ghoses of Polua-Magura came out of all this less battered than the Basus of Swarapur whose faith in the raj did little to save them from disgrace and ruin as described in *Neel-darpan*. For the play, its author's intentions notwithstanding, is a clear indictment of the futility of liberalism as a deterrent to tyranny. It is the story of the failure of a liberal government to shield its subjects from oppression and of the liberals to defend themselves.

III

Dinabandhu was himself a distinguished product of that historic hot-house of liberal culture, Hindu College, Calcutta. He exemplifies much of mid-nineteenth-century liberalism both in his own life and that of his characters. To identify a Hindu liberal of this period is easy enough. One has simply to find out where he stood on some of those great questions relating to Hindu society that separated the reformists from the no-changers. As a firm opponent of child marriage and kulin polygamy, and equally firm supporter of Vidyasagar's campaign for legalizing the marriage of widows, Dinabandhu ranks high as an advanced liberal. These themes figure prominently in many of his plays. What, however, is even more impressive is the radical sentiment, dating from his days in the Hindu College, that anticipates the reformism of the later years in much juvenile verse and prose. A poem written in 1853 calls for defiance of customary prejudice—*lokachar*—on these social issues and appeals to reason:

> Once upon a time people used to believe that the sun goes round the earth once every year, whereas it has now been established, thanks to the advancement of knowledge, that it is the earth which goes round the sun. [*15: 412-13*]

It is less than flattering for us to recognize that in order to justify Reason against Authority we have to recall Copernicus for the very first time in 1853 in a manner which in the West had already turned into a platitude by then! Yet, it is this very awkwardness which is so attractive about it. It is the fumbling and fall of a new scepticism that has just come into being and not found its feet yet. Its weakness is congenital and it will never qualify for Tagore's imagery of the Puranic fledgling shooting up through space to devour the sun. For Reason is born spastic in a colony. This is precisely why one must be prepared to acknowledge the pathetic dignity of its attempt to assert itself. In Dinabandhu's case we have his admiration for physics blasting tunnels for railroads, for museums as the store-house of scientific objects and literature, and for Cautley's engineering which in the teeth of Hindu superstition forced the waters of the Ganges into a canal [*24: 10*]—all to testify to a sense of wonder about science as an agent of change [*15: 342-43, 380, 404*].

If change, technological as well as social, is an achievement of science, the latter is itself a gift of education which, as Bankimchandra was soon to point out, had already become synonymous with western-style education by this time [*4: 680*]. In recording his own enthusiasm about it Dinabandhu speaks for the great liberal illusion of his age regarding the ability of a bourgeois intellectual culture to free society from feudal ideas all by itself. In *Neel-darpan* we have the response of an affluent Kayastha family to the new education. The father who apparently has had no western-type schooling himself, is still very proud of Bindumadhab, his younger son, going to college in Calcutta [*Act IV, Sc. 3*]. He has in fact considerably added to his own status by getting for his son a bride from a leading suburban Kayastha family who, normally, would not marry their daughter off to a lad from an obscure village, but were persuaded to do so when they found out how well educated Bindumadhab was [*Act V, Sc. 1*]. The young wife shows something of her relatively urban and emancipated upbringing when she soliloquizes about the dull domesticity of her existence thus:

> We are born women. We are not allowed to go for a stroll as far as into our own garden even when we are a company of five girls together. We are not allowed to go for a walk in the town. It is not possible for us to set up welfare societies. We have no colleges, no law courts, no Brahmo Samaj to go to. There is nothing to entertain a woman when she is sad in her heart. [*Act II, Sc. 2*]

She is obviously familiar with the fruits of western culture herself, for she has been pestering her husband for a volume of Shakespeare's works translated into Bengali. The young man, writing to her from the big city, enthuses about the benefits of literacy, for 'I am able to speak directly to you even over such a great distance!' But communication, alas, has to be a one-way traffic: his mother would not permit her daughter-in-law to write to him [*Act II, Sc.2*]. For female education is still at this time anathema even among the more advanced Hindus, particularly those who were country-based, and it is a measure of the still unimpaired feudal values that a girl of her social standing must not be allowed to make her sentiments so public as to write privately to her husband![4] The women of the family, however, are not against putting her literacy to good use and have her read out the tales of Vidyasagar's *Betala Panchavimsati* for their benefit during the afternoon break between chores [*Act I, Sc.4*]. Finally, note must be taken of the importance Navinmadhab, the principal protagonist of the play and elder son of the family, attaches to education. He hopes, when the planters' depredations are over, to set up a school within the grounds of his own homestead: for, 'nothing could make me happier than

---

[4] The attitude to female education continues, apparently, to divide Hindu middle-class parents and children during the second half of the nineteenth century much as it did in the eighteen-twenties and thirties. How a child-wife of nine, married in 1826 to a boy of fifteen, pursued literacy surreptitiously in the midst of her domestic chores and with the sole encouragement of her husband, is described thus in *The Life of Girish Chunder Ghose ... by One Who Knew Him* (Calcutta, 1911): 'In the intervals between her culinary operations she used to scrawl on the kitchen floor the letters of the alphabet with a piece of charcoal and thus learn to write. She would afterwards carefully rub out the letters lest she should be detected in the forbidden occupation. Her only preceptor was her husband who was then a student of the Hindu College. During his hebdomadal visits to Konnagar she used to learn from him to read and write at a late hour of the night when feeling quite weary after the day's hard work... In the course of a few years she was able to read all the best books then existing in the Bengali literature...' [*1: 63*].

to have our native boys schooled in my own house; this would indeed be a true fulfilment of my wealth and my exertions' [*Act II, Sc.3*]. Apart from projecting the author's own enthusiasm for the new education—he ran away from his village home, when still very young, in order to avoid old-fashioned schooling at the local *pathshala* and to seek 'English education' in Calcutta [*21: 249*]—this confirms how by the middle of the nineteenth century, schools had already joined the list of traditional public works such as temples and tanks as an index of a landlord's munificence as well as of his obligation to the rural populace. It was partly thanks to such patronage and official encouragement (Navinmadhab found the Inspector of Schools most responsive to his idea) that a network of colleges and secondary schools spread fast throughout the indigo districts during the eighteen-fifties [*14: 61*]—a development regarded by the planters and their barely literate minions[5] with utter hostility. 'My Lord', says the indigo factory's dewan to his boss in *Neel-darpan*, 'the establishment of schools in the countryside has made the peasants more turbulent than ever', to which the planter replies, 'Yes, I must write to our (Indigo Planters') Association to petition the Government on this issue. We shall fight to stop schools from being set up.' [*Act I, Sc. 3*].[6]

As the reader familiar with the text may well recall, these extreme sentiments follow immediately upon an exchange between Sadhucharan, a principal ryot, and the officials of the Begunberey factory. In a cleverly engineered dialogue Dinabandhu drops a word that explodes like a bomb. The ryot's use of an elegant, polysyllabic, sanskritized word, 'pratapshali' (meaning 'mighty'), sets off a series of interlocking reactions. The dewan, an uppercaste Hindu, asks him to shut up, because for a common peasant to use *sadhubhasha* is to overreach himself. The planter picks up the cue and (in a sentence which schoolboys have made all their own: *banchat bada pandit hoiachhe*) admonishes him for his pretension to learning. The amin is outraged that a man from a family that works on the land should use a language so far above his status. But he is not surprised. Such

---

[5] The *Tattvabodhini Patrika* (1850) had the following to say about the education of the planters' *karmacharis*: 'A little arithmetic represents the upper limit of their education; they have never tasted true learning, nor have they acquired a moral education' [*11 II: 129*].

[6] The European planters' objection to education for the peasantry in 1860 on the ground that this would increase their resistance to indigo, is echoed twenty years later by the Bengali zamindars of Bakharganj because 'they are afraid that education would make the ryots more obstinate in withholding payments of their rents' [*6: 289*].

insolence is only to be expected of a trouble-maker like this particular ryot who had the cheek to advise the peasants on points of law relating to indigo and thus stir them up.

IV

It is this link between an awareness of the law and what looks even remotely like education that the planters detested so much. The amin was furious because he suspected Sadhucharan of wanting to sue the planter. 'This bastard wants to take this matter to the Court', he reports to his master [*Act I, Sc. 3*]. The fear was not unjustified. In the very first scene we find Navinmadhab pressing his father to take the Basu family's dispute with the factory to a court of law. Throughout the indigo districts the victims of the planters had been trying to seek the law's protection. A Collector of Jessore observed early in 1862 how the ryots 'had acquired a fondness for litigation' [*14: 189*]. And it was true, too, that a man who was literate had a better chance of securing justice than one who was not. Sisirkumar reported a case from Jessore in 1860 to illustrate this point. A number of peasants falsely charged with assaulting a factory official were all convicted in a magistrate's court, whereas a zamindar's naib, accused at the same time of leading those very men into precisely the same affray, was acquitted, 'no doubt the Magistrate not daring to punish a man so unjustly, who knows how to read and write' [*2: 13*].

No wonder that the educated Bengali found in the law a great ally and an object of the highest admiration. Throughout much of the nineteenth century he went on propagating, defending and popularizing it. He brainwashed himself and others in favour of what he believed to be its impartiality and its powers to defend the poor. He queued up for jobs in the civil service as its junior custodian. He passed the law examination and set himself up to interpret and practise it in various roles as barrister, advocate, vakil, mukhtar and attorney. He suffered his *samiti*, *samaj* and *sammelan* with lawyers. He recruited lawyers to his *dawl*. He made the bar library the centre of local politics. He elected lawyers to represent him at the Indian National Congress. He sent lawyers to lobby for him in England, a mission greatly admired by our author [*15: 404*]. Above all, he made law into the most lucrative of the liberal professions. In some verses ('Asha' in *Dwadas Kavita*) about an unemployed youth frustrated in his attempts to find a job, Dinabandhu leads him on to conclude:

Oh, I have been quite mistaken... I shall no longer go on supplicating those men of affluence... I shall devote myself to the study of law, pass the law examina-

tion, set up as a vakil, make money on my own and share it with the poor... This
will make my family happy;

and of course, it all ends happily:

> He studied law, took the examinations, passed, enrolled as a vakil, became
> prosperous and all his hopes were fulfilled at last. [15.401]

This association between law and literacy is far from fortuitous. These
are two of the more essential components of the culture, respectively, of
the British ruling class and of their Indian collaborators. They mutually
sustain each other. The institutional network of education and that of the
law soon become interdependent. Education helps in schooling the bureaucracy that runs the colonial apparatus which it is the task of the law to
define (by jurisprudence), regulate (by the judiciary), defend (by penalties) and rationalize (with the aid of the legal profession). They both help
the violence of the State to express itself in a cultural idiom. The two work
fairly closely together throughout the entire period of British rule in India.
It is true that as we move towards the end of the nineteenth century, and into
the twentieth, certain non-antagonistic contradictions arising from the
progress of a section of our bourgeoisie from infancy to adolescence, begin
to make their appearance. But there can hardly be a doubt about the
thoroughness of their collaboration in the earlier period.

The Bengali lawyers' attitude towards the indigo peasants' revolt offers
some evidence in this respect. In *Neel-darpan* the author builds up the
lawyers as defenders of the ryots against the planters. Wood is very angry
about a law graduate who writes against the planters in the press [*Act III,
Sc. 1*], and Navinmadhab, his principal enemy, is said to have worked
closely together with some vakils and mukhtars in a recent law suit against
the factory [*Act I, Sc. 3*]. Dinabandhu Mitra thus helps to promote a myth
about the morality of the legal profession.

The truth of the matter is that with the exception of a few individuals like
the mukhtar Titu Chakravarti whose incarceration caused much noise at
the time [5: 183-4; 14: 152] the lawyers sided with the ryots only so far as
it was either *safe* or *profitable* to do so. Thus, they figure prominently in
the list of signatories in the anti-planter petition addressed to the Lieutenant-Governor in 1854 from Nadia while the agitation was still at a low key.
A similar petition addressed to the Lieutenant-Governor five years later by
the people of the same district when it is already poised on the brink of an
armed encounter between planters and ryots, carries not a single lawyer's
signature on it [*14: 65, 76*]. There were, indeed, a number of legal
practitioners—a 'small number', according to Kling [*14:86*]—who helped

the ryots by acting for them in the mufassil courts and reporting on their plight to the Calcutta press. Most of them had their services paid for by Harish Mukherjee [*14: 121*], and the young man who, in *Neel-darpan*, became the cause of the planter's displeasure, might well have been one such mukhtar-correspondent of the *Hindoo Patriot*. Some lawyers did better than others and were said to be receiving a monthly salary of a hundred rupees each from the British Indian Association for pleading for the ryots, an allegation which, it must be recorded, the Association was unable to confirm [*14: 119-20*]. In fact, when at the height of the revolt the planters went on a rampage and the local European officials in many areas swung into their support and venal rewards were not much in sight, legal advice for the ryots was not easy to mobilize. Even Harish Mukherjee with all his immense prestige barely managed to recruit three lawyers from Calcutta and send them up to defend some peasants hauled up before the law in Nadia. No mukhtar in that district outside the *sadar* station 'could be induced to take up a ryot's case', a fact remarkable enough to be noticed both by the editor of the *Hindoo Patriot* and the Lieutenant-Governor of Bengal [*5: 183-84; RIC 3878 e, 3880 e*]. And in Jessore, the scene of many clashes, the local lawyers were, as we gather from Sisirkumar Ghose's biographer, deterred by the fear of the planters' reprisal from holding brief for the peasants [*19: 404*].

Whatever the truth of the lawyers' attitude to the peasant revolt of 1860, Dinabandhu Mitra's enthusiasm for the legal profession was quite in order. By speaking well of it, he was speaking up for the law itself. For *Neel-darpan* is dominated by the idea of legality. The opposition between good and evil is represented throughout the play as a fairly straightforward opposition between those who are for the law and those against. Navin-madhab and his family are for the law, lawyers and legal processes; Wood and his men are against all these. What complicates matters is the undoubted existence of bad laws. Act XI of 1860 which gave criminal jurisdiction to magistrates in civil cases of breach of contract and thereby made indigo cultivation into a form of 'forced labour' [*14: 137*], was precisely such a bad law. The planter gushes over it and regards it as an improvement on the *shyamchand* [*Act III, Sc. 1*], the dreaded instrument of torture used so often to break the peasants' resistance to indigo.

The bourgeoisie, in the period of their ascendancy in the West, made defiance of bad laws into a political virtue of the highest order and invested the best of their heroes with it. By contrast, for the leaders of our up-and-coming liberalism the law, however harsh or wicked, is sacrosanct: not beyond question, it is still something that must not be actively defied. They would bend backwards to be considerate about it and explain away its iniquities in terms of maladministration. That they could adopt such an attitude even to an obvious monstrosity like Act XI—nicknamed 'ryot coercion act'—should be particularly revealing for those who retail the myth of a 'progressive' intelligentsia acting as a firm ally of the indigo peasantry in 1860. Even for Harish Mukherjee with all his involvement in the cause of the indigo cultivators 'it was not the law itself but its administration that was crucial' [*14: 137*].[7] And Dinabandhu Mitra expresses a central tenet of the political philosophy of his class in the words of his hero, Navinmadhab, thus:

> What a cruel Act has been introduced! But why blame the law or the law-makers? It would not have been so disastrous for the country had the custodians of the law handled it without partisanship. Alas, how many innocent persons are crying their eyes out in prison, thanks to this particular Act!... O Lieutenant-Governor, none of these ills would have occurred if only you had matched your laws with the appointment of equally good men to administer them. [*Act III, Sc. 2*]

The merit of this attitude is that it opens up for the liberal an immense hinterland of compromise and reformism into which to retreat from a direct contest for power with the colonial masters. Once the law and the law-maker—the rulers and the sanction by which they rule—are thus exonerated, one can settle for minor adjustments to make the State apparatus perform more smoothly and less offensively. The adjustments may be sought in the form of additions to the statute book or of administrative pawns pushed around from square to square. And, thus, 'improvement', that characteristic ideological gift of nineteenth-century British capitalism, is made to pre-empt and replace the urge for a revolutionary transformation of society. In *Neel-darpan*, Navinmadhab pleads, appropriately enough, for the addition of a *dhara* (meaning 'article' or 'clause' rather than 'regulation', as Madhusudan translates it) to Act XI to make it hot for the planters [*Act III, Sc. 2*]. Appropriately too, all the planters'

---

[7] Harish Mukherjee advised the ryots strictly to abide by the law, even by the wicked Act XI of 1860, and to go on begging for redress. 'I invariably advised them', he said, `to apply to the district authorities in the proper form for redress and to go to the next appellate authority, if they found no redress at the hands of the district authorities. I cautioned them against ever committing any breaches of the peace or committing themselves in any manner by acting illegally. I explained to them that the operation of the Act was temporary, and that better measures would be devised next year' (*RIC* 3873 e).

victims in the play, from landlord to day labourer, are seen to be discussing the relative merits of good and bad officials all the time and, taken together, making out a general case for the Government to substitute wicked magistrates by nice ones. Fair enough, if this is all that you want out of a massive, well-organized, armed uprising of the peasantry.

V

In other words, it is the function of *Neel-darpan* to generate illusion about British rule in India as a good thing with only a few minor faults here and there that can be easily mended. Put so bluntly, it might hurt liberal sensibilities. These had fed for over a hundred years on its reputation as predominantly a play of protest. Bankimchandra called it the *Uncle Tom's Cabin* of Bengal [*4: 834*]. Priests, professors and politicians have been unanimous in their description of the work as *exclusively* an indictment of the planters' tyranny. One has to turn to the text to see for oneself how partial and misleading such a description is. For the author's aversion to the planters is equalled by his reverence for the raj. One is a measure of the other. The blacker the planters, the whiter the regime.

There was nothing self-contradictory about the mid-nineteenth century liberal being anti-planter and pro-raj at the same time. A united front of the *sarkar* and the baboo against the planters was almost a historical necessity. For by 1860 the predatory phase of British rule in India was coming to an end. Not many princely vaults remained to be plundered. Many of the most important strategic annexations had been already made. Colonialism had by then found its social base in a neo-feudal class of its own creation and its cultural base in an emergent middle class capable of combining traditional values with a received, western-style enlightenment. Reform had struck roots. The 'classes' were queuing up for B.A. degrees and jobs. The 'masses' went on fighting until the Titu Mirs and the Dudu Mians, the Sidhus and the Kanus, the Nana Sahibs and the Tantia Topis had all been driven out, exiled, hanged, blown up. It was time now for the erstwhile conquistador to settle down nicely and respectably in his estate. In this process the indigo planters were of no help at all. On the contrary, they were an embarrassment. For the consolidation of Britain's power base in India it was essential that the Government should acquire the image of a well-run concern based on legality, order and responsibility. The planters undermined all these. The *Sambad Prabhakar* had noted as early as 1853 that the planters behaved as if they were above the law [*11 I: 200*]. To Edward De-Latour who, as a district official, had seen the Farazis of

Faridpur being goaded into rebellion by the planters, the latter seemed 'neither to recognize the existence of a Magistrate on earth nor a God in heaven' [*RIC 3917 e*]. They flouted the law whenever it suited them and openly perverted its processes. Their private armies weakened the standing of the official law and order at the local level. Their indulgence in torture, murder, rape and arson made the natives question the superiority of the white man's religion, civilization and morality. The long-term interests of the raj, therefore, demanded that the planters should be disciplined. To the extent that this perspective was not yet clear to many, if not most, of the junior European officials, they could still be seen to act in collusion with the planters. But at the higher levels of authority the planters were often regarded with contempt, although pro-planter pressure groups did succeed from time to time—as for instance, in the case of Act XI—in forcing the hands of the Government. On the whole, the anti-planter agitation had all the official wind in its sails. This is quite clear from the complicity of the Lieutenant-Governor and the Secretary to the Bengal Government in the translation, printing and circulation of *Neel-darpan*. In view of this it is difficult to understand Bankimchandra's somewhat belated tribute to its publication as an act of exceptional courage [*4: 825*]. The way the author advertises his loyalty throughout the play makes it quite clear that he had no intention of harming the prospect of his professional advancement.

In a foreword (left out of Madhusudan's translation) Dinabandhu defines what he wants his play to achieve. This is to influence the indigo planters to turn from 'self interest' to 'philanthropy' so that 'Britain's face could be saved', for their cruel ways 'have given the British a bad name'. He condemns their greed, rapacity and hypocrisy, but notes hopefully 'the signs of a new dawn of happiness for the ryots'. These signs are described as follows:

> The great Queen Victoria, compassionate mother of the projas, considers it improper that her children should be suckled by her wet-nurses. So she has taken them up in her arms and is feeding them at her own breast. The even-tempered, wise, courageous and liberal Mr. Canning has become the Governor-General. The high-minded and just Mr. Grant who punishes the wicked, protects the innocent, and shares with the ryots their weal and woe, has been appointed Lieutenant-Governor. And truthful, astute, non-partisan officials like Eden, Herschel and others are gradually coming to blossom as lotuses in the lake of the civil service. It must, therefore, be clearly evident from all this that we have an indication now of the great souls mentioned above taking up soon the *Sudarsan* disc of justice in order to end the unbearable misery of the

ryots who have fallen into the clutches of the wicked indigo planters. [*15: 1*]

This declaration of faith in the ultimate triumph of British justice is based on the illusion that the law is fair in absolute terms and all one needs is the fairy godmother of an impartial bureaucracy to make it available for all. It is a measure of this illusion that the Basu brothers in *Neel-darpan* remain unshaken in their faith in the Lieutenant-Governor and the Commissioner despite the altogether unwarranted imprisonment of their ageing father. When someone suggests that the magistrate who sent old Golok Basu to jail is no more wicked than the Commissioner, Bindumadhab flares up. 'Sir,' he says, 'you talk like this about the Commissioner because you don't know him well enough. The Commissioner is very impartial indeed. He cares for the advancement of the natives'. He and his brother are both convinced that the Governor is just on the point of quashing the sentence so that their father can be set free [*Act IV, Sc. 2*]. The curtain goes up on the next scene to reveal Golok Basu's corpse hanging in his cell.

In real life, too, we find some of the most radical of the contemporary liberals sharing the same sort of illusion. Even Harish Mukherjee appears to have been enthusiastic about the civil servants as late as February 1860 when the stage was already set for the first round of jacqueries [*14: 149*]. Sisirkumar Ghose's despatches to the *Hindoo Patriot* document this illusion in detail. In mid-August, 1860, he writes bitterly about the 'partiality' of M.G. Taylor, a magistrate recently posted in Magura. In another despatch the next week he talks about Mr. Taylor having 'considerably changed' for the better. His letter of 19 December 1860 is a long catalogue of Taylor's misdeeds and it concludes with the hope that the Lieutenant-Governor, 'highly spoken of here', might be less unheeding 'to the cries... of the indigo districts' than his local subordinates [*2: 30, 36, 42 ff*].

Thus, the liberal looking for perfect justice climbs, toehold by toehold, the edifice of colonial authority from its sub-divisional level to its gubernatorial summit. His faith in the morality of colonial rule survives all the evidence he has of collusion between planters and civilians and of the resulting perversion of the processes of the law. These latter are sought to be explained away as mere aberrations of an otherwise faultless system. Or, as Dinabandhu Mitra suggests more than once in his play, there are good sahibs and bad sahibs: the good ones run the Government and the bad ones run the indigo factories. Savitri, the Kayastha matron in *Neel-darpan*, spelt this out in terms of her own understanding of social distinctions when she described the planters as the white men's *chandals* [*Act I, Sc. 4*]. And in this she came very close to the distinction that the more sophisticated *Tattvabodhini Patrika* had made a decade ago between '*bhadra* Englishmen of good character' who never took up the indigo trade and 'the cruel, *abhadra* men' among them who did [*11 II: 129*]. Collaboration between the bhadra Bengalis and the bhadra Englishmen was, clearly, the need of the hour. In emphasizing this *Neel-darpan* was simply upholding an ideological tradition already well established among our liberals by the middle of the last century.

VI

The urge for collaboration runs through the entire range of Dinabandhu's literary works. The theme of the British conquest of India which for so many of our patriotic writers served as a cue for strident anti-British declarations, occurs in his *Suradhuni Kavya* as a vehicle of loyalism. The sad maiden haunting the grove at Plassey was for him not a symbol of the loss of Bengal's independence but the fall of the hateful Mughals [*15: 363 f*]. The careers of Sirajuddowla and Mir Kasim provided Girishchandra Ghose with material for seditious plays [*12: 39*] about the collusion between foreign invaders and native traitors against Bengali monarchs. Dinabandhu describes Sirajuddowla as a monster who richly deserved his cruel end [*15: 362 f*]. And he invests his account of the defeat of Mir Kasim's troops and the rescue of the Nawab's enemy, Krishnachandra Ray, by the British with the aura of a miracle comparable to the rescue of Bharatchandra's Sundar by the goddess Kali and of Mukundaram's Srimanta by the goddess Chandi [15: 360]. Still another idol of latter-day nationalism—Nana Sahib—was denounced by him as a brute and a coward, while he was no different from any other Bengali baboo of his time in representing the rebellion of 1857 as a senseless orgy of violence [*15: 345*]. For the permanence of the raj and the removal of all threat to it were for him highly desirable ends [*15: 443*]. And if these attitudes are about the same as those held by other stalwarts of the Bengal Renaissance, Dinabandhu excelled many of them in the obsequiousness with which he could put these into words. Anyone who wants this confirmed may do so by turning to his poem, 'Loyalty Lotus or Rajbhakti Satadal'. Rendered into English, it reads as follows:

Hail, Alfred, our brother, priceless object of our affection! The descendants of the Aryans are dancing with joy to-day. For, at this auspicious hour on this auspicious day they will all rejoice at the sight of royalty as they gaze upon your face which is handsome like the moon. Queen Victoria, our gracious mother, sheds lustre upon the realm as she manifests herself through you to-day.

O Scion of our Queen, take your seat on this mighty throne. The Lord of the Universe is thrilled by the beauty of this spectacle. It is after a lapse of a hundred years that the Queen, our mother, has been so kind as to send her beloved son to visit her Indian home. With all this display of affection for her Hindu children who can complain that our mother remembers us no longer?

A hope dawns irresistibly in our hearts that the Prince, our Emperor-designate, will, as a gesture of affection and in order to rear his subjects, bring along with him a venerable woman. An ocean of happiness will then flood this land of the Hindus. Long live the Prince! Long live the Prince! Long live the Prince!

And we hope that later on our mother Victoria who is so full of devotion to God, the Queen who has given birth to heroes and is by heroes much admired, will herself, together with her family, pay a happy visit to India with a joyous heart. The multitudes of her subjects are crying for their mother. She will, when she is here, take them up in her arms and kiss them.

Please be seated, brother Duke, among your Hindu brethren. Let us put a garland of white lotus on your neck. Let us in all affection feed you with milk, cream, cheese and some delicious *motichur* as well as with well-made and seductive *chandrapuli* cakes. And let us offer you even the tastier gift of love.

Strike up the music on *tabla*, flute, violin and *sitar*. When again shall we have a joyful day like this? Dress up in your *peshwaz*, put on your anklets and dance, O dancing-girl, matching your movements to the beat of the music. Sing, O singing-girl, such divine tunes as would excel those heard at Indra's[8] court in melody and in rhythm.

The Prince is holding court with Lord Mayo. Calcutta has been lit up by the sovereign's lustre. This city, bejewelled with lights, radiates a glow. To this has been added a glow radiating from the hearts of the subjects. And pious Hindu women with their moonlike faces are ululating (as they do on sacred occasions) and decorating their verandahs with earthenware lamps.

The women of Bengal are decorating their homes ingeniously with traditional designs (*alpana*) as an auspicious measure. With all their hearts they are engaged in worshipping the King with ceremonial offerings of fragrant flowers, paddy seeds and grass. Glory to you, the women of Bengal, in whom is stored up all that is beneficial; where else could one find females to equal you in chastity and piety?

The Prince has ascended the throne on this highly auspicious day. Who can now complain that India is unfree? You have now seen this land of India with your own eyes. You have seen how everything here has dissolved in an ocean of joy. On your return to England you can carry with you the good tidings that the sea of Indian loyalty has indeed flooded over.

[8] A god who, according to Hindu mythology, rules over the celestial kingdom. Many heavenly musicians are patronized by him at his court.

What is left to us to set down as an offering at our Queen's feet? For, all that we have already belongs to her. Please accept this loyalty lotus, the best thing we have in India. Let us melt in sentiments of loyalty. Let us cry, 'Victory to Victoria'. Let us applaud and rejoice. We have at last found ourselves in our mother's arms. Cry, 'Victory to Victoria'. Cry, 'Glory to Hari'. [15: 437 f]

We have quoted this poem *in extenso* not only as a representative sample of middle-class grovelling but also as a specimen of the canker that had eaten into elite nationalism at this early formative stage. Four out of the above ten stanzas talk of the Hindus and one of 'the descendants of the Aryans' as if they alone constitute the nation. India is, in fact, identified as the 'land of the Hindus', and correspondingly, in the poem 'Yuddha' [15: 398], the Musalmans are described as foreigners. This fragmented nationalism which excludes millions of Indians simply because they do not conform to the dominant religion, fits very well indeed with the parochial character of Dinabandhu Mitra's patriotism. At one point he appears to enthuse about the railways as a unifying factor. On a closer scrutiny, however, this turns out to be no more than a recognition of the way the railways *physically* brought together the peoples of the different parts of India across great distances [15: 404]. Far from embracing the whole of India, his image of 'the motherland' hardly takes in even the whole of his native Bengal. In 'Prabasir Bilap' [15: 392 f] his protagonist starts off his lament in exile promisingly with 'Oh my native land, the holy land of Bengal,' but ends up by leaving the reader in no doubt that this Bengal, is, for him, merely the sum of his own village and his family. It is not surprising therefore, that this immature sense of nationhood was gifted with an equally puerile idea of political independence. Since a member of the royal family favoured India with a visit, 'who can now complain that India is unfree?'

That a writer with such a communal, parochial and loyalist outlook should come to be regarded as a great nationalist, indicates that our nationalism has in it an ideological element with a fairly low anti-imperialist content. This element represents the contribution of that section of our bourgeoisie who are interested in opposing imperialism but cannot do so firmly and consistently owing to the historical conditions of their development. So their antagonism is compromised by vacillation and expresses itself spasmodically. Middle-class Bengalis in search of a radical tradition during one such spasm that culminated in the swadeshi movement, settled on *Neel-darpan* as a patriotic text [20: 274]. We thus got what we deserved: a loyalist play glorified as a manifesto of petty-bourgeois radicalism.

This had a rather sinister consequence for the historiography of the indigo rebellion. Adopted by many generations of baboos as a theme for their literary and artistic self-expression, this peasant revolt came to acquire the respectability of a patriotic enterprise led by benevolent talukdars like Navinmadhab with peasants like Torap following them with all the loyalty, strength and intelligence of a herd of cattle. (Torap is in fact likened to a buffalo in Act V, Sc.2). The emphasis has thus been laid on the unity of interest between the village poor and their native exploiters against a common, foreign enemy. This has helped to mask the truth about two important aspects of the upper class participation in this struggle— first, about the opportunism of the landed magnates and the fierce contest between them and the peasantry for the initiative of the struggle against the planters, and secondly, about the feebleness and defeatism displayed by the people 'of intermediate means', that is, the rich peasants and the lesser landlords.

VII

One can hardly expect to understand the character of the upper class response to the events of 1860 unless one is prepared to set aside the conventional view of the rebellion as an undifferentiated phenomenon. It is not enough to distinguish between the phases of the struggle in terms of its progress from lower to higher levels of militancy and organisation, as Suprakash Ray does in his most valuable work on peasant revolts [*19: 391 f*]. What is important is to recognize that in the course of this development the struggle went through significant changes in leadership and alignment, too. Ray does not seem to take these into account and the rebellion, as described by him, looks like an event with a uniform social content. Inevitably, as a result of this, he gets entangled in a number of confusing, if not mutually contradictory, statements about the nature of landlord and middle-class participation in the rebellion.

Since the zamindars and big talukdars were the leaders of rural society, 'it was inconceivable', according to him, 'that this class who were themselves created by the British, would be opposed to the class of indigo planters who enjoyed the patronage of the British rulers and ranked with the latter as exploiters. Furthermore, they used to make a great deal of money by leasing out parts of their estates in *patni* to the indigo planters at extremely high rates. It was, therefore, in their own class interest that they opposed the indigo rebellion' [*19. 402*]. Yet, the evidence of the clash of interests and of arms between zamindars and planters is far too vast and significant to be altogether ignored. Ray himself notes that 'some zamindars, urged by a spirit of revenge and by self-interest, went so far as to organise and lead the rebel peasantry' [*19: 394, 402*]. However, he regards this as exceptional, and concludes that 'but for a mere handful of the more humanitarian and vengeful zamindars and big talukdars the entire class of zamindars and talukdars stood by at a distance as silent onlookers' while the peasants were fighting [*19: 402*]. About the attitude of the middle classes, too, he is equally categorical. He regards the rural middle classes as 'reactionary' and 'degenerate', and finds it 'natural' that they should have opposed the rebellion, 'devoted their utmost efforts to the defence of indigo cultivation and of their masters, the indigo planters', and thereby brought upon themselves the wrath of the rebel peasantry [*19: 403, 404*]. The urban middle classes 'displayed indifference and satisfied their sense of duty by a little expression of sympathy from a distance' [*19: 404*], although there were among them some 'liberal and humanitarian' individuals like Harish Mukherjee and Dinabandhu Mitra who, according to him, were 'progressive' and came out 'more or less' in support of the peasant struggles [*19: 188*]. Here again, the burden of his conclusion appears to be that 'the middle classes, far from joining in the rebellion as a class opposed it and helped the indigo planters in various ways' [*19: 389*].

Two points thus emerge from these statements: first, that the landed and the middle classes were *in general* opposed to the indigo rebellion, and secondly, that such opposition was altogether consistent with the interests and outlook of these classes, so that any act or expression of sympathy for the rebels on the part of individual members of these classes must be regarded as exceptional. Both these points merit serious consideration not merely because they occur in what is unquestionably the most important work published so far on Indian peasant revolts, but also because they are representative of an influential and widely held view.

To turn first to the question of upper- and middle-class participation in the indigo rebellion, it would appear in the light of contemporary accounts that Suprakash Ray has grossly underestimated it. Eighteen-sixty was a well-documented year. Thanks to the passions that scored it, we have been left with a vast amount of evidence originating from parliamentary, administrative, missionary, journalistic and literary sources. Taken together it clearly testifies to the fact that the support offered by the zamindars, talukdars and the rural middle-classes to the rebel peasantry was far from fortuitous, rare or negligible.

Long before the rebellion of 1860 armed clashes between the zamindars and the white planters had come to be established as a notorious and recurrent feature of life in the indigo districts. The *Tattvabodhini Patrika* mentioned this in 1850 as a most familiar phenomenon and quoted an instance dating back to the middle of that decade when 'a fierce battle of staves' fought between the private army of a planter and that of a Muslim landlord in Krishnanagar had resulted in the killing and wounding of many on both sides and the drowning of four to five hundred heads of cattle belonging to the ryots [*11 II: 130 n*]. The *Sambad Prabhakar* in two of its leading articles in 1853 and 1854 regarded violent disputes between zamindars and planters as fairly widespread throughout Bengal [*11 I: 98, 200*]. These were common enough to be treated as typical by Peary Chand Mitra in his novel, *Alaler Gharer Dulal*, published in 1858. Its debauched and accident-prone hero has a pleasure trip to his zamindari estate rudely spoilt by an affray between the *lathials* of his own cutchery and those led by the manager of the local indigo factory [*16: 79-82*]. 'The planters are being increasingly oppressive in Jessore', says the author. Subsequently that year the *Hindoo Patriot* echoed the same concern and went on to record its appreciation of the zamindars siding with their ryots against the planters [*14: 118 & n 38*].

By 1860 as the revolt broke out in earnest, it appeared at least during its initial stages as if zamindari feud and peasant jacquerie had merged into a common resistance. A large number of the landed magnates—far in excess of the two families who come up for mention by Suprakash Ray [*19: 402*]—became actively involved in the struggle against the planters. The most formidable of them all was Ramratan Ray of Narail. His enmity towards the planters and the spectacular ways in which he expressed it, became legendary. He was said to have destroyed an indigo factory and covered up the traces by transplanting a coconut orchard onto its site—all done in a single night's work [*22: 14 f*]. He owned extensive properties in Jessore, Faridpur and Pabna, and in each of these districts his lathials fought the planters' mercenaries [*14: 87*]. In 1860 the rebel peasants of Kushthia enjoyed his active support and patronage. Insurgents in a Pabna village who chased away a party of policemen headed by a Deputy Magistrate, were reported to have been led by one of his agents [*14: 156 f*]. And later on that summer the lathials of the Narail zamindars fought a pitched battle against those of the Meerganj concern [*2:24 f*]. In most of these exploits Ramratan Ray had a powerful accomplice in his naib, Mahesh Chatterjee. It would be wrong, however, to regard the latter's opposition to the planters as merely derivative. He was a big zamindar in his own right and owned much property in the neighbourhood of the

Katchikatta factory in Nadia [*RIC 2921 e*]. By the middle of February 1860 we find him stirring up the Katchikatta ryots against the planters [*RIC 3111 e*]. Soon afterwards a complaint was lodged against him and his naib for inciting the villagers of Pachlia against the Poamaree concern and organising a murderous assault on one of its employees [*2: 12-15*]. By the end of March he had already been blacklisted by the District Magistrate as 'the most dangerous agitator in Nadia' [*14:87 n 11*]. Another officer of the same district testifying before the Indigo Commission accused Mahesh Chatterjee of holding nightly meetings where he exhibited parwanas hostile to indigo cultivation and of dissuading the ryots from sowing and 'encouraging them to refuse to allow the factory servants to enter the villages' [*RIC 3071 e*]. By April he was being blamed by the guardians of law and order for masterminding the Pabna riots against the planters on behalf of Ramratan Ray [*14: 157 f*]. The Pal Chaudhuris of Ranaghat, too, counted among the leaders of the landlord opposition. They had never been reconciled to the presence of European planters in western Jessore where they had their extensive zamindaris. Their Ranaghat home often served as a rendezvous for the anti-planter landlords of the region [*14: 89*]. Shyam Chandra Pal Chaudhuri had a long-standing feud with Robert Larmour of Mulnath, the notorious General Mufassil Manager of the Bengal Indigo Company. Lathials of the two parties had already clashed as early as January 1858. Two years later, in January and February 1860, a number of affrays occurred between them again [*14: 90*]. Another redoubtable enemy of the planters among the Jessore zamindars was Mathur Acharya of Shadooty (Sadhuhati). A report from Sisir Ghose to the *Hindoo Patriot* in October 1860 mentioned that 'the villagers with Mothoor at their head threw off the factory yoke during the late crisis' and that the planter was trying to persuade the authorities to send up to Shadooty a battalion of police for his own protection [*2:41 f*]. According to Satishchandra Mitra, the Acharyas 'sided with the peasantry and incited and organized them' into a force of thirty thousand men who easily overpowered the planter's retainers and plundered the houses of the factory employees [*19: 394*]. Yet another zamindar who allied himself for a while with the indigo ryots was Brindaban Sarkar of Sibnibas. The Sarkars had been involved in a long-drawn litigation with Larmour over the ownership of Sibnibas. During the very early stages of the rebellion they incited the ryots against the planter and helped them with advice and money to the extent of paying off the fines imposed by the courts on those found guilty of violation of Act XI of 1860 [*14: 90 f*; RIC 2172-2174 *e*]. Then, there was Jagadbandhu Ghose, a zamindar, who, in the spring of 1860, was directly responsible for provoking a revolt among the ryots of the Aurangabad concern in a manner that had some unforeseen consequence as noticed later on in this paper. Subsequently that summer Sisir Ghose sent up to the *Hindoo Patriot* a story about 'a great battle... fought at Mullickpore between the owner of the villages and John MacArthur of Meergunge' [2: 16]. The owner mentioned above was Haranath Ray, zamindar of Mallikpur. Srihari Ray of Chandipur, too, a proprietor of seven villages, took a conspicuous part in the rebellion, according to Satishchandra Mitra [19: 391]. So, it appears, did Prankrishna Pal, identified by James Forlong of the Nischindipur factory as an influential zamindar and an enemy of the planters, who had been trying 'to induce the ryots not to sow indigo and to sow a large breadth of rice cultivation' [RIC 56 e]. Two other landlords, Ramnidhi Chatterjee and Nabokisto Pal, both big talukdars, are also named by Forlong as active and powerful opponents [RIC 2924 e, 2925 e, 3505 e]. This roll-call is necessarily incomplete. But I do not think it is possible to regard this as indicative of anything but a fairly large participation of the landed magnates in the rebellion during its earlier stages.

As for the middle classes, the distinction between an urban section 'somewhat sympathetic' to the rebellion and a rural one, hostile to it, is perhaps a bit too neatly drawn. Physically as well as culturally, town and country still formed a smooth continuum during the middle decades of the nineteenth century: physically, because the metropolis shaded off into the deep country by degrees along a gradient of suburban villages and rural towns; and culturally, because the metropolitan society itself was still 'subject to a strong rural pull' as the *nouveaux riches* in Calcutta and its suburbs continued to cherish all the more important traditional values, while the ancestral village 'continued to exert a spiritual influence even on the new generation which had grown almost entirely on urban soil' [22: 12, 63, 79, 80, 85]. It is, therefore, difficult to agree with Suprakash Ray that the division of the middle-classes between those who were 'permanently resident in the country' and the rest, 'predominantly resident in towns', corresponded to a general division between a 'relatively reactionary' section and a 'progressive' one [19: 188, *passim*]. And this was true of attitudes relating to social reform as well as to politics. The middle classes, in town and country, were all equally attached to the raj. If anything, the Calcutta baboos, judging by the noisy display of their solicitude for their 'guardians' in 1857, appeared to be quite capable of outshouting their mufassil class brethren in protestations of loyalty. Again, if sympathy for a peasant revolt (not to be confused with pity for the victims of repression after a rebellion has been crushed) is any measure of a progressive outlook, the urban middle-classes could hardly be credited with it. Thus, the revolt of the Santals, the other big uprising of the decade just ended, proved to be a cause for much concern to the *Somprakash*, identified by Benoy Ghose as 'an organ of the educated and liberal Bengalis of intermediate means during the second half of the nineteenth century' [*11 IV: 25*]. One of its correspondents reported how a great fear spread among the rich rural gentry as in a particular area the rebels approached the neighbourhood, while, editorially, the *Somprakash* pleaded with the authorities to garrison the small towns so that the insurgents were contained in the jungle until the cold season when the troops might enter it and deal with them [*11 IV: 790, 791*]. The *Sambad Bhaskar*, edited by Gourishankar Bhattacharya who in his youth had learnt the art of liberal journalism as an assistant at the *Jnananveshan*, the organ of the Young Bengal 'ultras', was worried that 'the Muslim infidels of Bihar were spreading a rumour to the effect that the Santal insurrection would soon force the British Government out of power', and that the Kols, too, felt encouraged to take up arms against the Government, thanks to 'the mild manner' in which the latter had treated the Santals [*11 III: 34, 298, 300*]. Residence in towns could, thus, hardly be an influence on middle-class opinion in favour of a peasant revolt, particularly if the latter bore even the least tinge of disloyalty to the raj. For, as the milkman in *Neel-darpan* put it, 'The city baboos are pro-British' [*Act V, Sc. 1*].

This, however, is not to deny that some elements of rural society who had, customarily, stood between peasant and landlord, fed as parasites on the latter's resources and been the agents of his tyranny over the ryots, were now in the planters' employment and sided with them against the enemies of the factory. Already in 1850 the *Tattvabodhini Patrika* had noticed the presence of 'a community of ruffians engaged at many places in the countryside in oppressing the people' under the planters' patronage [*11 II: 129*]. They were still there, a continuing scourge, when the *Somprakash* wrote bitterly about them fourteen years later [*11 IV: 80*]. The characters, Dewan, Khalasi and Amin, represent them in *Neel-darpan*. The khalasi has considerably added to his authority by allowing the junior planter to sleep with his sister [*Act III, Sc. 1*]. But the amin who also made a gift of his sister to the sahebs, has not been promoted to the post he had expected as a fair return [*Act I, Sc. 2*]. The dewan laments that his services are not appreciated enough, although he has allowed himself to sink, morally, to the lowest depths in order to do his master's bidding. Yet he is determined not to shrink from anything, however base, for 'once I have taken up this job', he assures his boss, 'I haven't been the least swayed by any consideration of fear, shame, timidity, honour or prestige; incendiarism, cow slaughter and the slaying of brahmans and women have all become habitual with me' [*Act I, Sc. 3*].

Suprakash Ray has quite rightly emphasised the role of these corrupt elements as the planters' henchmen [*19: 403 f*]. But here again one must be careful not to oversimplify. There were, indeed, some outstanding cases of the planters' own managerial staff turning against them and actively participating in the rebellion or even leading it. Morad Biswas who led the Aurangabad rebels up to a point, had for some time acted as a rent collector for the factory which his men were to attack later on [*14: 94*]. The Mallik brothers, Ramratan, Rammohan and Girish, who led the insurgents in Jessore, had all been factory *karmacharis* [*14: 96*]. And, Digambar and Bishnucharan Biswas, whose reputation as leaders of the rebellion became legendary, had also served as senior employees of the Bansberia Concern [*14: 95*]. What is even more important to recognize is that the rural middle class was by no means coterminous with the group of factory karmacharis. It consisted of two very large sections of the rural population. First, there were many small landholders including jotedars, talukdars and mahajans who obviously belonged to this category. Not all of them were friendly to the planters. It was they who petitioned the Lieutenant-Governor during his tour in 1859 seeking his intervention against the tyranny of William White of the Bansberia Concern [*14:76 f*]. It was the opposition of these small landholders which the planters, Maurice Tweedie and Adam Hume Smith, complained of in their evidence before the Indigo Commission [RIC *3409, 3494*]. Some of the most formidable leaders of the rebellion came from this class. There were the Biswases of Poragachha, minor landlords and moneylenders[9] who organised, armed and financed the insurgents [*14: 95 f; 19: 386, 390 f*]. Then there were the Malliks of Jayrampur whose anti-planter sentiments received much encouragement from the Pal Chaudhuris of Ranaghat [*14:89*]. The manner of their militant opposition to the planters earned for Ramratan, the head of the family, the heroic sobriquet, 'Nana Saheb of Bengal' [*14: 96f*]. Sisir Ghose, too, came from a middle-ranking talukdari family. Unlike the other talukdars mentioned above he wielded the pen rather than the lathi against the factories at the height of the rebellion [*2: 22 f, 28, 30, 36; 19: 220, 389, 391, 402*]. He was, indeed, a member of that small but increasingly important section of the rural *madhyabitta* who combined landed interests and country

---

[9] Suprakash Ray describes them once as 'middle class' [*19: 389*] and then, again, as 'primarily peasants' [19: 390].

connections with a western-style education and a liberal professional training acquired in the cities. Operating from the rural towns and the larger villages, they constituted a vital link between the indigo ryots and Calcutta public opinion. It is they who acted as Harish Mukherjee's eyes and ears, for the Editor of the *Hindoo Patriot* had never been to any of the indigo districts himself except Barasat and Hugli [RIC *3876 e*]. No wonder that they were not popular with the planters and their friends among the local bureaucrats. According to Sisir Ghose, members of 'the anti-indigo party' blacklisted by the Jessore authorities in July 1860 included, among others, a court official, a post-master, a teacher and three other educated natives for all of whom the gravest consequences were predicted if the law could lay its hands on them [*2: 22 f*].

Apart from these junior representatives of the rural gentry endowed with talukdaris and/or a little western-style education—the *bhadrasantans* whose distress had provoked a protest in the correspondence column of the *Sambad Prabhakar* in the summer of 1859 [*11 1: 105 f*]—there was still another section of the non-urban middle classes whose attitudes must be taken into account. These were the rich peasants[10] who, acting as modols (or, more elegantly, mandals i.e., headmen) occupied a strategic position both in indigo production and in village society. Benoy Chowdhury draws upon a contemporary account of the role and resources of these elements to conclude that the planter found them 'eminently useful in enforcing his system', and the modols, thanks to their triple function as principal tenants, moneylenders and leaders of the local community of ryots, 'misappropriated the peasants' surplus' as the reward of their collaboration [*5: 169 f*]. A pucca house adorned with a string of *golahs* (granaries) for rice was the visible symbol of a headman's affluence in a village in the indigo districts. But the collaboration was short-lived and the planters by their rapacity alienated this useful section of the middle classes, too. A modol family mentioned in the opening scene of *Neel-darpan* owned not so long ago ten ploughs, forty to fifty bullocks, a massive cowshed, a big courtyard, a large annual crop of monsoon paddy and money enough to provide sixty meals a day for its members and dependants. But within less than three years after the planters had acquired a

---

[10] To make doubly clear what should be quite obvious to the reader at this point, the demographic difference between the total rural population and the sum of all big zamindars, big talukdars, middle peasants, poor peasants, landless peasants, agricultural labourers, bandits, vagabonds and *lumpen*, constitutes for us the 'rural middle classes' in 1860.

lease of the village, the family was altogether ruined: two of the three brothers who were severely beaten up by the men of the factory for refusing to change over from rice to indigo, had fled the village, and the other too, was about to do so. It was oppression of this sort that drove the rich peasants throughout the indigo districts to make a common cause with the rest of the peasantry against the planters in 1860. More often than not it was they who emerged at the flash point of a local insurrection as its grass-root leaders inciting, leading and organising the ryots. In Lal Behari Day's *Bengal Peasant Life* it is the intervention of the mandal 'who seemed to be somewhat better dressed than the rest and to exercise some sort of authority over the assembly', that swayed village opinion decisively in favour of action against the planter, Mr. Murray. '*Mari Salake maro!* (Strike the scoundrel Murray)—that should be our battlecry', he exhorts. 'The indigo planters have been the ruin of our country. Before those *salas* came, this country was as happy as Ayodhya in the time of Rama. But now everything has gone to wreck and ruin. They oppress us; they beat us; they imprison us; they torture us; they dishonour our wives and daughters. Down with the blue monkeys! *Mari salake maro!*' [*10:247*]. All contemporary observers testified to the rich peasant's oppositional role as indicated by Day in his novel. Herschel, the highly knowledgeable Nadia official, claimed to know of 'hundreds' of village headmen who had acted as the local leaders of the insurgents [RIC *2832 e*]. And for Blair Kling, a modern historian of the 'Blue Mutiny', the evidence is ample and eloquent enough to allow him to state 'that the village headmen or mandals whose names appear as leaders in the indigo disturbances are too numerous to recite' [*14: 97*].

VIII

It should be clear from all this that the indigo rebellion which was primarily the work of the great majority of the villagers ranging from middle peasants to the rural proletariat, was still not without participants, and occasionally leaders and organizers, drawn from the classes above that level. But those upper-class elements consisting of zamindars, talukdars and rich peasants, although they opposed the planters up to a point, did not do so all for quite the same reason as the rest of the rural society. Their resistance differed from that of their poorer neighbours in its aims as well as in its quality.

Let us turn to the landed magnates again. Rabindranath Tagore retailed a myth about the zamindars rescuing the ryots from 'the usurious noose' by which it was the planters' policy to try and hang them [*23: 342*]. The truth is that the big landlords were quite happy to let the white planters do what they liked with the ryots so long as they did not themselves feel threatened. The poet's grandfather, Dwarakanath Tagore earned some at least of his princely fortune from bargains made with the planters [RIC *3997 e*]. He was emphatically in favour of 'the cultivation of indigo and the residence of Europeans' as a result of which, he claimed, 'the ryots materially improved in their condition'. And it was as a token of the white seigneurs' coexistence with the brown ones that he was quoted by the former in their memorial to the Secretary of State in defence of the indigo system [*17: 44 f*]. But the era of mutual tolerance, never altogether free of feuds, was not destined to last long. Dwarakanath's death in 1846 may be said to have symbolized its end. The Union Bank crashed the following year. As capital dried up, the rush to acquire more land for indigo, which had already started in 1829 and considerably intensified since 1837, developed into a scramble. It was the ryots, inevitably a marauder's first kill, who were hurt most as the wicked weed fought, burnt and litigated its way into paddyfields, residential plots, country paths and the borders of village ponds—lands no tyrant had ever laid his hands on. The zamindars cashed in on the peasants' distress. 'The landholders knew that the planters could not do without their land. This immensely strengthened their bargaining position'. This comment, by a modern historian [*5: 166*], on the venality of the zamindars is confirmed by the views of many contemporary observers including some who testified before the Indigo Commission. Among these were Maurice Tweedie of the Loknathpur factory, who was no friend of the native landlords [RIC *3407 e*], Mahesh Chatterjee who, as noticed above, was no friend of the planters [RIC *3526 e*], and James Forlong, a liberal planter friendly to both sides [*14: 27,59 f*]. 'While I have been in this district', said Forlong after thirty years' residence in Krishnanagar, 'I have never yet found a zamindar hesitate in handing over his ryots to the planter as soon as his terms are complied with' [RIC *2902 e*]. It is the near unanimity of all on this point which led the Commission to conclude 'that the only difficulty experienced by the planter has been that of settling the pecuniary terms' with the landlord concerned in a dispute, and that 'in any case there is usually but one termination to these disputes', that is, an agreement on 'the price demanded for a *putni*' or 'the *bonus* demanded for a lease' [RIC *42-43 r*].

But an agreement was often not so easy to reach. The landlords would try to exact 'the largest possible rental and the largest possible price' [RIC *2917 e*], anything up to four times the annual rental as the price of a *patni*, amounts ranging from 25 to 100 per cent as the *selami* for an *ijara*, and so on [5: 176f]. Only the larger indigo concerns backed by the agency houses could afford to buy on such terms, and even they found their resources badly strained when the failure of the Union Bank made funds hard to come by. And thus as the normal procedure of buying and selling was hindered by a shortage of credit, both parties were tempted to use non-economic measures of persuasion. The zamindar would try to force his own terms for a lease on the planter by inciting the ryots against him. The planter, on his part, would use the obliging and sinister authority of the local police and the courts in order to bully the landlord into a settlement favourable to himself. If such devices failed, they would directly resort to violence and allow the question of prices and perquisites to be decided by combats between their respective bands of lathials and spearmen.

The use of such non-economic measures was, in a certain sense appropriate for these transactions. For, what was at issue here was not just a price regulated by demand and supply, but feudal rights of access to the peasant's land and labour. The zamindars of Bengal already enjoyed these rights which they used, until 1829, to stop the planters from contracting directly with the ryots for indigo [14: 52], and after that date, to boost their bargaining power in deals over patnis and ijaras in the manner mentioned above. Inevitably, therefore, during the three decades preceding the rebellion and particularly after 1837 when legal restraints on the acquisition of property were all removed, the planters intensified their drive to acquire and expand their *ilaka* (proprietary) estates where they could get the ryots to grow indigo without having to go through the zamindars as in be-*ilaka* (non-proprietary) areas. An ilaka offered the planters a two-fold advantage. First, they could develop the cultivation of indigo here on a traditional *ryoty* pattern based on small peasant holdings worked by family labour. This considerably reduced the managerial costs and the uncertainties of labour supply which made the plantation type of agriculture in *nijabad* lands unattractive for them [5: 124-30]. Their preference showed up clearly in the fact that by 1860 for every bigha of nijabad growing indigo in Lower Bengal there were almost as many as three bighas of ryoty [RIC 1 a]. Secondly, ilaka cultivation equipped the European planters with socio-political powers over the peasantry on a par with those held by the native landlords. This was recognized by all concerned—by the Government, by the planters themselves and, of course, by their rivals the zamindars. A Midnapur official pointed this out in 1855 when he observed that by securing 'the proprietary right in the land where his future operations are to be carried on' the planter 'obtains power over the ryots' [5: 166]. The planters' view was represented, among others, by James

Forlong of Nischindipore factory. He preferred ilaka because, he told the Indigo Commission, 'the authority of the zamindar seemed to me to be the only authority possessing any practical character whatever as to rights over the people'. Pressed further by the Commission to state whether these 'rights' constituted 'feudal power', he answered in the affirmative [RIC 54 e, 65 e]. The consensus in favour of such power had, by 1860, led the planters to acquire vast proprietary estates throughout the indigo districts. A little over seventy-seven per cent of all land growing indigo for the Bengal Indigo Company was ilaka; so was nearly seventy-four per cent of the area under James Hills' concerns [14: 53 f]. It was inevitable, therefore, that expansion on this scale should end up in clashes between planter and zamindar. The zamindar felt that the planter operating independently of him in the neighbourhood, was a challenge to his own authority over the peasantry. Disputes between their servants exacerbated feelings on both sides. The ryots often sought the zamindar's protection against the planter's brutalities, while the zamindar would not hesitate to use armed bands of ryots to discourage a planter from intruding into lands he owned or coveted. All such conflicts, as noticed in the *Report of the Indigo Commission* [RIC 41 r], were characteristic of the rivalry between an established feudal magnate and an aspiring one. No wonder that these were widespread in areas where the planters had expanded most, such as Nadia, where they set up as landlords over two-thirds of the district [14: 55].

This acute and often violent contest for 'feudal power' explains much of the landlord opposition that merged *up to a point* with the resistance of the peasantry in 1860. Yet this opposition was by its very nature limited. W.J. Herschel, Magistrate of Krishnanagar, was quite accurate in his observation that the principal zamindars did 'on the whole' throw their weight into the scale against the planters, but by no means to the extent they could have done [RIC 2833 e]. The limitations of the landlords' involvement in the rebellion showed up in a number of ways. There were those who, like Srigopal Pal Chaudhuri, intrigued and sympathized with Ramratan Mallik and other rebel leaders against the planters, but would not refuse them leases or defy them openly in any other way for fear of expensive and protracted litigation [RIC 3551 e]. It is this prospect of harassment generated by long-drawn judicial processes and by the partisan intervention of the local officials that made even such inveterate foes of the planters as Ramratan Ray and Mahesh Chatterjee weaken sometimes [RIC 3526 e] or withdraw temporarily from a confrontation [RIC 3516 e]. Even Mathur Acharya who once led a miniature peasant army against the factories, chose after some time to act as a mediator between planters and ryots and persuaded the latter to disarm [*19: 394*]. And we have it again on the authority of the knowledgeable Mr. Herschel that Robert Larmour, once the *bete noire* of the Pal Chaudhuris of Ranaghat, owed 'a great deal to the ultimate assent of the two principal zamindars, Sham Chunder Pal Chowdhury and Habib ul Hossain' [RIC *2834 e*] for the suppression of the rebel peasantry.

Common to all these retreats and betrayals was a logic of proprietary interests which led the native zamindars to go along with the ryot so long as he refused to grow indigo but back out the moment he decided to defy the planter-landlord by withholding rents. The development of the struggle against indigo into a rent strike is thus of immense importance for the history of the rebellion. It was, however, the inevitable consequence of the planters' accession to zamindaris. This equipped them with the power to impose advances on the ryots and enforce the cultivation of indigo in a manner they could never hope to do in be-ilaka villages thanks to the 'constant and capricious interference' of the native landlords [RIC *2928 e*]. As a district official observed in 1859, it was 'compulsory' for the ryots in ilaka areas to accept advances for indigo [*5: 166*]. The procedure for compulsion consisted, first, of a planter's 'use of the power of summoning ryots to increase his indigo cultivation' [RIC *2607 e*]. Once forced into the planter's presence, the ryot could be coerced in a variety of ways ranging from blackmail to bastinado to contract for indigo. The pressure that often told most was the threat of a rent increase [*5: 168*]. It was stepped up after the failure of the Union Bank in December 1847 when in conditions of a reduced supply of capital the planters became 'more careful in collecting arrears of rent from the peasants' [*5: 194*]. In the absence of a law enabling them to choose crops for the ryots to grow the planters fell back on their right to increase rents as the next best inducement in favour of indigo. An unobliging tenant could always be charged a higher rental and be thrown out for not paying it. It is not that enhancement and eviction actually took place on a large scale [*5: 182*]. The threat was often enough to break down an individual peasant's opposition. To press a ryot for rents was to press him for indigo. Or, as the planters would say, 'When we settle for rent we settle for indigo' [*5: 189*]. The intensity of their feeling against the Rent Act of 1859 (Act X) which appeared to curb their power of enhancement (though in reality it did not, as the planters were soon to recognize), was indeed a clear indication of the value they put on it as an instrument of their 'practical authority' in the ilaka areas [*5: 187*; RIC *2266 e, 2932-33 e*].

Thus, the increasing use of rent as a lever to impose indigo on the ryots forced the latter to react to this, inevitably, as a concomitant evil. As a result, what had been so far a resistance to indigo began to take on the shape of a rent strike too. Thus began the 'rent disturbances' which, as Kling explains, was the official description for 'the refusal of the ryots who were tenants-at-will to submit to eviction from lands claimed by planters' [14: 173]. Although this aspect of the rebellion became most pronounced after September 1860, that is, during the period immediately following the publication of Neel-darpan, it had already made an impact earlier that spring as a growing and potent element of class antagonism which could hurt all landlords, white as well as brown. Some of the latter who knew how to read the signs, started immediately to decelerate on their collision course. Jagadbandhu Ghose made his peace with the Aurangabad concern by March 1860; Mahesh Chatterjee, too, withdrew to Calcutta that spring; Shamchandra Pal Chaudhuri negotiated a truce with his enemy Robert Larmour by the summer of 1860 [14: 88, 90, 92]. For by then a tocsin had been sounded in the ilaka areas for all to hear. As early as March the peasants of northwestern Jessore had gone on a rent strike against the planter-zamindars of the Joradah concern. And within three months it spread to the Salmagudia concern in southern Pabna and from there to the northern part of that district and to Jessore, Nadia and western Faridpur [14: 173]. By autumn rents emerged as the 'central issue' of the conflict between planters and ryots, and the rebellion assumed the character of a struggle fought more closely than before on class lines.

An important feature of this later and more determined phase of the rebellion was the increasing solidarity of the poorer sections of villagers. It is not merely the tenant-cultivators who were tyrannized by the planters. Poor and landless peasants, hired for an assortment of jobs connected with the transport of the indigo crop and with the making of the dye [14: 29], were also subjected to iniquities of all kinds. They had, of course, their share of the physical brutalities, the regime of the Shyamchand, which the planters and their myrmidons inflicted on all and sundry.[11] Equally, if not more, galling was the fact that they were underpaid for their labour. An editorial in the Sambad Prabhakar of 12 March 1860 noted how a rising cost of living had been pushing up agricultural and industrial wages for all except those employed by the factories. Consequently, 'the peasants of Nadia have united in a strike (ekatra hoia dharmaghat sthapan kari-

---

[11] The Tattvabodhini Patrika wrote in 1850: 'It is not merely the peasants who are subjected to coercion and punishment by the planters and their henchmen; the same treatment is meted out also to those who transport the indigo leaves by cart or boat or on their heads and do any other work of this kind' [11 II: 128].

achche) and determined not to work for the planters for unremunerative wages'. The *Prabhakar* went on to describe the intensity of local feeling thus: 'In certain villages the peasants have built up such a solidarity among themselves that even the servants of the factories, however armed with lathis, dare not confront them' [*11 1 112 f*]. A most remarkable fact about this solidarity is that it appears to have cut across the ethnic division between the local poor and the tribal poor—the so-called *buna* coolies brought over annually in large numbers from the districts of Bankura, Birbhum or the South-West Frontier Agency [RIC *24 r*] to serve as a cheap labour force for the planters. We have nothing on record to show that the tribal labourers participated in any armed action during the rebellion. But there are some clear indications that they, too, were caught up in the spirit of the *dharmaghat* against ill-paid work. A letter to the *Englishman* (7 June 1860) from a Jessore planter complained of the 'ingratitude' of the tribal workers for demanding an advance of five rupees and a monthly wage of five rupees as against the customary advance of two rupees and an equally low wage per month. Less than two-thirds of the usual number of tribal labourers had turned up to work for him [*14: 166 f*]. That they were no great friends of the planters, but, on the contrary, ready to sympathize with their opponents, is also indicated by Navinmadhab in *Neel-darpan* when, after rescuing Kshetramoni he decides to take the path that leads through the tribal quarter of the village as the safest escape route, because the tribals, he is sure, 'will do us no harm if they come to know what has happened' [*Act III, Sc. 3*].

Solidarity of this kind between the poorer sections of the rural population contrasted sharply, at this particular point of the rebellion, with the increasing dissociation between the militant peasants and the landed magnates. The latter had already, as noticed above, started back-pedalling. This had sometimes a most disastrous consequence for the rebels, particularly when they failed to sustain a struggle all on their own and looked up to the big landlords for leadership. What could happen in such cases, is illustrated by a story published by both the *Hindoo Patriot* and the *Sambad Prabhakar* in January 1860 [*11 1: 109-12*]. The peasants of Goahpota, Shyamnagar and Badochuluri, three small villages in Nadia district, were subjected continuously to a series of harassments by the gomasta of a local indigo factory until their patience ran out and they banded together, beat up the planters' henchmen and rescued from their custody a village headman and an aged ryot who had been quite unjustly put under arrest. On hearing about these incidents, the European chief of the Bhajanghata factory decided to try and bully the villagers into submission. So he paraded through the area with a posse of lathials and ordered the modols of each village to report to him at the local factory, which they refused to do. Defied thus, the planter immediately lodged a false complaint to the local magistrate alleging that the ryots had raided the houses of the Europeans and robbed them. Not content with invoking the law in his favour, he also hired a contingent of fifty spearmen from Jessore and began systematically to spread terror in the neighbourhood. 'But what could the solidarity of the villagers hope to achieve by itself?' asks the *Sambad Prabhakar*. 'For they were all poor and of course, no one ever comes forward to help the poor. They became very scared indeed for having been involved thus in a dispute with the planter. So they conferred among themselves and decided that they must have someone who was wealthy enough to be able to protect them'. The person whom they begged for help was Brindaban Pal Sarkar, a big zamindar, who owned a large number of villages in the area. He refused to help because, he said, he had far too many quarrels of his own with the planters to take over other people's disputes. 'Thus ended the hope the villagers had of securing help from affluent quarters and as a result, they lost heart.' They appealed to the magistrate, but to no avail. And yet the planters had no difficulty in persuading the same guardian of the law to sanction the employment of two dozen armed guards for the protection of the gomasta whose misdeeds had actually triggered off these troubles. Eventually, 'the ryots who had by this time lost all hope and could no longer bear up with oppression, presented themselves in a body to the planter one day and asked for his forgiveness.' The planter responded by imposing on the three villages a collective fine of three hundred rupees in addition to an equal amount they were obliged to pay to the gomasta as the price of peace. The story ends by noting that these villagers 'are now altogether subservient to the planter and bear up, heads bent, with all his ukases.'

All this was in 1858. When, subsequently, the rebellion breaks out, we find the peasants in a more aggressive and resolute mood. W.J. Herschel's summary of violent incidents, appended to the *Report of the Indigo Commission*, offers some insights into the insurgents' temper. *Case no. 62, Government vs. Muniruddin Biswas and 22 others*, may be cited as a typical instance:

Resistance to the Police. A petition was sent to the Darogah of Bagoda by one of Mr. Larmour's people to say that, on their going to Barakhanpore, they were driven out by the villagers; the Darogah went to the spot, saw the Naib reading the Parwana and Proclamation of March 29 and April 9 to some 25 ryots. They laughed at it and swore not to obey it; the ringleaders, four, were arrested and presently rescued by a party of lathials, two hundred. Subsequently the Military Police under the Magistrate (Mr. Macneile) arrested 18 men and one more was next day arrested and identified'. [RIC *11 a*]

This militancy was informed with a sense of independence which the *Hindoo Patriot* had already noted, disapprovingly, in 1854 as a product of the 'growing estrangement' between ryots and zamindars—'their natural heads' [*5: 200*]. When by the spring of 1860 the rent question began to assume a critical importance, this spirit of independence contributed much to strengthen the peasants' morale. For, 'their natural heads' had by then started ganging up against the ryots with their not-too-natural heads, the white planters. A common proprietary fear cut across ethnic distinctions. A zamindar would often call it a day after a mere round or two of sharp encounter and hasten to settle with a hostile concern for a small consideration, such as an ijara or a job for a kinsman. The ryots, left in the lurch by such a sordid compromise, would, at this point, need all the initiative they could summon in order to continue independently their struggle against the planters. Blair Kling has put together detailed first-hand evidence about one such ongoing struggle [*14: 91-93*] in a manner that proves how beyond a certain point the indigo rebellion became the peasants' own show.

In this case again the insurrection broke out when in February 1860 the ryots of the Aurangabad concern in Murshidabad district were pushed beyond their customary threshold of forbearance by a factory gomasta's oppressions. Neither the manager nor his assistant, both Europeans, did anything to stop this wicked man from screwing out of the ryots large cash payments as the price of their freedom from sowing and weeding indigo. His retainers bound up some of the headmen for refusing to oblige him with more than half the sum he demanded on one occasion. This was a signal for the villagers to rebel. They rescued the captives, and chased, seized and beat up the gomasta himself. As tempers rose, three thousand peasants gathered and marched on the factory. All this provided the big local zamindar with a handle to advance a cause he had long cherished: he wanted the *gomasta's* job for his brother. He now set to exploit the situation by using some of the lesser landlords to work on the peasants' wrath for his own benefit. These leaders put up a spurious fight 'by calling out the ryots for further attacks and lathiyal battles and then dismissing them before the battles materialized,' until the planter, scared out of his wits, decided to buy his peace by dismissing the unpopular *gomasta* and appointing the zamindar's brother in his place.

That was the end of round one. The zamindar, pleased with his bargain, withdrew from the conflict. But the peasants were no longer in a mood to give up resistance. Having got rid of an oppressive employee of the factory, they now demanded a reduction of the amount of indigo to be sown. And as the struggle developed, they raised their sights still higher and resolved to do away with the cultivation of indigo once and for all. This radical enhancement of the aims of the insurrection corresponded to an increase in its thrust as it spread across the borders of Murshidabad into Malda district. The spontaneity of the initial jacqueries, too, matured into a conscious drive for unity and organization. Hindus and Muslims swore together never again to sow indigo. Villagers subscribed to common funds to finance the struggle. Drum-beats and other agreed signals were used by them to warn each other of an imminent raid by the planters' men and to mobilize large numbers of armed villagers in defence of their homes. 'In fact, they had it all their own way', observed an official investigator and added ruefully: 'The Police were afraid'.

The lesser landlords who, earlier, had acted as the zamindar's instruments of incitement, wanted nothing better than to draw in their horns at this point and opt out of the struggle. But the peasants would not let them. 'At this stage', says Kling, 'the ring-leaders found themselves borne along by the torrent which they had set loose; they were no longer in control of the ryots, but prisoners of a movement which they still appeared to direct'. Failing to stop it, they tried to use it for their own ends. Morad Biswas, the leader of this group, instigated the ryots to concentrate their attack on another factory in the neighbourhood in order to force its manager to employ him as its *gomasta*. As he was about to clinch his deal with the planter, the peasants, betrayed by the zamindar not so long ago in precisely the same way, refused to let Morad withdraw from the campaign and forced him to commit his son to join in an assault on the factory scheduled for the following morning. The assault failed and Morad, a reluctant rebel, was taken prisoner.

The history of the struggle against the Aurangabad concern illustrates much of what was typical of the landlords' involvement in the rebellion of 1860. It shows that they did participate in it and even led it sometimes, but that this participation was inspired by opportunistic and limited aims which took it no further than the point beyond which their interests were antagonistic in absolute terms to those of the peasant masses. Hence, the emergence of rents as a central issue in the struggle against the planters constitutes a historic turning point. Landlord leadership weakens considerably thereafter and is increasingly characterized by vacillation, compromise and betrayal. Not to acknowledge the fact of the landlords' participation in the rebellion amounts, therefore, to exonerating them for all their capitulation and treachery during the subsequent phase of the rebellion, and correspondingly, to taking the edge off the sharp contest between them and the poorer sections of the village for the initiative of the struggle.[12]

IX

If the landed magnates proved to be opportunist in their support for the rebellion, the rural middle classes proved to be defeatist. Their opposition to the planters was inspired by a relatively greater degree of self-interest. More often than not they had to oppose in order to survive. Forced into a conflict they were almost exclusively dependent on the ryots for combat power, if only because they had no private armies of their own to match those of the cutcherries or the kuthis. Yet with all the backing they received from the poorer sections of the village, they were tempted to throw up their arms far too soon in any encounter that promised no easy victory.

It is this inclination to run away from a fight that made such middle-class leadership particularly short-lived. This is indicated in a statement made by a contemporary observer who was second to none in his knowledge of the indigo districts. Asked by the Indigo Commission if he knew of 'any head ryots who have sufficient resolution and knowledge to stir up their own ryots and also to communicate with other ryots in other parts of the district and thus create a comibination among themselves', W.J. Herschel of Nadia said: 'I could point out hundreds such. *But the village leaders in this case, with few exceptions, had a strong interest in the question themselves.* Leaders have sprung up in one village, who have, in an incredibly short space of time, gained an enormous influence in

---

[12] The betrayal of the upper classes became even more frequent and blatant as the rent struggle intensified after October 1860. As Kling points out, the planters began to play, with remarkable success, on the native landlords' proprietary fears and enlist their support as allies [*14: 173 f*]. But they were not the only class to feel threatened by the increasing militancy of the ryots. A section of the rich peasants, too, appears to have gone over to the planters in some areas, generating, wherever this happened, jacqueries followed by the mass exodus of the ryots. A report from Pabna published by the *Somprakash* in 1864 said: 'Some planters... have appeased the principal and wealthier *projas* of the villages by offering them employment at the factories, and consequently, the latter are now gathering up the helpless mass of the projas as fuel for the flames of indigo. Wherever the mandals are yielding to such temptation, affrays and associated acts of violence are invariably breaking out there. It is precisely because of this that many projas have deserted Khadampur and other villages' [*11 IV: 79 f*].

numbers of adjacent villages, and have lost it almost as quickly' [RIC 2832 e]. Suprakash Ray quotes this statement [19: 388] minus the twenty-five words we have italicized above. Shorn thus of much significant information about the rich peasants' motivation and the limited character of their influence, the statement serves for Ray as evidence of 'a collective leadership of the rebel community of peasants' [19: 387]. Read in full, however, Herschel's testimony can mean only one thing; that is, the class interests of the rich peasants acted as a brake on their participation, and consequently, the leadership of the village headmen proved to be far too transient for the poorer ryots to develop any abiding trust in it. The fragility of rich peasant leadership is well illustrated by Lal Behari Day in his account of the taming of a loud-mouthed modol who had, earlier, incited the ryots to a fight against the local factory, but agreed, after a little third-degree treatment, to accept advance for indigo and lie to the police in order to shield the planter from criminal charges [10: 252-59]. No wonder that the influence gained by some rich peasants 'in an incredibly short space of time' was lost 'almost as quickly' for want of a sustained combativity.

This feebleness was characteristic of the other groups of middle-class leaders, too—the talukdars and the lesser landlords above the rich peasant level. *Neel-darpan* is the story of such a minor landowning family, the Basus of Swarapur. It has as its hero the elder son of the family, Navinmadhab. Before the planters clouded his horizon, his annual income amounted to seven hundred rupees in rents alone and his other assets to fifteen warehouses for grain, sixteen bighas of garden land, twenty ploughs and fifty day-labourers. He could afford to celebrate the Pujas in great style, throwing large banquets, distributing gifts, entertaining his guests with music and *jatra* [Act III, Sc. 2]. They grew enough food to feed the family for the whole year and still had a good deal left over [Act I, Sc. 1]. They made a bit of money by selling a part of their own produce of mustard and tobacco. And all this income from land was supplemented by usury [Act III, Sc. 2]. It is an index both of Navinmadhab's culture and of his resources that he could afford to have a siesta in the middle of the working day [Act I, Sc. 4].

His affluence impresses by contrast. Sadhucharan, described as a *matabbar*, i.e. principal ryot of the village, has only one and a half ploughs and twenty bighas of land [Act I, Sc. 3] of which about half has been rendered unproductive by saltwater [Act I, Sc. 2]. He cannot afford to have his house protected by a fence, so that his wife and daughter are quite defenceless once the men have left for the fields [Act I, Sc. 4]. Talk of security under the raj, it does not apply to chasha families like his own. Old

Golok Basu can boast of growing enough mustard to meet his family's annual requirement of oil and still have sixty to seventy rupees worth of seed to spare [*Act I, Sc. 1*]. The *matabbar's* wife must beg for a little oil from the *kolus* (oil-men) before she can light up for the evening [*Act I, Sc. 4*]. And to complete the picture of the relative prosperity of the Basus, we have in the play an agricultural labourer, a Tikiri by caste, who, representing as he does the very lowest depth of village society, comes from a family that never possessed a single plough and he has no land, no cattle, no cowshed [*Act IV, Sc. 1*].

Neel-darpan is a play about this Basu family. But it is not a play that invites us to witness the authentic aspects of the economic and social operations of these not inconsiderable exploiters in a village of impoverished peasantry where even a matabbar ryot is not better off than a middle peasant of the poorer sort. The author appears to be primarily concerned to emphasize the paternalistic element in the given agrarian relationship. Navinmadhab is so utterly dedicated to the welfare of the ryots that the planter's dewan is tempted to jibe, 'That bastard talks like a missionary' [*Act I, Sc. 3*]. Even in the midst of his worst trials our hero sticks to his principle that 'to do good unto others is the highest virtue' [*Act II, Sc. 3*]. And the ryots persecuted by the planters turn to him as their provider. An innocent peasant dragged away from the fields by the guardians of the law accompanied by the planter's hirelings, cries out to Badobaboo (as Navinmadhab is called by the villagers) to save his two children from starvation during his involuntary absence from home [*Act II, Sc. 3*], and sure enough, in the next scene, we have the dewan complaining to the planter how Badobaboo was indeed providing food for the families of four of the arrested ryots and having their lands cultivated at his own expense with his own ploughs, cattle and hired labour [*Act III, Sc. 1*].

Far from exposing the less benevolent side of the Badobaboo's transactions as a landlord, the play goes so far as to make a virtue of an even more sinister aspect of his role in the village economy—that of a moneylender. We gather from Act III, Scene 2, that usury is, for the Basus, a subsidiary, though not meagre, source of income. We gather, too, that even the matabbar's family—not to mention others poorer still—have to share their hard-earned income with the *mahajans* every year [*Act I, Sc. 2*]. Yet in a play supposed to uphold the cause of the peasantry, moneylending, that scourge of the rural poor, appears as the theme for a panegyric [*Act V, Sc. 1*]. 'The moneylender is a well-wisher of his debtors', we are told (Madhusudan Datta misses out on this key sentence in his translation). Or, again, 'The mahajans never sue their debtors'. The two speeches where.

these nuggets occur, are quite explicit in their praise of mahajani not merely for its alleged superiority to *dadni* favoured by the planters, but in absolute terms for the advantages it is supposed to have for the peasants. The speaker, in both instances, is the dewan who with all his wickedness is still endowed by the author with enough sympathy for the ryots to make him suspect in the planters' eyes. In choosing such a relatively positive character to plead in favour of mahajani Dinabandhu Mitra set himself up as a defender of the contemporary landlord-money-lender as typified by Navinmadhab, his hero. And he was by no means the only liberal of his age to hold this particular brief. Lal Behari Day does precisely the same thing in his *Bengal Peasant Life* [*10: 219-23*]. Here we find the village mahajan insisting on 'interest at the rate of two *payasa* per *taka* a month' that is, 37.5 per cent per annum, from a heavily indebted peasant whose house has been burnt down by the landlord's men. Yet the author describes Golaka Poddar, the moneylender, as 'a most respectable man' who 'never cheated anyone and was honest and upright in his dealings'. Should the reader be tempted to credit men of Golaka Poddar's class with anything but the purest of motives, Day warns: 'The reader must not suppose that all mahajans of Bengal are as hard-hearted and inhuman as Shakespeare's model Jew... we do not believe that, in Bengal at least, the moneylender is so much detested by the peasantry as a portion of the Indian press represents him to be.[13] Indeed, but for the good offices of the mahajan, many a Bengal raiyat would have to cool his heels in the cells of some prison-house'. In real life, however, few mahajans would come as clean as that. Take, for instance, the case of Sri Hari Rai of Chandipur who testified before the Indigo Commission. It was his custom to charge the ryots interest at an annual rate of 24 per cent on loans made up of money, 37.5 per cent on composite loans of money and grain, and anything up to 50 per cent on loans made up of grain alone [RIC *3481 e*]. Operating on these lines he got a decree against a ryot called Selim Biswas, attached his property and arrested him. But Biswas happened to be a ryot who grew indigo for the Khalbolia factory. As the peasant was thus sent 'to cool his heels in the cells' of the moneylender's cutcherry, the planter retaliated by seizing Sri Hari Rai's *gomasta* and imprisoning him in a godown [RIC *3477 e*].

This story contradicts not merely the liberal missionary Lal Behari Day, but the arch-liberal Rabindranath Tagore himself. For Sri Hari Rai, the mahajan of Chandipur, was also 'a zamindar of moderate substance'

---

[13] For a specimen of the hostile press the moneylenders got, see an editorial published in the *Sambad Prabhakar* on 23 November 1863 [*11 I: 113-15*].

who owned seven villages [RIC *3473 e*]. And his record is a part of the massive evidence which shows that contrary to Tagore's claim about the zamindars saving the peasantry from the planters' 'usurious noose', the landlord-usurers often acted as the hangmen themselves. Moneylending was a common practice with landowners of all kinds—large, medium and small. It was so with magnates like Jaykrishna Mukherjee of Uttarkpara (to whom Lal Behari Day dedicated his famous work as to 'one of the most enlightened zamindars in Bengal') who had vast estates in Hugli and the Twenty-four Parganas, paid an annual sadar jama of ninety thousand rupees and had, on his own admission, nearly a hundred thousand rupees 'floating' as money lent on 'interest from twelve to twenty-four per cent'. No wonder that half of his ryots were in debt [RIC *3807 e, 3821 e, 3821 e, 3825 e*]. There were others like Prankrishna Pal of Latoodaha and Sri Hari Rai of Chandipur who, though not of the magnate class, still combined their extremely large zamindari incomes with mahajani [RIC *2918 e, 3473 e, 3562 e*]. And there were, of course, those 'numberless small talukdars' mentioned by James Forlong in his evidence before the Indigo Commission, who—like the Basus of Swarapur—had 'very strong mahajani interest to promote' [RIC *2910 e*]. It is these large usurious interests of the rural gentry which were directly threatened by the planters. An exchange between the President of the Indigo Commission and Prankrishna Pal put the nub of the matter thus:

> *President*: You are a mahajan as well as a zamindar, does Indigo cultivation interfere with your lending money?
> *Prankrishna Pal*: It does interfere, because the ryots are not allowed to sow rice until the Indigo is sown and afterwards weeded, and it becomes too late to do anything for the crop, and consequently poor ryots are not able to pay off their rice as well as cash debts to me [RIC *3562 e*].

The mahajan's sense of loss was particularly heightened by the fact that 'rice debts' made up of grain for the ryots' use as food as well as seed, which normally fetched a rate of interest ranging from twenty-five to fifty per cent [RIC *3481 e, 3482 e*], tended to be more lucrative than ever before precisely at this time when the price of rice was rising steeply [5: *197*; RIC *3824 e*]. It was inevitable, therefore, that in 1860 leading planters like Larmour and Forlong should complain about the mahajans trying 'to induce the ryots not to fulfil their indigo engagements' and doing 'all in their power to induce the ryots not to sow Indigo and to sow a large breadth of rice cultivation' [RIC *2024 e, 2910 e*]. Thus, in spite of Jaykrishna Mukherjee's hypocritical plea for a peaceful co-existence of planters and landlord-usurers [RIC *3822 e, 3823 e*] a direct clash between them could hardly be avoided under these circumstances. For what was at issue here was neither the zamindar's urge to protect the ryot from the money-lender, as Tagore wanted us to believe, nor, as Maurice Tweedie of Loknathpur Factory righteously claimed [RIC *3407 e*], 'the opposition shown to him (i.e. the mahajan) by the planter when he endeavours to screw too high a rate of interest out of the ryot'. The truth simply was that the poor ryot was caught in a cross-fire between dadni and mahajani, two contending systems of usury patronised respectively by the planter and the landlord, both of whom were equally interested in appropriating the peasants' surplus [*7: 229 & n*]. The two systems, as the *Somprakash* perceptively observed, bore a close family resemblance [*11 IV: 89*]. In upholding the landlord-usurer against the planter-usurer, Dinabandhu Mitra was, as the Bengali proverb would put it, siding with the crocodile in the water against the tiger on the bank. In doing so he, a 'progressive' writer, was at the same time striking a blow, however unconsciously, for an emergent class who, as Kling points out, benefited most from the indigo rebellion: 'Ultimately they snatched the fruits of victory from the peasants, and the indigo disturbances mark the transfer of power from planter to moneylender in Lower Bengal' [*14: 75*]. In this, too, he anticipated by many decades an element of liberal-nationalism which insisted on the unity between the rural poor and their feudal and semi-feudal exploiters as a condition of success in the struggle against the raj. Represented in all its maturity in the political philosophy of Gandhism, it is expressed with characteristic vigour by Vallabhbhai Patel during the Bardoli satyagraha. 'The Government wants to divide you and the shahukar (money-lender)', he said in an address to Harijans, Dublas and artisans in the spring of 1928, 'but for you your shahukar is everything. You should laugh at and consider him to be a fool if somebody says that you should change your shahukar. It is just like saying to a pativrata (a chaste and dutiful wife) that she should change her husband. How can you leave the shahukar who has helped you in your difficulties?'[14]

X

We have already noticed how so many of the political beliefs and social

---

[14] I am grateful to D.N. Dhanagare for drawing my attention to this extract from *Satyagraha Patrika* quoted in his D. Phil. (University of Sussex) dissertation [now published as *Peasant Movements in India, 1920-1950* (Delhi, 1983)].

attitudes of the Basu family were almost identical with those cherished by Dinabandhu Mitra himself. This is of course quite appropriate in view of the author's affinity to his protagonists in class terms. It adds greatly to the authenticity of the Basu family's ideological portrait as presented in the play. What, however, comes through as less than authentic is the ideological portraiture of the peasantry. A close look at Dinabandhu's characterization of Torap should make this clear.

Torap is largely responsible for Dinabandhu's fame as a 'progressive' writer. For each successive generation of middle-class Bengali radicals throughout the twentieth century—and this includes even such a perceptive historian as Suprakash Ray [*19: 500*]—Torap has been the symbol of peasant insurgency. That this has been so reveals much about the baboo's mental image of peasants and rebels. For, the author of *Neel-darpan* has endowed this agricultural labourer with so much of his own virtues of liberalism and loyalism that he has, in fact, turned him into a perfect petty-bourgeois. One is, therefore, not surprised to learn from Sivnath Shastri how, on the publication of the play, Torap immediately endeared himself to the readers. What could, indeed, be more endearing to the Calcutta intelligentsia of 1860 than to read about someone who thought like themselves and was helped by his fictive existence to perform such brave and noble deeds as beating up a wicked white planter and saving a pregnant peasant woman from his lust?

The opening scene of Act II provides us with some details of Torap's ideology. He is seen as a prisoner here in the warehouse of the Begunberey indigo factory. Four other ryots are his fellow captives. Of these at least one is an agricultural labourer, the Tikiri whom we have already met. The rest, too, are either labourers or poor peasants. Torap dominates the conversation as one who obviously is a leader of the village poor and deals with the doubts and questions of the others with a certain amount of knowledgeability and authority. One of the ryots seems to have little faith in the sahebs. He has been the victim of false criminal charges twice. The second time his accuser was the white planter of the Bhabnapur factory whom some people regard as such a nice fellow. Torap, however, jumps to the saheb's defence. The ryot must have done something wrong to merit his punishment, he says, for 'the saheb of Bhabnapore wouldn't cause trouble unless there was good reason for it... Had they all been like him, none would have spoken ill of them'. The ryot retorts by pointing out that this nice saheb has in fact been found to have illegally detained seven persons including a small child and that he still continues to rob the peasants of their cattle. Torap changes the subject, but not the theme. 'As soon as they come across a saheb who is really a good chap,' he says, 'they want to destroy him'. Which then raises the interesting question of the distinction between good and bad sahebs—a distinction the other ryot is unable to grasp. Torap explains this in terms similar to those used by the matron of the Basu family: the magistrates are scions of respectable families (*bandonoker chhawal*) while the planters are the low-caste people of England (*belater chho-tonok*). 'Then', quips another captive peasant, 'how come that our former Governor went around the factories being feasted like a bridegroom just before the wedding?' The much-harassed ryot appears to have shared the view Sisir Ghose found common among the villagers in the indigo districts that the British officials lived off (*patramara*) the planters' bounty [*2: 35 f*]. Unable to shake off his suspicion about the mutuality of interest between the Government and the Indigo Establishment, he suggests that the late Governor Saheb—that is Halliday—judging by his cordiality towards the kuthis, must have been linked with the planters as a business partner. Torap dismisses this as absurd, but not being in a position to say much in favour of Halliday, passes quickly on to the latter's successor who provides him with another toehold for his undaunted loyalty to the raj: 'If by the grace of God our present Governor lives long enough, we shall have all we need for two square meals and the spectre of indigo will no longer press on our shoulders.'

This Torap has nothing in common with a peasant up in arms against his oppressor. He, like the baboo who created him, is full of a sweet reasonableness which is ready to exonerate the colonial regime for all its crimes against the peasantry. For every bad planter, bad magistrate, bad Governor named by the harassed and embittered ryots, our so-called rebel has a good planter, a good magistrate and a good Governor to name. This tendency to lean backwards in order to accommodate a 'moderate' point of view essentially hostile to that of the insurgents is clearly illustrated by Torap's attitude to the crucial question about the future of indigo cultivation in Bengal. A decisive swing away from indigo appears to have begun already in the summer of 1860 in parts of Jessore, where, according to one of Sisirkumar Ghose's reports sent in as early as May, the peasants 'with one voice... said that they would no longer cultivate indigo'. Pressed by the Joint Magistrate, Mr. Skinner, to carry on with indigo, they firmly declared, 'We won't sow indigo any longer' [*2: 10*]. Later on, in August, in another part of the district they defied still another official plea for indigo by saying, 'No, Saheb, if you cut our throats, even then we won't, if we die for this Indigo, the majority of our countrymen will in future live happily' [*2: 27*]. Mahesh Chatterjee told the Indigo Commission on 18 July 1860 that 'according to their present temper and feeling' the ryots were unlikely to sow any indigo again [RIC *3529 e*]. And here is a selection of answers from the peasants themselves to the Commission's question whether they would want to cultivate indigo in future.

'No, not at a rupee for two bundles, nor at a bundle the rupee.'

'Not for two bundles a rupee, not for a bundle a rupee nor for 100 rupees a beegah.'

'I would rather go to a country where the indigo plant is never seen or named.'

'Rather than sow indigo I will go to another country; I would rather beg than sow indigo.' [RIC *1132 e, 1163 e, 1180 e, 1216 e*]

The authorities appear to have taken this defiance seriously enough. 'Reports that the ryots would oppose the October sowings', wrote O'Malley, 'led Government to strengthen the military police in the indigo districts and to send two gunboats to the rivers of Nadia and Jessore, and Native Infantry to the headquarters stations of these two districts' [*18: 106*]. It is clear thus that by the summer of 1860 masses of armed, angry peasants were fighting to end the cultivation of indigo once and for all, and the struggle had by then nearly gone beyond the bounds of the immediate economic issues involved. Yet it is precisely at this apical hour that the rebel peasant, thought up by a luminary of Bengal Renaissance, is busy trying to reform the planters. He still seems to be hoping that they would give up their predatory ways and take up indigo cultivation as a regular agricultural pursuit in which, Torap assures them, they can depend on the cooperation of the peasantry [*Act II, Sc. 1*]! And he continues in this vein— backed by a nodding assent from Podi Moirani, the white man's procuress [*Act II, Sc. 3*]—right up to the rescue scene when after he has beaten up Mr. Rogue, he still finds it useful to try and persuade him to 'carry on your business by mutual consent' [*Act III, Sc. 3*]. This is not an angry, insurgent peasant's voice addressing an enemy; it is a baboo's voice begging the saheb to come to terms with the turbulent chashas before they get out of hand.

Which brings us to the question of violence. The widespread use of violence by the ryots against the combined forces of planters, police and troops is a fact about the indigo rebellion recorded in all contemporary evidence. A few words from a local observer, such as those of Sisirkumar Ghose from Jessore in August 1860, can even at this distance in time sum up for us a quickening situation: 'The planters are collecting revolvers, ammunition and lathials... while the villagers are gathering clubs and spears...' [*2:26*]. *Neel-darpan*, published within weeks of this despatch, has nothing in it even remotely approaching the coiled tension of these lines. This is so because the author is simply not responsive to the music of a clash of arms between the peasant and his oppressors.

It is not that there is no violence in the play. There is indeed a lot of it, in the form of the planters' terror backed by the official engines of repression. But the theme of a retaliatory violence on the victims' part is kept firmly under control throughout the text. The author does not allow his realism (for which he is so highly rated by the literary pundits) to get the better of his philosophy. He is ready to douse with moderation every surge of people's anger. When two hundred ryots, armed with lathis, are poised for an attack on the factory, Sadhucharan is made to pacify them [Act V, Sc. 2]. When Torap succeeds at last in laying his hands on the wicked junior planter, Navinmadhab, our hero, tries to reason with him: 'Why beat him, Torap? We don't have to be cruel to them even if they are so themselves' [Act III, Sc. 3]. And it is a measure of Navinmadhab's sense of values that he allows himself to get into a fight with one of the planters who insults his dead father's memory but not with the other planter who tries to rape a pregnant peasant woman. In any case, his advice is not lost on Torap who, after he has bitten off Mr. Wood's nose in the affray involving Badobaboo, says that he would have inflicted further physical punishment on the planter if he had a chance, 'but I would not have killed him, as he is a creature of God' [Act V, Sc. 2].

This highly sanctimonious tone, this neatly drawn distinction between chastisement and annihilation can come only from one who can afford not to have his hatred boiling over, not to indulge in 'excesses' when the planters are looting, burning, raping all around. This cannot be the attitude of an indigo peasant of 1860 involved, by all accounts, in a most sanguinary battle for survival. Torap is a pseudo-peasant and a pseudo-rebel.

Note how, in the first place, he displays none of the initiative with which all contemporary observers credit the peasant rebels. In fact, he hardly ever confronts the planters *on his own*. On each of the two occasions when he actually puts up a fight, he does so as the Badobaboo's strong-arm man. The author endows him with only as much militancy as would be needed to highlight the overriding quality of all—that is, his loyalty to Navinmadhab. And it is precisely because of this—and here is a second important thing to note about him—that his militancy has little in common with that of the rebels of 1860. All contemporary accounts agree on the highly organized and steady combativeness of the insurgent peasantry, such as was exhibited in that minor epic of a peasant war fought against the

Aurangabad Concern. By contrast, Torap's combativeness appears to be spasmodic. It is not disciplined because it is not informed by the consciousness of the rural proletariat. Gifted by his liberal maker with a petty-bourgeois consciousness his militancy explodes in brief, intermittent bursts. What goes well with this is the politics of the bomb and of middle-class terrorism, and not the politics of a revolutionary peasant war. Finally, it should be noted that he is not even all that brave: he confesses to being mightily scared at the sight of Navinmadhab being hit on his head by the planter, and in his very last speech he says: 'Let me now hide myself inside the barn. I shall give them the slip after dark. The scoundrel will let hell loose on the village to make up for his lost nose' [Act V, Sc. 2]. This does not strike one as exactly the sentiments of a peasant hero, a fish in water, at the height of a popular rebellion sweeping the countryside.

The defeatism of Torap's parting words represents the spirit of the play as a whole. Forced into a confrontation with the power of the planters, he wants to run away as soon as possible. He is not the only one to do so. At one point Sadhucharan, too, thought of leaving the village with his entire family in order to take refuge in a neighbouring zamindar's estate [Act I, Sc. 2]. And although in the opening scene Golok Basu rejects all advice in favour of desertion, the idea occurs to his son soon afterwards when he finds the old man threatened with imprisonment under the new law [Act II, Sc. 3]. This temptation to run away is only equalled by all the principal protagonists' efforts to placate the planters, reason with them and arrive at a compromise—and failing all this, to try and defend themselves by litigation. The only character to take a stand of total defiance is a poor ryot who, even as he is forcibly led away from the fields, asserts that he 'would rather rot in jail than grow indigo for that skunk of a planter' [Act II, Sc. 3]. But this one firm voice of a genuinely rebellious peasant is drowned in the chorus of petty-bourgeois wailing of the rest of the cast. In the end there is nothing in all their faith in the law, the civil service, the Lieutenant-Governor and the Queen that can save the Basus of Swarapur from being utterly ruined. Neel-darpan, written by a liberal in the midst of a peasant revolt, shows where the liberal stands at the time of a peasant revolt: he stands close to the power of the state seeking cover behind the law and the bureaucracy. It also shows what happens to him if he does so: he is destroyed.

REFERENCES

1. Anon., *The Life of Girish Chunder Ghose... by One Who Knew Him* (Calcutta, 1911).
2. Bagal, Jogesh Chandra (ed.), *Peasant Revolution in Bengal* (Calcutta 1953).
3. Buckland, C.E., *Bengal under the Lieutenant-Governors*. Vol 1 (Calcutta, 1901).
4. Chattopadhyay, Bankimchandra, *Bankim Rachanavali*. Vol. 2 (Sahitya Samsad Edition. Third Impression. Calcutta, 1964).
5. Chowdhury, Benoy K., *Growth of Commercial Agriculture in Bengal* (1757-1900). Vol. 1 (Calcutta, 1964).
6. Ibid., 'Agrarian Relations in Bengal, 1859-1885' in N.K. Sinha (ed.), *The History of Bengal, 1757-1905* (Calcutta, 1967).
7. Ibid., 'Growth of Commercial Agriculture and its Impact on the Peasant Economy' in *Indian Economic and Social History Review*, vol. 7, no. 2 (June 1970).
8. Dasi, Binodini, *Amar Katha O Anyanya Rachana* (Calcutta, 1969).
9. Datta, Michael Madhusudan, *Madhusudan Rachanavali* (Sahitya Samsad Edition. Calcutta, 1965).
10. Day, Lal Behari, *Bengal Peasant Life* (Reprint of the 1878 edition incorporating the two-volume edition of *Govinda Samanta* published in 1874. Calcutta, 1955-66).
11. Ghose, Benoy, *Samayik Patre Banglar Samajchitra*. 4 vols. (Calcutta, 1955-66).
12. Ghose, Girishchandra, *Girish Rachanavali*. Vol. 2 (Calcutta, 1971).
13. Indramitra, *Karunasagar Vidyasagar* (Calcutta, 1969).
14. Kling, Blair B., *The Blue Mutiny. Indigo Disturbances in Bengal 1859-1862* (Philadelphia, 1966).
15. Mitra, Dinabandhu, *Dinabandhu Rachanavali* (Sahitya Samsad Edition. Calcutta 1967).
16. Mitra, Peary Chand, *Alaler Gharer Dulal* in *Tekchand Granthavali* (Hitavadi Edition. Calcutta, 1912).
17. Natarajan, L., *Peasant Uprisings in India, 1850-1900* (Bombay, 1953).
18. O'Malley, L.S.S., *Bengal District Gazetteer: Murshidabad* (Calcutta, 1914).
19. Ray, Suprakash, *Bharater Krishak Bidroha O Ganatantrik Samgram* (Calcutta, 1966).
20. Reisner, I.M. & Goldber, N.M. (ed), *Tilak and the Struggle for Indian Freedom* (Indian Edition. New Delhi, 1966).
21. Shastri, Sivnath, *Ramtanu Lahiri O Tatkalin Bangasamaj* (New Age Edition. Calcutta, 1955).
22. Sinha, Pradip, *Nineteenth-century Bengal. Aspects of Social History* (Calcutta, 1955).
23. Tagore, Rabindranath, 'Rayater Katha' in *Rabindra Rachanavali*. Vol. 13 (Centenary Edition. Calcutta, 1961).

24. Whitcombe, Elizabeth, *Agrarian Conditions in Northern India*. Vol. 1 (Berkeley, Los Angeles, London, 1972).

GLOSSARY

Abhadra not *bhadra* (see below).

Amin a native employee on the supervisory staff of an indigo factory.

Banchat bada pandit hoiache literally, 'The sister-fucker is showing off his learning'.

Bhadra pertaining to the *bhadralok; bhadralok*: a general term used to indicate the elite status shared by the three highest ranking Hindu castes of Bengal; *santan*: scion of a *bhadralok* family.

Bigha a land measure which, in Bengal, stands for 1600 square yards or a little less than a third of an acre.

Blue Monkey translated from the Bengali words 'neel bandar' used in a contemporary popular verse to describe the white indigo-planters.

Chandal one of the lowest and least pure castes.

Chasha peasant.

Dadni a system of cash advances made out to peasant cultivators in order to induce them to grow indigo and sell it to the factories.

Dawl faction.

Dewan a native employee on the supervisory staff of an indigo factory.

Dharmaghat strike.

Dudu Mian principal leader of the Farazi peasant insurrections in eastern Bengal in the nineteenth century.

Gomast a native employee on the supervisory staff of an indigo factory.

Ijara lease.

Jatra a form of folk theatre.

Joteda a class of rich peasant farmers.

Kanu one of the two brothers who led the revolt of the Santal peasantry in 1855.

Karmachari employee, official.

Kayastha one of the elite castes of Bengal.

Kulin a term denoting some of the 'purer' groups among the elite castes like Brahmans and Kayasthas.

Kuthi the headquarters of an indigo-planter.

Lathial a mercenary armed with a heavy bamboo cudgel.

Madhyabitta those belonging to middle income groups.

Mahajan moneylender.

Mufassil the country, as distinct from Calcutta.

Mukhta a legal agent or attorney who, in most cases, is not allowed to plead.

Naib manager of a landlord's estate.

Nana Sahib one of the principal leaders of the Great Rebellion of 1857.

Parwana an order; a written command.

Pathshala a primary school of the traditional, as against western, kind.

Patni lease of lands with zamindari rights.

Proja tenant.

Pucca brick-built.

Puja ritual worship often accompanied by festivities.

Putni same as *Patni* above.

Sada a principal seat of government in a province or district, as distinct from subdivisional or other secondary administrative centres.

Sadhubhasha elegant, somewhat Sanskritized vocabulary associated with elite status in Bengali society.

Sala wife's brother; here used as a term of abuse.

Samaj an association usually (though not always) based on the identity of its members' social status.

Samiti an association usually (though not always) based on the identity of its members' professional or occupational interests.

Sammelan conference, meeting, gathering, etc.

Sarkar government, regime.

Selami a gratuity or offering on receiving a lease.

Shyamchand an instrument of torture used by landlords against their tenants and by the planters against the indigo peasants.

Sidhu one of the two brothers who led the revolt of the Santal peasantry in 1855.

Sudarsan a mythical instrument of war, shaped as a circular blade, used by the god, Vishnu, to cut down his enemies.

Swadeshi the national protest movement (1903-1908) against the partition of Bengal.

Talukdar a hereditary landlord whose rank, in Bengal, is usually inferior only to that of the zamindar.

Tantia Topi one of the principal leaders of the Great Rebellion of 1857.

Titu Mir leader of the famous peasant revolt of 1831 in Barasat near Calcutta.

Vakil a lawyer who is allowed to plead in a lower court.

Zamindar a Bengali landlord of the highest denomination whose hereditary title to property was confirmed by the Permanent Settlement of 1793; *zamindari*: (n.) a zamindar's estate (adj.), pertaining to a zamindar.

Chapter Two

The Agrarian League of Pabna, 1873[*]

KALYAN KUMAR SEN GUPTA

The revolt of the occupancy ryots of Pabna[1] against the Permanent Settlement landlords in 1873, was one of the great events in the socio-economic history of Bengal in the nineteenth century.[2] This unique revolt conducted under the aegis of a powerful and well-organised agrarian league created the conditions for the outbreak of similar revolts in other parts of East and Central Bengal. Patterned on the model of the Pabna revolt these revolts gradually took the aspect of a general protest against absolute landlordism in cash-crop producing areas of East and Central Bengal in the late nineteenth century. These agrarian disturbances demonstrated the great power of unionism which was fast developing during this period in Eastern Bengal among certain sections of the peasantry, underlined the basically unstable nature of the landlord-tenant relationship during the seventh and the eighth decades of the nineteenth century and sharply pointed to the inadequacy of the existing law which regulated these relationships. In fact, the agrarian disturbances in Pabna and other areas

[*] I am very grateful to Dr Barun De, Professor of Economic and Social History, Indian Institute of Management, Calcutta and currently, Fellow of Indian Institute of Advanced Study, Simla, for reading an early draft of this article and making valuable comments.

[1] A district in Central Bengal lying at the South-East corner of Rajshahi Division. Originally it formed part of the great district of Rajshahi but became a separate district in 1832.

[2] Even the ultra-orthodox Chief Secretary, C.E. Buckland, in the official history of late nineteenth century Bengal has accepted this characterisation. See *Bengal under the Lt. Governors*, Calcutta 1901, Vol. I, pp. 546-548.

'were really the origin of the discussion and action which led to the enactment of the Bengal Tenancy Act of 1885.'[3]

Unfortunately, however, no comprehensive study of this important movement has appeared so far. Scholars like Dr Benoy Choudhury or Dr P. Sinha have in passing used the Pabna disturbances as a case study in their published work on nineteenth-century Bengal.[4] But these authors have only dealt with some aspects of the movement and a principal deficiency is that they have not adequately stressed the importance of the Pabna Agrarian League. Some Marxist scholars[5] have also studied the Pabna revolt but though the writings of Narahari Kabiraj and Dr Sunil Sen have been quoted and used extensively by E.N. Komarov,[6] these studies are based on insufficient data and represent preliminary outlines which present an agenda for further enquiry. Consequently certain important questions pertaining to the Pabna agrarian league have remained unasked or have not so far been satisfactorily answered. The purpose of the present paper is to determine the nature of the Pabna agrarian league—to examine the causes which led to the formation of the league, to analyse its internal structure, leadership and methods of agitation, to assess some aspects of the character of the movement organised by the league and to discuss the methods adopted by the landlords to meet the challenge of the league.

I. Why the League Came into Existence

The agrarian disturbances of Pabna originated in *parganah* Yusufshahi of the sub-division of Serajgunge. This parganah, formerly the property of the Raja of Natore, was purchased by five different families, the Tagores of Calcutta, the Banerjees of Dacca, the Sanyals of Sallop, the Parrasis of Sthal and the Bhaduris of Porjona,[7] during the years 1796-1815 when a

---

[3] Buckland—*op. cit.*, p. 548.

[4] N.K. Sinha Ed., *History of Bengal*, Calcutta 1967, pp. 288, 291, 294; P. Sinha, *Nineteenth Century Bengal*, Calcutta 1965, pp. 22-31.

[5] Suprokash Roy—*Bharater Krishak Bidroha 0 Ganatantrik Sangram* (in Bengali), Calcutta 1966, pp. 416-432; Narahari Kabiraj—*Swadhinatar Sangrame Bangla* (in Bengali), Calcutta 1957, pp 152-153; S. Sen— Peasant Struggle in Pabna 1872-73 , *New Age*, Vol. 2, Bombay 1954; Hiren Mukherjee—*India's Struggle for Freedom*, Calcutta 1962, pp. 74-75.

[6] I.M. Reisner and N.M. Goldberg, eds., *Tilak and the Struggle for Indian Freedom*, New Delhi 1966, pp. 221-223, pp. 236-237.

[7] File 448 Nos 111-113, B.J.P.P. July 1873; *Hindoo Patriot* July 7, 1873; *Friend of India* July 24, 1873; *Englishman* July 26, 1873.

major portion of the Natore estate was brought to the hammer for arrears of revenue under the sale law of 1793.[8] The new landlords belonged to a section of the *nouveaux riches* grown rich by taking advantage of trends which were accelerated by the early British occupation of Bengal.[9] Armed as they were by law[10] with very large powers to cancel all engagements entered into by former landlords with the ryots, these people were most relentless in their demands.[11] They bought the estates as speculative investments and expected to make the most of the bargains. They had, moreover, not the social position of the previous proprietors and consequently their emergence as land-owners added an element of social tension to the economic exploitation. Precisely for this reason the relations of these new-comers with their tenants were unfriendly from the very beginning.[12]

The new landlords from the very outset tried to rack-rent maximum profits out of the landed property which they purchased by using various extra-legal methods. Thus they demanded enhanced rent by deliberately lowering the standards of measurement,[13] exacted from time to time heavy and unauthorised cesses of various sorts[14] and subjected the tenantry to physical torture if their demands were not conceded.[15]

The rapacity of the new landlords did not however lead the tenantry to combine prior to 1873, though on occasions they resisted individually the landlords' demands.[16] In fact, during the years 1858 to 1873, the tenantry met the landlords' demands for enhanced rent without any serious protest because they felt that such demands would not be repeated in future.[17] This led the *Pioneer's* special correspondent to remark:[18] 'It is not high or

[8] Radharaman Mukherjee—*Occupancy Ryot*, Calcutta 1919, p.74, Reg. VIII of 1793.

[9] Radharaman Mukherjee—*op. cit.* p. 74.

[10] Reg. XLIV of 1793, Sec. 5.

[11] Radharaman Mukherjee—*op. cit.*, pp-69-70.

[12] File 448 Nos. 111-113, B.J.P.P. July 1873; *The Bengalee* Oct. 25, 1873; *The Friend of India* July 24, 1873.

[13] Misc. coll. 14 Nos 15-16; B.L.R.P. Jan. 1874; *The Bengalee*, July 26, 1873.

[14] No. 44 B.L.T.P. July 1874.

[15] File 448 Nos. 111-113 B.J.P.P. July 1873; *Friend of India* July 24, 1873; *Pioneer* July 29, 1873.

[16] File 448 Nos. 111-113 B.J.P.P. July 1873; Molony to the Govt. of Bengal dt. 16 July, 1873.

[17] File 448 Nos. 111-113, B.J.P.P. July 1873; Misc. colln. 14 Nos 2-4 B.L.R.P. Jan. 1874; *Bengalee* July 26, 1873; *Englishman* July 10, 1873; July 26, 1873.

[18] *Pioneer* July 24, 1873.

enhanced rates the ryots object to; they are willing to pay what would be considered rack-rent, if they were only assured that the rent would be all that they would have to pay and that the amount would be permanent for a length of time.' This shows clearly enough that the root of agrarian discontent at least in Pabna was not the question of enhancement of rent which is what Dr. P. Sinha and Dr. Benoy Choudhury would like us to believe.[19]

The root of agarian discontent was, however, the tendency of the landlords to tamper with the tenants' right of occupancy. This became evident when some of the leading Zamindars forcibly extorted from the tenants written engagements or *Kabuliyats* which if enforced would have made the occupancy ryots mere tenants at will.[20] The commencement of the agrarian disturbances in the estates of the Banerjees[21] who had taken from their tenants the maximum number of such Kabuliyats clearly showed that this particular issue actually precipitated the crisis. It is thus evident that mere enhancement of rent was not the primary objective of the Pabna landlords. Their principal concern was to take away all the rights which Act X of 1859 had conferred on the occupancy ryots in order to clear the way for the future ejectment of the latter and the subsequent re-allotment of their holdings at competitive rental based on the current market value of land.

Apart from this, the landlords displayed a general tendency to convert the illegal cesses into legal rent. This became apparent when, following the general enquiries into the nature and the extent of illegal cesses in 1872 and the enactment of the Road Cess Act of 1871, the zamindars in a desperate bid to settle the rents at a higher rate, consolidated the rents and the *abwabs* into one single sum without taking any return or formal consent from the ryots.[22] This illegal action of the landlords coupled with their efforts to do away with the legal privileges of the occupancy ryots sharpened the latent

---

[19] P. Sinha—*op. cit.*, p 22; N.K. Sinha, ed., *op. cit.*, p. 291.

[20] Report on the Administration of the Registration Department, Govt. of Bengal for 1872-73; B.J.P.P. Sept. 1873; *Bengalee* Aug. 9, 1873; *Indian Daily News*, quoted in *Pioneer* July 17, 1873; *Friend of India* July 17, 1873, July 24, 1873; *Englishman* July 26, 1873.

[21] Nolan's Report dated 23.4.1874. Settlement Reports of Pabna and Bograh quoted in Radharaman Saha—*Pabnar Itihas* vol.III (in Bengali) Pabna 1330. B.S. p. 98.

[22] File 448, Nos. 111-113, B.J.P.P. July 1873; Nolan to Tayler, Campbell to Northbrook, dated July 4, 1873; Reel 317, A.P. Buckland—*op. cit.*, p. 489; the *Spectator* quoted in *Bengalee* Sept. 27 1873; *Friend of India* July 10, 1873.

class antagonism and created the climate for the formation of the agrarian league.

The formation and the rapid spread of the league throughout the district was facilitated by certain objective factors. A ruling given in a rent case in 1872 by the District judge of Pabna in favour of certain ryots of Yusufshahi enthused the latter to a large extent,[23] created among them a belief that the government was sympathetic to their cause and initially encouraged them to combine.[24] This spirit of combination then developed quickly in the preponderantly Muhammedan district of Pabna, because combination was easier among the Muslims than among the Hindus who were divided into innumerable castes, jealous and distrustful of each other.[25] The turbulent and the lawless character of the Pabna landlords, who had always the means to ruin a poor cultivator, also contributed to the growth of a spirit of combination among the tenantry who realised that individually they were no match for such landlords.[26] Besides, the Indian Revenue law which allowed as a legitimate cause of enhancement the fact that a higher rate than the one in question was paid by neighbouring ryots of the same class for similar lands indeed led to extensive union among the former as it became the interest of every ryot to prevent enhancement in any land near his own though at first sight he might seem to have nothing to do with the matter.[27] All these factors together contributed to the growth of a general spirit of unionism among the Pabna tenantry and directly led to the formation of the agrarian league.

II. *The League's Internal Structure, Methods of Agitation and Leadership*

The League, which formally came into existence in May 1873 solidly en-

---

[23] File 448 Nos. 111-113, B.J.P.P. July 1873.

[24] Nolan's Report dated 23.4.1874 *loc. cit*; *Hindoo Patriot* July 14, 1873.

[25] Misc. colln. 14 Nos 26/27. B.L.R.P. Jan. 1874.

[26] File 448, Nos 111-113, Letter no. L.C.T., Nolan to Tayler dated 1st July 1873, B.J.P.P. July 1873, Annual Report Rajshahi 1874-75, File 149/1 B.G.P. Oct. 1875; *Indian Mirror* quoted in *Pioneer* July 23, 1873; *The Indo-European Correspondence* quoted in *Bengalee* Aug. 9, 1873; *Indian Daily News* quoted in *Indian Observer* June 28, 1873.

[27] Nolan to Tayler, Misc. colln. 14 Nos 14-15 B.L.R.P. Jan. 1874; *Hindoo Patriot* July 28, 1873; *The Indian Mirror* quoted in *Pioneer* July 23, 1873; *Friend of India* July 24, 1873.

trenched itself throughout Yusufshahi by the middle of June.[28] By the end of the month, it not only engulfed the entire district but its internal organisation and methods of agitation were also perfected.[29]

Initially, the League's primary concern was to create a reserve fund to meet the expenses of civil litigation against the landlords. This fund was gradually raised by collecting subscription not only from its active members but also from persons inimical to the landlords.[30]

The methods devised by the League to rally the ryots to its cause were very effective. By sounding buffalo horns or musical instruments like drums etc., its members usually assembled the villagers and exhorted them to resist the unlawful demands of the landlords.[31] A contemporary journal thus vividly described the manner in which the league organised the villagers: 'Those who were in India during the Indigo Disturbances often describe the peculiar effect of the night cry in the villages which were opposed to the planters. It is a shout given by all the inhabitants of a village at night in chorus and taken up by hamlet after hamlet. In some of the places where the agrarian question is most hot, this shout may now be heard at night and very impressive effect it has, showing more clearly than anything else the numbers and unanimity of the ryots'.[32] Those who have lived through the communal riots of 1946 or later in Calcutta might find this to be a fairly accurate description of the cry of 'Nar-e-Takbir-Allah-u-Akbar' raised from one *mahalla* to another in urban areas to keep the Muslims watchful and defensive.

Messengers were also sent by the League to those villages which kept aloof to hold out threats of armed bands being sent.[33] These messengers, however, were not themselves the leaders of the villages which joined the League because they generally withdrew from all further action after issuing instructions and left the villages to appoint their own leaders.[34] On some pretext or other, the League's messengers used to enter the houses of the ryots who had remained attached to the landlords and conveyed to

---

[28] Nolan to Tayler dt. 1st July 1873, File 448 Nos. 111-113 B.J.P.P. July 1873, File 448 No. 77 B.J.P.P. Aug. 1873.

[29] Tayler to Nolan, File 448 No. 48 B.J.P.P. Aug. 1873.

[30] *The Indian Daily News* quoted in *Indian Observer* June 28, 1873.

[31] Tayler to Molony, File 448 Nos. 111-113 B.J.P.P. July 1873; *Amrita Bazar Patrika* quoted in *Hindoo Patriot* July 7, 1873; *Hindoo Patriot* July 14, 1873.

[32] *The Indian Daily News* quoted in *Indian Observer* June 28, 1873.

[33] Tayler to Molony. File 448 Nos. 111-113 B.J.P.P. July 1873.

[34] *Ibid.*

them the messages of the League couched in the form of circulars.[35] The *Amrita Bazar Patrika* published what it claimed to be an exact copy of a circular issued by the unionists to induce all ryots to come over to their side. 'So and So *Projas*!! As soon as you see this circular hasten over to the side of the insurgent party. If you fail to come over within this day, rest assured that we go to fish in the *beel* close by your village. We have already fished in the beels of so and so villages. Know this order is peremptory.'[36] Either dissatisfied with the effect of such messages or to make sure that the effect should not die out, the unionists formed large bands of men, visited the villages along the line of the route over which they travelled, at some places being content with threats alone, at others compelling the ryots to join the procession.[37]

The League however, never instructed the ryots to defy the authority of the British Government. On the contrary, its members declared their intention to become the ryots of the queen of England.[38] The unionists were under no delusion that under the new regime of their heated imagination they were to be exempt from the payment of rent as they knew perfectly well that the tenants of the Government estates had to pay their rents fully and punctually. The real secret of their wish to become the queen's ryots was their belief that the queen's governnment would collect from them only a fixed and reasonable rent.[39]

The local officers[40] and a section of the Calcutta press[41] initially thought that the original intention of the League was to pay no rents at all. Professor Hiren Mukherjee and Dr Pradip Sinha, have also accepted this point of view.[42] But as the League gradually consolidated its position, it appeared

[35] *Education Gazette* 27 June 1871 (R.N.P. Beng. 1873); *Amrita Bazar Patrika* quoted in *Hindoo Patriot* July 7, 1873; *Hindoo Patriot* July 14, 1873.

[36] *Amrita Bazar Patrika* quoted in *Hindoo Patriot* July 7, 1873.

[37] Tayler to Molony, File 448, Nos 111-113 B.J.P.P. July 1873; B*engalee* Sept.6, 1873, *The Spectator* quoted in *Bengalee* Sept. 27, 1873; *Englishman* July 10, 1873; *Amrita Bazar Patrika* 18 July, 1873. Bholanath Chandra—*Life of Digambar Mitra*—Vol. II, Calcutta 1906, pp. 73-74.

[38] File 448 Nos 111-113 B.J.P.P. July 1873. *Hindoo Patriot* July 14, 1873, Raja Narendra Krishna Deb's speech on the 20th Sept. 1873 (B.I.A.R.); *The Spectator's* article "The agrarian riots in India" quoted in *Bengalee* Sept. 27, 1873; *Friend of India* July 10, 1873; *Pioneer* July 24, 1873.

[39] *Pioneer* July 24, 1873.

[40] File 448 Nos 111-113 B.J.P.P. July 1873.

[41] *The Englishman* July 26, 1873.

[42] Hiren Mukherjee—*op. cit.*, p. 74; P. Sinha—*op. cit.*, p. 25.

that the non-payment of rent was only a tactical move initiated by the League to keep the zamindars out of all rents till they settled the questions in dispute.[43] Generally speaking, the ryots had no disposition to refuse the rents but on the contrary, they offered rents which the landlords considered inadequate and in many cases the preferred rents were actually deposited to the collector whom the ryots considered the representative of the queen.[44] Withholding of all rents was, therefore, not the basic aim of the League which had a positive economic programme based on certain definite demands such as the measurement of land by a large cubit, the reduction of rates to those in existence at the time of the Natore Raja, the abolition of the illegal cesses, and the granting of pattahs on the above terms.[45] In other words the programme of the Agrarian League was not merely a non-constructive resistance as Chaudhury has suggested[46] but did have a constructive aspect.

The leadership of the League was provided by men of considerable means such as the petty landlords like Ishan Chandra Roy of Daulatpore,[47] the village headmen like Shambhunath Pal of Meghoolla[48] and *jotedars* like Khoodi Mollah of Jogtollah.[49] Generally speaking the village headmen who belonged to a lower section of the rural gentry led the movement, collected the subscriptions and organised the villages with a view to offering physical resistance to *lathiyals*—men armed with heavy sticks hired by the landlords.[50] This particular section of the rural gentry including certain *amalahs* of the landlords was all along opposed to the big landlords for various personal reasons. Taking advantage of the prevailing agrarian tensions those men tried to consolidate their hold on the peas-

[43] Misc. colln. 14 No. 6. B.L.R.P. July 1873.

[44] File 448 Nos 111-113 B.J.P.P. July 1873; *Hindoo Patriot* July 14, 1873; *Friend of India* July 10, 1873.

[45] File 448; Nos 111-113 B.J.P.P. July 1873; *Englishman* July 26, 1873; *Friend of India* July 24, 1873; Report of the Progs. 22nd Annual Meeting of the British India Association dt. 7th April 1873 (B.I.A.R.).

[46] N.K. Sinha, ed. *op. cit.*, p.291.

[47] File 448 No. 44 B.J.P.P. July 1873, File 448 Nos 111-113 B.J.P.P. July 1873; *The Pioneer* Aug. 6, 1873.

[48] Misc. colln. 14, Nos 12-13 B.L.R.P. Jan 1874; *Amrita Bazar Patrika* July 18, 1873; *Englishman* July 26, 1873; Nolan's Diary dt. 1.7.1873, File 448 Nos. 111-113 B.J.P.P. July 1873.

[49] *Hindoo Patriot* July 14, 1873; *Amrita Bazar Patrika* July 18, 1873.

[50] *The Pioneer* July 14, 1873.

antry.[51] In fact among those who were convicted in connection with the disturbances there were three Hindus of respectable caste who were not cultivators but lived by service alone.[52] This disgruntled section of the rural gentry, because of the eminent social position which they occupied in rural society, became really prosperous through the cultivation of jute, a cash crop, which only the substantial ryots could afford to cultivate.[53] This was proved by the heavy demand for immigrant labour in Pabna in this period and the corresponding absence of emigration, notwithstanding the spectacular population growth.[54] These rich peasants were sufficiently well-to-do to offer an effectual resistance to the landlords.[55] The occupancy ryots led by a section of the rural gentry and in alliance with substantial peasants thus constituted the hard core of the League.

Other sections of the peasantry, namely, the tenants-at-will, the share croppers, and the agricultural labourers, were not directly affected by the rent disputes but in Pabna the rapacity of the landlords threw even these sections of the peasantry into the arms of the occupancy ryots. The Agrarian League, in consequence, managed to enlist the overwhelming support of the lower orders.[56] The entire district was thus divided into two rival camps; one espousing the cause of the League and the other supporting the landlords.[57] This led Peter Nolan, the Joint Magistrate of Serajgunje to remark: 'This class feeling was so universal that the opinion of any native to the agrarian question may be told to a certainty by looking at his dress. If he wore a light 'chaddar' on his shoulder, used shoes on his feet and carried an umbrella, one could make sure that he was a zamindar's man. If merely clad in the Dhoti and Gamcha (a towel) he was at heart a unionist'.[58] The arraying of the ill-dressed against the well-dressed, of poor against rich, comes out even from Nolan's words.

[51] File 448 Nos. 111-113 B.J.P.P. July 1873; Englishman July 15, 1873.
[52] File 448. No. 77 B.J.P.P. Aug. 1873.
[53] Bengalee Dec. 20, 1873.
[54] Bengal Administration Report 1871-72, p. 34.
[55] Misc. Colln. 14 No. 11 B.L.R.P. Jan. 1874.
[56] Nolan's Report dated 23.4.1874, loc. cit.
[57] Ibid.
[58] Ibid.

III. *Some Aspects of the Character of the Movement Organised
by the League*

As the movement organised by the League gradually assumed the character of righteous protest against high-landlordism, the landlords of Bengal became conscious of the fact that a further extension of the movement would adversely affect the overall position of the landlord class. It was apprehended that the force of the movement would eventually compel the government to review the entire question of landlord-tenant relationship in Bengal which might in its turn lead to a new tenancy legislation in the interest of the tenantry. Consequently the landlords of Bengal who had practically the entire communication media on their side, made a deliberate attempt to confuse the issues. Spearheaded by the redoubtable *Hindoo Patriot* the pro-landlord press frantically tried to convince the government of Bengal that the movement organised by the Pabna League was not an agrarian movement but really a communal movement of the Muhammedan tenantry against their Hindu landlords.[59] The *Halishahar Patrika* even went to the extent of suggesting that by pursuing a soft policy in Pabna the government was actually placating the Muslims.[60] Thus we see how class conflict aroused by the exploitation of the landlords led fortuitously to feelings later classified as communalism since the exploiters were Hindus and the rank and file of the peasantry who combined against the landlords were Muslims. It was perhaps this social character of class conflict in Eastern Bengal in the pre-Independence days which led the Muslim peasantry in East Bengal to welcome the partition in 1947.

Those who sought to impute a communal character to the movement had, however, completely overlooked the fact that the two top leaders of the League, Ishan Chandra Roy and Shambhunath Pal to whom the League owed its very existence were caste Hindus who would certainly not have associated themselves with the League if it had been directed against the entire Hindu community. On the contrary, the members of the League did not themselves make any distinction on grounds of religion and equally punished all those who refused to join the League.[61] Besides, the statistical survey conducted by the Government of Bengal in 1871-72 revealed that the majority of the inhabitants of Pabna were Muhammedan converts from the lower castes of Hinduism, most of whom were completely ignorant of even the elementary doctrines of the *Koran* and were rejected by the true

[59] *Hindoo Patriot* July 14, 1873; *Halisahar Patrika* July 4, 1873 (R.N.P. Beng. 1873).

[60] *Pioneer* July 28, 1873.

[61] *Amrita Bazar Patrika* 25/26 June 1873.

Muhammedans as unfaithful.[62] Such men could not consciously rise against the landlords merely because the latter were Hindus. A contemporary journal thus remarked, 'though the majority of the ryots are Muhammedans and the landlords and the upper class are chiefly Hindus there is nothing like a *Faraizi* rising.'[63]

The organs of the Bengal landlords also made a conscious attempt to create an impression that law and order had broken down in Pabna where the ryots were committing all sorts of atrocities.[64] Thus Dwijendranath Tagore,[65] one of the famous cultural figures of the Bengal renaissance drew the attention of the Lt. Governor of Bengal to the acts of wanton violence committed by the Pabna peasantry upon the 'inoffensive people'.[66] Tagore's complaint was corroborated by a group of Pabna *muktears*,[67] the agents of the local landlords, who in a petition to the government of Bengal painted a lurid picture of mob-violence in the district.[68] The organs of the Bengal landlords, *Hindoo Patriot* and *Amrita Bazar Patrika*, also sent from Pabna reports of wanton plunder, rape and arson committed by the enraged peasantry.[69]

The enquiry conducted by the local Pabna officials following the publication of these reports in the press, however, established the fact that most of the reports were either false, or exaggerated.[70] This was confirmed by the results of the Pabna trials, which showed that the ryots arrested on charges of rioting and plunder were either acquitted or let off with light punishment.[71] Most of the influential newspapers also thought that the reports of mob-violence deliberately circulated by the landlords were grossly exaggerated since the outrages committed by the ryots were remarkably few.[72]

[62] Appendix A being No. 85 of the B.R.P. Aug. 1872, p. 79.

[63] *Friend of India* July 10, 1873.

[64] *Hindoo Patriot* July 14, 1873.

[65] Eldest son of Maharshi Debendra Nath Tagore. Born 11th March 1840: Died 19th Jan. 1926.

[66] File 418 No. 37 B.J.P.P. July 1873.

[67] An agent, attorney.

[68] File 448 No. 44 B.J.P.P. July 1873.

[69] *Hindoo Patriot* July 14, 1873; *Amrita Bazar Patrika* July 18, 1873.

[70] File 448 Nos. 111-113 B.J.P.P. July 1873.

[71] File 448 Nos. 78-79 B.J.P.P. Aug. 1873.

[72] *Indian Daily News (Bengalee* July 19, 1873); *Bengalee* Aug. 9, 1873; *The Spectator (Bengalee* Sept 27, 1873); *Friend of India* July 17, July 24 1873; *Englishman* July 5, July 12, 1873; *Pioneer* July 8, 15, 1873; *Indian Mirror (Pioneer* July 10, 1873); *Indian Observer* July 12, 1873.

Basically, the Pabna movement was a non-violent and legalistic agrarian rising. Instances of violence were rare because the peasant leaders were too wise to take the law into their own hands and actually advised the ryots to keep themselves within the bounds of the law.[73] In fact, in parts of Yusufshahi where the influence of Ishan Chandra Roy was greatest and most direct, the movement was accompanied with the least excesses and carried on in a legal manner, but in the Sadar sub-division where his influence was vague and more general there had been far more plundering and rioting.[74]

Illegal violence, however, never formed an essential part of the agrarian union.[75] The cases of violent crime, apart from occurring in conflicts between the followers of rival landlords, were also due to the criminal classes who took advantage of the excitement and were definitely responsible for violence in certain areas.[76] Even then the outrages committed by people who were considered 'bad characters' by the district authorities fell far short of the outrages habitually perpetrated by the landlords of the district.[77]

The members of the League, on their part, strictly confined themselves within the bounds of the law as their organisation was essentially defensive.[78] It is also significant that in spite of the great class antagonism prevailing in Yusufshahi during this period, the landlords were not molested, the roads and ferries were not beset and the landlords' agents could freely come to the magistrates to lodge complaints or to present their petitions.[79] Thus no one was killed or seriously injured and there was no significant damage to property even when the miasma of class hatred was fast spreading throughout the district.[80] The Pabna peasantry thus did not imitate the crimes of their landlords taking all the vengeance that makes a *jacquerie* and their movement at no stage showed any tendency to

---

[73] *The Hindoo Patriot* July 14, 1873.

[74] *Pioneer* Aug. 4, 1873.

[75] *Ibid.*

[76] Nolan's Diary, File 448 Nos. 111-113, B.J.P.P. July 1873; *Friend of India* July 24, 1873; *Pioneer* July 28, 29, 1873, *Indian Daily News (Bengalee* July 19, 1873); *Sadharani* 18 *Kartick* 1280 B.S. (Nov 2, 1873); Nolan's Report dt. 23.4.1874, *loc. cit.*

[77] *Pioneer* July 29, 1873.

[78] File 448 No. 40 B.J.P.P. July 1783.

[79] Letter No. 321 File 448 Nos 111-113 B.J.P.P. July 1783.

[80] File 448 Nos. 78-79 B.J.P.P. Aug. 1873.

gravitate towards the criminal courts.[81] This led Sir William Hunter to remark: 'They (the rural population) have fought with keen persistence but with few ebullitions of violence the struggle between landlord and tenant and are conducting before our eyes an agrarian revolution by due course of law.'[82]

IV. *The Landlord Reaction*

Following the formation of the League, the landlords' initial reaction was merely to watch the course of the events. They held off in the hope that the ardour of the League would cool down or its members would get into trouble in the criminal courts and so would lose spirit.[83] The subsequent events in Pabna, however, belied these expectations because in spite of the efforts of the local officers to bring about a compromise, the Pabna peasantry did not relent and carried on its struggle with determination.[84] This forced the landlords to reorient their strategy for combating the League.

At first the landlords endeavoured to harass the ryots in the civil courts,[85] whose dilatory procedure and the imperfections of the law which these courts administered, provided the former with ample opportunities for concocting evidence against the members of the League.[86] Besides, the landlords tried to meet the challenge of the League by granting perpetual leases or *putnis* to their under-tenants in large numbers.

[81] *Pioneer* July 8, 1873; *Friend of India* July 24, 1873.

[82] Statistical Account of Bengal (1875-76), Vol. IX Preface. The facts stated above clearly show that Hunter was not 'over simplifying the situation' as stated by P. Sinha, *op. cit.*, p. 25.

[83] File 448 Nos 111-113 B.J.P.P. July 1873.

[84] *Bharat Sanskarak* March 26, 1875 (R.N.P. Beng. 1875); Divisional Report Rajshahi 1875-76, File 122/1 B.G.P. 1876; *Ibid* for 1877-78; File 144/I B.G.P. Sept. 1878, Paras 73 and 109.

[85] *Bharat Sanskarak* March 26, 1875 (R.N.P. Beng. 1175); Divisional Report Rajshahi for 1875-76, File 122/1, B.G.P. Aug. 1876; *Gyan Bikashini* (Pabna) Jan. 5, 1874 (R.N.P. Beng 1873) Buckland, *op. cit.*, I, p. 548.

[86] Article in *Friend of India* quoted in *Hindoo Patriot* Sept. 22, 1873.

PERPETUAL LEASES GRANTED BY THE LANDLORDS OF PABNA
SINCE 1873–74[87]

Year	Number of deeds registered	% of Increase since 1873-74
1873-74	627	
1874-75	1406	124%
1875-76	1579	151%
1876-77	1633	160%

It is evident from the above table that since 1874 the putni system was fast developing in Pabna in all its economic starkness as a measure by the Permanent Settlement landlords to create the sub-intermediary as an effective means to forestall the challenge of the League. This led to the creation of a new type of rich peasant bulwark within the tenant class which in its turn brought about a very material change in legal relationships and economic alignments.

The system, however, was a flimsy bulwark against the powerful Agrarian League which existed in Pabna. This was corroborated by the reports of the Divisional Commissioners of Rajshahi which showed that notwithstanding the pressure of civil litigation and the development of the putnis, the League's solidarity could not be broken during all these years.[88] Consequently the situation remained inconclusive in Pabna which continued to be a sort of hot-bed of smouldering discontent and distrust between the landlords and the tenants.[89] In the circumstances the landlords had to strain their resources to the utmost to meet the mounting expenses of civil litigation. Precisely for this reason, the landlords who had enough resources in 1875[90] to harass the ryots in the civil courts had become by 1878 quite impoverished.[91] They had also learnt that they could no longer obtain the illegal cesses and that the ryots were able and determined to fight for their rights.[92] The landlords were thus brought face to face with the

[87] Reports on the Administration of the Registration Department Bengal for these years.

[88] Divisional Reports, Rajshahi for 1876-77, 1877-78, 1878-79, 1880-81, 1883-84 and 1885-86 B.G.P.

[89] *Ibid.* for 1875-76 B.G.P. Aug. 1876.

[90] *Bharat Sanskarak* March 26, 1875 (R.N.P. Beng. 1875).

[91] Divisional Report Rajshahi, 1877-78, File 144/1 B.G.P. 1878, para 73 and 109.

[92] *Ibid.*

question of giving up all demands in excess of actual rent. It appeared that this question was likely to become a general one in the years to come.

V. *Conclusion*

It is apparent that the causes, character and consequences of the Pabna disturbances are open to conclusions which have not been spelled out by previous Marxist or social and economic historians. The grievances of the ryots were not narrowly concerned with regard only to the enhancement of rent; they felt that high landlordism struck at the very roots of their occupancy status and their uprising had a more general character than is usually ascribed to it. The organisation and activities of the combinations which they formed to carry on the movement clearly show the arraying of the poor against the well-to-do. The communal character which was attributed to the movement by the mouthpieces of the property owners and the imputations of widespread arson and violence reflect the growth of communalism among the propertied class but are not an accurate description of the complex factors which shaped the general character of the disturbances. The disturbances had a wider effect than is commonly supposed in shaping the reactions of landlords towards the continuance of agrarian combinations but the innate strength of the Agrarian League finally foiled all the efforts of the landlords to destroy the solidarity of the peasants.

ABBREVIATIONS

B.J.P.P.	Bengal Judicial (Police) Proceedings.
B.L.R.P.	Bengal Land Revenue Proceedings.
B.R.P.	Bengal Revenue Proceedings.
B.G.P.	Bengal General Proceedings.
B.L.T.P.	Bengal Local Taxation Proceedings.
B.I.A.R.	British Indian Association Reports.
R.N.P. (Beng.)	Report on Native Press, Bengal.
A.P.	Argyll Papers (N.A.I. microfilms).

Chapter Three

Peasant Revolt: An interpretation of Moplah violence in the nineteenth and twentieth centuries

CONRAD WOOD

Definition

> Not mere riots or affrays, but murderous outrages, such as have no parallel in any other part of Her Majesty's dominions.[1]

The violence periodically manifested during the nineteenth century by the Muslims of Malabar, the Moplahs, was a perpetual source of horrified fascination for British officials in the Madras Presidency. The wonder consisted in the configuration of the violence, styled the Moplah 'outbreak' or 'outrage'. Characteristically, the preparations for an outbreak involved the intending participants donning the white clothes of the martyr, divorcing their wives, asking those they felt they had wronged for forgiveness, and receiving the blessing of a *Tangal*, as the *Sayyids* or descendants of the Prophet are called in Malabar, for the success of their great undertaking. Once the outbreak had been initiated openly, by the murder of their Hindu victim, the participants would await the arrival of government forces by ranging the countryside paying off scores against Hindus they felt had wronged them, burning and defiling Hindu temples, taking what food they needed, and collecting arms and recruits. Finally, as the government forces closed in on them, a sturdy building was chosen for their last stand. Often the mansion of some Hindu landlord (frequently the residence of one of their victims) was selected, but Hindu temples, mosques, and other buildings were also used, the main criterion being, apparently, to avoid being captured alive. As a Moplah captured at Payanad temple in 1898 put it, it was decided to die there 'as it was a good

---

[1] Minute by J.D. Sim, member of the Madras Executive Council, nd, Government of Madras Judicial Proceedings (hereafter MJ), 1606-A/5A, 28 August 1874.

building and we were afraid lest we would be shot in the legs and so caught alive'.[2] By the time the government forces surrounded them, the outbreak participants had worked themselves into a frenzy by frequent prayers, shouting the creed as a war-cry and singing songs commemorating the events of past outbreaks, especially that of 1849 in which fifteen Moplahs armed mainly with 'war knives' scattered two fully armed companies of sepoys. The climax of the drama came when they emerged from their 'post' to be killed as they tried to engage in hand-to-hand combat.

Divergences from this ideal pattern were frequent but the essence of the Moplah outbreak, demarcating it from other forms of violence, lay in the belief that participation was the act of a *shahid* or martyr and would be rewarded accordingly. As one participant (who recanted at the last moment and was captured) said in explanation of why he and his associates 'went out' (i.e. participated in the outbreak): 'I have heard people sing that those who... fight and die after killing their oppressors, become shahids and get their reward. I have heard that the reward is "Swargam" [Paradise].'[3] The pattern of a Moplah outbreak was dictated by the fact that participants had no intention of evading the heavy hand of justice. On the contrary their objective was to compass their own destruction by hurling themselves in a suicidal charge against the forces sent to deal with them. In the words of a wounded Moplah captured at Manjeri temple in 1896: 'We came to the temple intending to fight with the troops and die. That is what we meant to do when we started.'[4] The defining characteristic of the Moplah outbreak was devotion to death.

Analysis

The Moplah outbreak was a more or less regular occurrence from 1836 to 1919, between which dates twenty-eight separate occasions can be distin-

[2] Statement of Trasheri Unni Ali, 4 April 1898, in the Court of the Special Assistant Magistrate of Malabar, MJ 1737-40, II November 1898, p. 23.

[3] Statement of Ambat Aidross, 16 March 1896, India Office Public and Judicial Department Papers (hereafter IOP & JDP), 996/96, p.15. See also statements of captured outbreak participants Pottanthodika Alevi (MJ 1737-40, II November 1898, pp. 24-5) and Moideen (MJ 794, 1 December 1849, p.4774), as well as the statement of an outbreak leader's father quoted in the Report of T.L. Strange, 25 September 1852, MJ 483, 23 August 1853, p. 4549 (hereafter Strange Report).

[4] Statement of Aruvirallan Muttha, 13 March 1896, taken by Mr Winterbotham, member of the Madras Board of Revenue, deputed to report on the 1896 outbreak, IOP & JDP 996/96, p. 12.

guished in which Moplahs actively sought their own death. In all twenty-eight cases except three, the number of participants ranged from a single Moplah to nineteen. The final total depended partly on how quickly the outbreak was suppressed, because usually the initiators were joined by recruits as the outbreak took its course. Thus, in the case of the 1896 outbreak, which was prolonged, exceptionally, for several days, the number of participants grew to a record ninety-nine, and others appear to have been on their way to join when the gang was rapidly destroyed, only five being taken alive. Of the three hundred and forty-nine participants in the twenty-eight outbreaks only twenty-three failed to achieve their end, and this includes the twelve forced to surrender in the very exceptional and significant case of the 1898 outbreak.

The *type* of man who participated in outbreaks is summed up in a report on the 1896 affair as 'field-labourers, porters, timber-floaters, mendicants, and others of the lowest class, living from hand to mouth'.[5] In fact report after report on Moplah outbreaks indicates that the great majority of participants were wage-workers, poor tenants, and the like, with a sprinkling of *mullas* of barely-distinguishable economic standing, criminals on the point of having their careers cut short by authority, the chronically diseased, and men who were rather more comfortably-off but who, often, had experienced economic decline. It was calculated that in the 1896 affair more than three-quarters of those involved were 'more or less really poor', 2 or 3 per cent 'comfortable' and the rest youths living with, and more or less supported by, their parents.[6] In fact, although men of almost any age might be found in the ranks of participants, young men predominated. In the 1896 gang about 30 per cent were in their teens, 40 per cent in their twenties, 20 per cent in their thirties and only 7 per cent between 40 and 60.[7]

All the victims of the twenty-eight outbreaks were Hindus, with the exception of two Collectors of Malabar, H.V. Conolly, who was murdered in 1855, and C.A. Innes, who narrowly escaped the same fate in 1915. (The incident in 1851 in which a Moplah, discontented with an arrangement concerning family property, stabbed five of his own relatives and announced his intention of dying as a martyr is not included in the twenty-

[5] Preliminary Report of H.M. Winterbotham, 10 April 1896, *ibid.*, p.6.

[6] Report of F. Fawcett, Superintendent of Police, Malabar, 5 June 1896, IOP & JDP 2060/96, pp. 101-3 (hereafter Fawcett Report).

[7] Report of J.T. Gillespie, Acting Special Assistant Magistrate, Malabar, 26 April 1896, *ibid.*, pp. 38-47.

eight outbreaks since I have no evidence that he 'actively sought his own death'.[8]) Of the eighty Hindu victims the caste status of seventy-six is determinable. Of these sixty-two were members of high castes (thirty-four Nairs, twenty-two Nambudri Brahmins, and six non-Malayali Brahmins) and the other fourteen of castes ranking below Nairs in the hierarchy, eleven being Cherumar, traditionally agrestic slaves in Malayali society. Something of the class background of seventy of the Hindu victims is known. Eighteen were rich *jenmis* and or moneylenders, four were *kariastans* and advisers of jenmis, seventeen guests, retainers, servants, and similar dependants of jenmis, ten members of jenmis' families, five village headmen (where not definitely known to be jenmis as well, as headmen in Malabar villages usually were) and four other government officials, making a total of fity-eight known to be themselves powerful figures in the Malabar countryside, or directly associated with such. Of the twelve remaining victims whose class status is known, three were labourers, five members of one labourer's family, one priest, two 'cultivators', and one teacher.

Of the twenty-eight jenmis and moneylenders and their families who fell victim to Moplah outbreaks nineteen were Nambudri Brahmins and eight Nairs, with one of unknown caste, while the eleven agents and officials comprised one Nambudri, nine Nairs, and one unknown. This restriction of those victims classifiable as 'powerful' to the high Hindu castes was not fortuitous. Malabar, and especially that part where outbreaks occurred, was throughout this period pre-eminently a land of the big Nambudri jenmi and the Nair official. Nearly all the big jenmis were in fact high-caste Hindus. In 1915 Collector Innes gave figures showing that the eighty-six biggest landlord families, owning many hundreds of thousands of acres and paying about a fifth of the total land revenue, were all high-caste Hindus except two Moplahs, one Tiyyan, and one Goundan.[9] The powerful[10] and often hereditary post of *adhigari* was usually in the hands

---

[8] See report of H.V. Conolly, 23 July 1851, MJ 29 July 1851, pp. 2504-5.

[9] Strictly confidential note by C.A. Innes on the Malabar Tenancy Legislation (hereafter Innes Note), Government of Madras Revenue Proceedings (hereafter MR), 3021, 26 September 1917, p. 25. See also the great predominance of high caste Hindus among the signatories of the papers in which the principal landholders acquiesced in the settlement principles of 29 June 1803: W. Logan (ed.), *A Collection of Treaties, etc.,* 2nd edition (Madras, 1891), Part II, p. 355.

[10] See the details of revenue, magisterial, and police duties of the Malabar Adhigari in B.H. Baden-Powell, *Land System of British India*, 3 vols (Oxford, 1892), III, pp. 88-9.

of the biggest local jenmi, and therefore frequently of a Nambudri,[11] while the ranks of the administration and the judiciary were heavily weighted with Nairs and other high-caste Hindus.[12]

One striking feature of the Moplah outbreak was its virtual restriction to only one part of the area of Malabar inhabited by Moplahs. With only three exceptions, every outbreak took place in the rural parts of interior south Malabar (and of the remaining twenty-five all but one in Ernad *taluk* or northern Walluvanad). The three exceptions include two in rural north Malabar and one in which Moplahs from interior south Malabar made their way to the residence of District Magistrate Conolly in Calicut town to butcher him and initiate their outbreak. This phenomenon of geographical restriction was not the least of the enigmas the Moplah outbreak presented to perplexed British administrators, for whom the fact that Pandalur Hill happened to be located roughly in the centre of the outbreak zone came to assume obsessive proportions. 'I have puzzled for twenty-five years why outbreaks occur within fifteen miles of Pandalur Hill and cannot profess to solve it,' was the lamentation of H.M. Winterbotham in his report on the 1896 outbreak.[13]

I *Official Rationale*

For decades this startling phenomenon of the Moplah outbreak presented British administration with the most taxing problem of interpretation, a problem it seemed necessary to solve if appropriate policy measures were to be adopted. Since so many of the attacks involved the selection of victims who were rich landlords or their agents and since so many of their assailants were men in social positions vulnerable to the adverse exercise of their economic and social power, it seemed obvious to ascribe Moplah outbreaks to antagonism between landlord and tenant, or landlord and labourer. What strengthened the assumption was the fact that on the occasions when the grievances of outbreak partipants were recorded, alleged oppression by landlords, buttressed by the courts and local administration, figured prominently. Thus the participants in the 1843 Pandikkad outbreak, which was directed against overbearing local notables, com-

[11] C.A. Innes, *Malabar Gazetteer* (Madras, 1908), I, p. 105.

[12] See, for example, H.V. Conolly to Secretary, Judicial, 26 February 1853, MJ 379, 2 July 1853, p. 3671, and Minute of J.F. Thomas, 3 November 1852, MJ 483, 23 August 1853, pp. 4696–97.

[13] IOP & JDP 2060/96, p.62.

plained that 'it is impossible for people to live quietly while the Atheekarees [village headmen] and Jenmies... treat us in this way'.[14] Again, the leader of the 1849 gang, Athan Gurikal, left behind a document in which he claimed that the behaviour of landlords in collusion with public servants, 'the majority... being of Hindoo caste', was a source of grievance to 'the Mussal-men inhabiting the inland part of Malabar'.[15]

One of the main forms in which landlords were felt to be behaving oppressively was their use of powers of eviction, recognized by the courts. One specific method of exercising these powers which was especially resented, and which became popular with jenmis in the course of the outbreak period, was by granting of *melcharths*, by which the jenmi virtually sold (sometimes by literally putting it up for auction[16]) to a third party the right to oust and replace one of his tenants. Thus a Moplah captured in 1896 gave as a reason for the outbreak the 'fact' that 'poor folks who have only two or three paras of land are ejected and put to trouble by the grant of melcharths over their heads'.[17]

Almost without exception every British official concerned with interpreting the Moplah outbreaks was prepared to concede that all was not well with landlord-tenant relations in Malabar, and the grievance over insecurity of tenure was repeatedly stressed by them. Explaining outbreaks as anti-jenmi manifestations, however, posed thorny problems which those Malabar Collectors most sympathetic to tenant grievances grappled with more or less unsuccessfully. In particular, since *Hindu* tenants and labourers admittedly suffered quite as much, if not more, from the great power of the big jenmi, why were outbreaks confined to the Muslim community? Moreover, why should some of the assaults have been directed against Hindus who were not only *not* landlords, but members of the slave caste at least as vulnerable to the exercise of jenmi power as many of the assailants themselves? Failure by those who stressed the agrarian explanation for outbreaks adequately to answer such questions undermined their case for legislation to grant occupancy rights to tenants, a measure they urged as essential if the Moplah problem were to be solved.

[14] 'The Wurrola chit [anonymous writing] written for the perusal and information of Walluvanaad Tahsildar', no date or signature, left behind in the house in which they had 'taken post', MJ 69, 27 January 1844, p. 286.

[15] 'Writing of Syed Assan, Manjery Athan and all the others who have taken possession of Manjery temple', MJ 794, 1 December 1849, p. 4784.

[16] Innes Note, p. 21.

[17] Statement of Puzhutini Kunyayu, 14 March 1896, IOP & JDP 996/96, p. 12.

The shortcomings of the case presented by the 'pro-tenant' school of Malabar Collectors were seized on by the government of Madras which, in this period at least, was most reluctant to intervene in agrarian relations in Malabar in favour of the tenant. Would-be reformers were fully conscious that before any meddling with the powers of the jenmi could be considered it was incumbent on them 'to show some political necessity for interference'.[18] The failure of Collector Innes to show any such thing when he presented his case for legislation in 1917 met with this response from the Board of Revenue:

> Mr. Innes speaks of the janmis of Malabar as a 'political force on the side of Government'. In the Board's opinion there can be no doubt that tenancy legislation of the kind now suggested would be a grave political mistake, as it would alienate this force from the Government, and the Government could not count on receiving from the tenants anything in the way of gratitude to replace this loss.[19]

The Board strongly recommended that the question of tenancy legislation for Malabar should be dropped and the government of Madras readily agreed.

As early as 1852, when the outbreak situation had become so serious that a Special Commissioner, T.L. Strange, had been appointed to ascertain the cause of outbreaks, the local government had hinted at another interpretation of the phenomenon and a different policy approach when it issued Strange with his instructions. He was to bear in mind that his 'grand object' should be 'to secure to the Nair and Brahmin population the most ample protection and safety possible against the effect of Moplah fanaticism'.[20]

With such a direction, perhaps it is not surprising that Strange came to the conclusion that outbreaks were not due to agrarian grievance but that 'the most decided fanaticism... has furnished the true incentive to them', adding that the 'pride and intolerance fostered by the Mahomedan faith, coupled with the grasping and treacherous, and vindictive character of the Moplas in these districts drawn out to its worst extent have fomented the

[18] Report of Madhava Rao Commission to advise the Government of Madras on legislation on Malabar land tenures, 17 July 1884, in Government of Madras, *Malabar Land Tenures* (hereafter *MLT*) (Madras, 1885), p. 122.

[19] Reference from the Board of Revenue (Land Revenue) 2105-A/Gt. 15-2/C. 30, 1 May 1917, MR 3021, 26 September 1917, p. 6.

[20] Minute 123, 17 February 1852, MJ p. 678.

evil and it may be said to lie at the root thereof'.[21] As an 'explanation' for the Moplah outbreak this posed more questions than it answered. In particular, why should the Muslims of two Malabar taluks have reacted fanatically to their religion when so many Muslim communities, including those in the rest of Malabar, did not? The claim that it was the 'character' of these particular Muslims to react in this way in itself explained nothing.

Moreover, the attributing of outbreaks to religious fanaticism itself posed policy problems for the government. British administrators in India tended to believe that interference in the religious affairs of the people was more likely to stir up trouble than allay it. As was said of an 1896 proposal to regulate the teaching of the *ulema*:

> Any real attempt to control religious teaching and preaching would be viewed as persecution, and we should have sedition preached on the hill-tops, in the depths of the jungle, and in dens and holes in the earth.[22]

Even so, the government did act on Strange's report. Its representatives in Malabar had already persuaded one of the leading Ernad Tangals, Syed Fazl, the Tirurangadi Tangal, who was suspected of fomenting outbreaks, to remove himself to Arabia. But its main policy instruments were the repressive Moplah Acts of 1854. Formulating a repressive policy to deal with men whose very aim was death was not found to be an easy task. The Moplah Acts, however, provided for the banning of the Moplah war knife, the deportation without trial of anyone suspected of intending to participate in an outbreak, the confiscation by the government of the property of participants, and the levying of fines on the inhabitants of localities involved in the disturbances. This last provision was especially significant. It reflected the conviction of all government servants in Malabar that the great majority of Moplahs were sustaining outbreaks by their sympathy with them. The remark of Collector Conolly in 1843 that it seemed evident that the Moplahs' 'real sympathies were always enlisted on the side of the Criminals'[23] was typical, while in 1849 Conolly remarked that seldom did 'a Moplah of the lower order' pass the grave of any participant in earlier outbreaks 'but in silence and with an attitude of devotion, such as is usual in this district in passing a mosque', adding that 'despite the prohibition of

[21] Strange Report, pp. 4587 and 4591.

[22] Final Report of H.M. Winterbotham, 5 May 1896, IOP & JDP 2060/96, 30 September 1896 (hereafter Winterbotham Report), p. 64.

[23] Conolly to Secretary, Judicial, 4 November 1843, MJ 69, 27 January 1844, p. 216.

the authorities, ceremonies are from time to time secretly performed in their remembrance to an admiring audience'.[24]

It would be natural to enquire, in view of this frequently noticed sympathy on the part of the great mass of Ernad Moplahs, whether the Moplah community of the outbreak zone, or at least an important section of it, was gaining in any way from outbreaks. Although there is no record of any thorough investigation of this question by the government of Madras, it was clear that it was hoped the policy of fining would mean that the community as a whole, and more particularly its richer and more influential members who were to be the special target of the fines,[25] would come to believe it must *lose* whenever an outbreak occurred, and act accordingly. In fact, the record shows that the frequency of outbreaks *did* decline after the passing of the Acts. From 1836 to 1854 when the measures were enacted, sixteen outbreaks occurred. From 1855 to 1887, when a change of policy was effected, the total was only seven. In the first period (of 18 years) outbreaks occurred at a rate of rather less than one every year, in the second period (of 32 years) one every four or five years.

The administration began to feel that their aim of rendering outbreaks 'comparatively unimportant and unfrequent'[26] was being achieved. Even so outbreaks *did* continue, and when in 1880 an anonymous petition was received setting out tenant grievances, especially regarding eviction, and threatening that 'the severity of the oppression of the Malabar land lords will lead to great disturbances, at which a great number of people will lose their lives... disturbances and bloodshed of a kind unknown in Malabar'[27] a second Special Commissioner, W. Logan, was appointed to investigate Malabar land tenures.

Logan argued that the Moplah outbreak was the outcome of administrations imposing through the courts their British agrarian preconceptions on a state of society fundamentally different to that of their homeland. By recognizing the jenmi as the absolute owner of his holding and 'therefore free to take as big a share of the produce of the soil as he could screw out of the classes beneath him' the British had, Logan claimed, presented him

[24] Connolly–Secy, Jdcl, 28 July 1949, MJ 503, 7 August 1949, pp. 2574-8.

[25] Strange Report. See his comments on Section 6 of his draft Moplah Act No. I, p. 4626.

[26] Minute 123, 17 February 1852, MJ, p. 678.

[27] 'Petition purporting to be addressed by certain Mussulmans, Nayars, Tiyyans and men of other castes inhabiting Malabar', 14 October 1880, MLT, pp. 14-15.

with powers which were not customary in Malabar.[28] Logan pressed in vain for occupancy rights to be conferred on certain categories of tenant, to curb jenmi power. All the government would concede was a measure, the Compensation for Tenants' Improvements Act of 1887, providing for payment by jenmis, in the event of their resorting to eviction, of the 'full' value of improvements made by tenants at a rate determined by Court Commissioners. The hope was that this requirement would impose 'a check on the arbitrary exercise of the power of eviction'.[29] Even generally pro-tenant administrators like H. Bradley and C.A. Innes[30] agreed (though with much qualification) that the Act at first worked favourably for the tenants, especially after it was strengthened by amendment in 1900. Meanwhile, after two bad disturbances in the earlier 1890s, outbreaks came to an end for an unprecedented seventeen-year period with the remarkable outcome to the abortive affair of 1898.

Like most outbreaks that of 1898 departed from the ideal pattern in several ways. In particular the initiators probably had no settled intention of becoming shahid in the period of preparation for what at first was simply an assault on a big Nambudri jenmi by timber carters aggrieved at the payment they had received from him and the beating he had had meted out to them when they protested. When the jenmi was actually killed in the course of the assault, the gang appears to have decided to die as martyrs and the affair then took the course of a normal outbreak. The late decision to become shahid was probably not unique and may well have been a feature of outbreaks on previous occasions, especially those of November 1841 and August 1849. What was very significant, however, was the way the affair ended. When the government forces arrived at the temple selected for the last stand, they found it surrounded by three or four hundred Moplahs, led by a local Tangal, in the process of trying to induce the gang to surrender, which it in fact did.

Quite apart from the persuasive powers of the Tangals who parleyed with the gang it is clear that the continuation of the outbreak by means of the final suicidal charge became impossible once several hundred Moplahs had separated the gang from its objective, hand-to-hand combat with

---

[28] W. Logan, 'Report on Malabar Land Tenures' (hereafter Logan Report), in *Report of the Malabar Special Commission*, 3 vols (Madras, 1882), I, p. xvii.

[29] Government of Madras, *Malabar Land Tenures Committee Report* (hereafter *MLTCR*), 22 February 1887, p.5.

[30] Report on the working of the Act, 31 January 1894, MJ 2374 (Confidential), 1 October 1894, p.2. Innes Note. p.9.

the government forces. The British administration, whose servants in Malabar had openly despaired of ever being able to devise a method of capturing outbreak participants alive,[31] had been witness to a pacific demonstration of the power of numbers of which Gandhi himself might have been proud. As G.W. Dance, the Collector, pointed out in his report, such a thing had never before been known in the history of Moplah outbreaks and yet this had happened in Payanad, an *amsom* notorious for its outbreak record and this after the gang had murdered an unpopular Hindu jenmi.[32]

The British had for decades been familiar with the widespread support, both overt and covert, for outbreaks on the part of Ernad Moplahs. That this striking demonstration in the heart of the outbreak zone indicated a shift in attitude within the community seems even more likely when the following seventeen years of freedom from outbreaks are considered. As one British official perceived in 1896, 'until outbreaks become unpopular with the Moplahs as a body, they will not cease'.[33]

II *Thesis*

The manifest support for outbreaks on the part of the great mass of Ernad Moplahs, more especially 'among the lowest orders',[34] seems to have been rooted in resentment at the exercise by high-caste Hindus of massive power based ultimately on a virtual monopoly of land-ownership. Malabar in this period was a country in which land-ownership was not only restricted to the high Hindu castes (see figures given by Innes.[35]) What was described by Raja Sir T. Madhava Rao (by no means an opponent of the principle of landlordism)[36] as 'an extraordinary... a stringent and systematic monopoly of land' by the big jenmis, was 'well fortified by law on all

---

[31] H.V. Conolly to Secretary, Judicial, 23 December 1843, MJ 69, 27 January 1844, p. 249. See also Winterbotham's preliminary report cited in footnote 5, p.3.

[32] Report of G.W. Dance, 19 April 1898, MJ 1737-40, 11 November 1898, p.18.

[33] Winterbotham Report, p.66.

[34] Report of W. Robinson, Head Assistant Magistrate, Malabar, 18 October 1849, MJ 794, 1 December 1849 (hereafter Robinson Report), p. 4951.

[35] 6.8 per cent of the total population actually working in agriculture according to the 1911 census (i.e. 30,455 landowners out of a total of 449,719 landowners, tenants, farm servants and field labourers), *Census of India*, 1911, XII, Pt. 2, pp. 140-142.

[36] See 'Additional Remarks of Madhava Row', nd, in *MLT* pp. 190, 202-5, in which he spoke strongly against any abolition of landlordism.

sides', including the law of primogeniture and the system of joint families, the law providing for adoption in the case of failure of heirs and that debarring the jenmi from making gifts of land.[37] Above all it was fortified by the jenmi's strong traditional aversion to the sale of land, noted by all officials from the earliest period of British rule.[38] This withholding of the smallest portions of the major source of subsistence, of power, and even of the means of religious practice, was sometimes manifestly resented by Moplah outbreak participants. As the spokesman of the Kolatur gang of 1851 bitterly remarked about the difficulty local Moplahs were experiencing in trying to obtain a site for a mosque from their Nair landlord, 'what is the loss to the Nairs and Numboories if a piece of ground capable of sowing five Parrahs of seed be allotted for the construction of a Mosque? Let those hogs [the soldiers] come here, we are resolved to die.'[39]

But it was the power to manipulate the British legal system to the disadvantage of the tenantry that the jenmi land monopoly conferred that was most consciously resented by the Ernad Moplah. T.L. Strange in his report of 1853 noted that five of the relatives of the 1851 outbreak participants when examined subsequent to the outbreak said 'they had been taught to believe that if a poor man had been evicted from land it was a religious merit to kill the landlord.[40] W. Robinson, in his 1849 report, went so far as to say of the total destruction of one big jenmi's papers (a frequent event in Moplah outbreaks) that it was 'so natural a step for a set of ignorant Moplah had [lads?] taught from the [sic] childhood to look on these as the weapons with which the Nair and Rajah Jenmis... were ruining their Caste in the Courts and elsewhere, that their preservation had been to me unaccountable'.[41]

Where reports on Moplah outbreaks are detailed enough, time after time it is found that participants (or their families) had suffered from the attempt of a jenmi to exploit his powers under the British legal system at the expense of his tenants. The case of the Nair jenmi who changed the terms of one of his Moplah tenant families from a rent of 59 per cent of the

[37] 'Note by Rajah Sir T. Madhava Row, being a brief statement of the exceptional circumstances of the Malabar district, demanding exceptional treatment', nd, enclosure to Appendix F(c), *MLTCR*, pp. 29-30.
[38] See, for example, Major A. Walker, *The Land Tenures of Malabar, Report of 20 July 1801* (Cochin, 1879), p. II.
[39] 'Memorandum of the conversation between the Walluvanad Tahsildar and the fanatics', 28 August 1851, MJ 700, 20 November 1851, p. 3774.
[40] Strange Report, p. 4560.
[41] Robinson Report, p. 4939.

net produce to one of 77 per cent for a doubling of the annual expenditure of labour and found himself as a result the object of an outbreak led by one of the family[42] is merely one of the best documented instances. As Commissioner Logan pointed out, British administrations by permitting the jenmi to take as much as he could from his tenants had introduced a concept of landlord rights conflicting sharply with what the Malabar tenant considered legitimate. During his investigations Logan found 'how familiar even the most illiterate of the agriculturalists' were with the shares of the produce due respectively to landlord and tenant according to custom, and that they could usually 'say at once what the shares are of any particular bit of land... though the shares now actually paid as '*rent*' are very much greater'.[43]

British administration in Malabar, having conferred on the jenmi what the Moplah tenant saw as unjust powers of land-ownership, was extremely reluctant to legislate to curb the exercise of those powers (despite the advice of many of its own servants most familiar with the district) and on the contrary was prepared to use the most exceptional repressive legislation to suppress the challenge to that power. Little wonder then that almost throughout the nineteenth century the Ernad Moplah evinced little inclination to rely on appeals to the administration for curbing landlord power. A British District Superintendent of Police with much experience of Malabar towards the end of the century noted that the Ernad Moplahs had 'an insane idea that Europeans hate them and want to destroy them'.[44] The countervailing force against jenmi power had to be one generated by the tenants and labourers themselves.

The creation of such a force would seem necessarily to involve making the main resource of these groups, their superior numbers, a reality by combination. In Malabar circumstances made this especially difficult. Not only was the bulk of the rural population engaged in work which confined relations with others within a very narrow compass but also Malabar was a district of isolated houses, so that even the degree of human interaction which life in a nucleated village normally fosters would, no doubt, be absent here. Moreover, the isolation would be increased by the notoriously bad means of communication in the district (especially in the wild, less

[42] H.V. Conolly to T.L. Strange, 30 September 1852, MJ 716, 6 November 1852, pp. 4661-62 and 4667-69.

[43] Logan Report, p. xxviii.

[44] Fawcett Report, p. 112.

thickly populated country in the heart of the outbreak zone)[45] as well as by poverty.

For the Muslim section of the population, however, an important compensating factor for these obstacles to combination would have been their congregational form of worship which must have been partly responsible for the greater 'independence' of the Moplah as compared with the Hindu, often remarked on by the British in Malabar.[46] Certainly, among the many uses made of the mosque by outbreak participants, its utility as an aid to confederation emerges clearly in many outbreaks. In the case of the abortive affair at Malappuram in 1884 the plotters for a month before the attack took to sleeping in their mosque, constantly praying together and planning the project in a room above the mosque.[47]

The opportunities for Moplahs to achieve a greater degree of independence from the power network linking Hindu jenmis and officialdom would be increased by the existence of a Moplah ulema capable of playing a sanctioning and even a leadership role, to the extent that they derived their income from the contributions of the faithful. Indeed some British officials went so far as to ascribe an originating or manipulative role to the Moplah Tangals who were alleged by one administrator in 1896 to 'have been at the bottom of most of the outbreaks'.[48]

There is certainly plenty of evidence of Moplah divines sanctioning and sometimes (as in 1849) actually leading outbreaks. Apparently, it was the favourite text of Syed Fazl in his Friday orations at Tirurangadi mosque that it was no sin but a merit to kill a jenmi who evicted.[49] It was clearly necessary for outbreak participants to have religious sanction for taking the path of martyrdom, and it was deemed most important to receive the blessing of the Tirurangadi Tangal (or sometimes any Tangal) before

[45] C. Collett, Joint Magistrate, Malabar, to T. Clarke, Malabar Magistrate, 31 January 1856, MJ 552, 22 May 1856, p. 3342.

[46] See, for example, A.F. Pinhey, Malabar Magistrate, to Chief Secretary, 8 May 1904, MJ 1067, 12 July 1904, p.72.

[47] Reports of J. Twigg, Acting Special Assistant Collector, Malabar, 9 July 1884 and C.A. Galton, Acting Malabar Magistrate, 16 September 1884, MJ 2776-81, 1 November 1884, pp.3 and 5.

[48] Minute of J. Grose, member of the Executive Council, 18 August 1896, IOP & JDP 2060/96, p.131.

[49] Information given by C. Kanaran, late Deputy Collector of Malabar to W. Logan, Government of Madras, Malabar *Special Commission, op. cit.*, II, p. 48. Kanaran had had immediate charge of the negotiations resulting in Syed Fazl's departure for Arabia.

'going out'.[50] Even so, where their source of livelihood was from the Moplah community, the dependence of the Tangals on the population supporting outbreaks was as great as that of the intending martyr on the Tangals. Significantly, this was the case with the Tirurangadi Tangals, who had 'no apparent property' but professed to be *fakirs* and were supported by 'the voluntary oblations of their followers'.[51] If it really was true that the doctrine of the Moplah outbreak had originated with the Tirurangadi Tangals,[52] they were merely catering for their followers' needs.

Nevertheless, there existed important limits to the facility with which even the Moplah tenantry could combine without external assistance. Most ulemas, especially in interior south Malabar, had a level of economic standing and a range of experience and contact which little different from that of the rural mass. On the other hand, nearly all those who *did* stand out in these respects and who might therefore be more likely to attract a considerable following, the important Tangals of the towns and religious centres in the coastal zone, were primarily in contact with the coastal Moplahs who were a more prosperous commercial community having, the British were sure, 'no sympathy with the Moplah agriculturalist'.[53] The one really important exception was Syed Fazl who, with his ability to attract thousands of devoted followers, had caused the administration much concern until he was exiled in 1852.

The second important limitation to the extent to which Moplahs were likely to combine was probably the size of mosque congregations. Figures for 1831 and 1851 seem to indicate an average of the order of four hundred Moplahs per mosque for Malabar as a whole.[54] Probably the figure would be smaller for rural areas like the outbreak zone. With possibly no more than a hundred or so adult males per congregation, it would seem that the religion of the Moplahs normally promoted a degree of regular intercourse sufficient for combination on only a limited, localized scale, and combination on such a scale would present a strictly limited range of alternative forms of action against jenmi power. There is evidence of occasional

[50] H.V. Conolly to Secretary, Judicial, 12 October 1849, MJ 794, 1 December 1849, p. 4750.

[51] H.V. Conolly to Secretary, Judicial, 4 November 1843, MJ 69, 27 January 1844, p. 197.

[52] T.L. Strange to H.V. Conolly, 29 July 1952, MJ 154, 16 March 1853, p. 1610.

[53] H. Wigram, Officiating District Judge, South Malabar, to Chief Secretary, 8 November 1883, MLT, p. 17.

[54] Strange Report, p. 4593.

attempts at combination by Ernad Moplahs to withhold rent and to prevent the releasing of land from which tenants had been evicted,[55] but this mode of resistance may well have been of limited efficacy in the face of competition for leases from land-hungry outsiders. Under normal conditions perhaps the forms of anti-jenmi action generated by the Ernad Moplahs themselves had, of necessity, to be those involving the mobilization of relatively small numbers.

Combination on this limited scale produced forms of action, varieties of 'normal' crime especially, which are common to perhaps any rural society in which mass organization is not feasible. Apart from common murder of local notables by small groups, such as the three Moplahs who in 1865 killed an importunate moneylender,[56] Moplahs supplied most of the *dacoits* of the district.[57] In times of dearth *dacoities* would be committed by starving people for food, as in 1897 when one British official observed that 'the Mappillas, who suffered the most by far, are not those to sit down under that kind of adversity which consists in starvation while fat Hindu landlords had plenty'.[58] At least one outstanding example of what E.J. Hobsbawm has called 'social banditry'[59] is recorded with the case of Athan Gurikal, an idle young man (of a Moplah family of status in greatly reduced circumstances) who secured a living for his gang by levying 'protection money' from rich jenmis, but who also set himself up as a champion of the oppressed Moplahs, among whom he enjoyed great prestige.[60]

But throughout the period Moplah resistance to the rural establishment constantly gravitated towards the form of the outbreak. This persistence over many decades of the outbreak as the chief form of action may well have been because it entailed the wreaking of the maximum degree of terror with the minimum resources, for nothing was more chilling to the local Hindus than the thought of frenzied fanatics for whom death not only

[55] Logan Report, I, p. iv.

[56] Judgment in High Court of Judicature at Madras, 21 October 1865, Case 166, 1865, MJ 1550, 31 October 1865, p. 1903.

[57] Memorandum of H.V. Conolly, 25 March 1852, MJ 154, 16 March 1853, p. 1506.

[58] F. Fawcett, Superintendent of Police, Malabar, to Inspector-General of Police, Madras, 28 January 1898, MJ 819, 25 May 1898, p. 90.

[59] E.J. Hobsbawm, *Primitive Rebels* (Manchester, 1959), p.5.

[60] For Robin Hood-style 'righting of wrongs' by this Gurikal see Robinson Report, pp. 4867 and 4873.

held no fears but was eagerly sought. Perhaps it was partly because 'the Moplah [was] only formidable when under the effects of fanaticism'[61] that the despised coolie and abused tenant was attracted by such a suicidal form of action in which even war-like Nairs in possession of arms 'rushed into the jungle, climbed trees, and... descended into wells leaving their wives and children and their property at the mercy of [a] gang'[62] of outbreak participants.

The main strength of a outbreak form, however, was that the inevitable death of all the participants meant that direct retaliation by the powerful was impossible. As long as men could be found who preferred the rewards of paradise to life, a means of resistance could be employed which was not subject to the main disadvantage of other forms of action, like common murder, dacoity, and 'social banditry', in which relatively small, and therefore vulnerable, numbers were involved. Thus the Moplah outbreak seems to have been essentially a peculiar form of rural terrorism which functioned as what, in the circumstances, was probably the most effective means of curbing the enhanced power of the jenmi, for the earthly benefit of Moplahs who themselves did not become participants.

It has been argued that the outbreak was merely a form of action instigated by the more influential elements among the Moplahs for their own ends. In particular it has been claimed that the outbreak was 'only the culmination of the constant struggle of the wealthy kanamkar to get possession of the land'.[63] Kanamkar, variously regarded as tenants or mortgagees, held land from jenmis for terms of twelve years, making a loan to the jenmi at the beginning of this period and recovering the interest on the loan from the produce of the land, the net profits being paid over to the jenmi.[64] Some *kanamkar*, such as the family of one outbreak participant in 1851 which held one and a half acres on *kanam* and also worked as coolies,[65] were clearly of a status little different from the usual 'fanatic'.

---

[61] H.V. Conolly to Secretary, Judicial, 29 November 1851, Government of Madras, *Moplah Outrages Correspondence*, 2 vols (Madras, 1863), I, p. 223.

[62] Report of H. Bradley, Acting Magistrate, Malabar, 16 May 1894, MJ 2186-2192, 8 September 1894, p. 109.

[63] Sir Charles Turner, *Minute on Malabar Land Tenures Draft Bill* (Madras, 1885), p. 60.

[64] Proceedings of the Court of Sudr Udalut, 5 August 1856, in *ibid.*, appendix XVI, p. 100.

[65] C. Collett to H.V. Conolly, 20 September 1851, MJ 558, 8 September 1851, p. 3732.

Others however were substantial, powerful men, frequently Moplahs in the outbreak zone, jealous of the jenmi land monopoly which was an obstacle to their ambition and, indeed, often the means of their downfall. Moplah kanamkar who had gone from relative riches to destitution in a matter of years sometimes became participants in outbreaks.[66] The kanamkar certainly had every reason to look with favour on any movement directed against jenmis.

Moreover, the evidence for the key role of kanamkar in the Moplah outbreak is strong. They clearly provided the leaders of a number of outbreaks, such as those in April 1841 and December 1841 and were strongly suspected of instigating or directing others, as in August 1851 and September 1880. As Logan realized, the conception of the outbreak as a weapon used by kanamkar in their struggle with the jenmi also provides an explanation for the apparent fact that outbreaks did not begin until 1836, several decades after the beginning of British rule (1792).

Before 1792 Malabar had been subject to invasion and control by the Muslim rulers of Mysore, and large numbers of the Hindu jenmis had fled. Before doing so, they had had to make what bargains they could with their Moplah kanamkar who for small sums secured deeds assigning to them large kanam claims.[67] On their return in the wake of the British in 1792 the jenmis found themselves with the rights of absolute landlords on paper, but in practice so heavily in debt to the kanamkar that their position was weak. Only when prices began to rise in 1831-32 did the jenmis find it possible to begin to pay off their debts and to exploit their rights which had lain dormant.[68] The actual origin of the Moplah outbreak is a matter of conjecture, but the phenomenon certainly seems to have appeared at a time when the kanamkar found it useful, it undoubtedly continued with the support of such men, and it would seem to have disappeared temporarily in 1898 when it no longer served the interests of this group.

Even so, it cannot be said that the outbreak was 'nothing but' a weapon of the substantial kanamkar, and that the men (of whatever social and economic status) who actually devoted themselves to death were mere tools. The outbreak was far too popular with the great mass of Ernad Moplahs, and especially the poorest, for this to be true. As District Magistrate Conolly said of the doctrine that the murderer of an evicting jenmi would be entitled to paradise, such notions were 'far easier sown

[66] *Ibid.*, pp. 3736-7.
[67] Logan Report, II, p.43.
[68] *Ibid.*, p. 46-7.

than rooted out amongst a wild people and that exhortation from superiors, other than spiritual, are of little avail against a popular faith'.[69]

Nor is this surprising when the material gains to be derived from 'fanaticism' were so widespread. Apart from the plunder of Hindu mansions by large numbers of Moplah neighbours and the mass destruction of jenmis' deeds and moneylenders' accounts, 'the tendency of an outbreak [was] to benefit the Moplahs as a class' since, as one British official pointed out, it 'keeps their name up; deters many landlords from enforcing their legal rights; and supplies temporary employment on easy terms to many hundreds of Moplah "guards"'.[70] (When 'fanatics' were on the rampage, many Moplahs quartered themselves as 'guards' on the terrified inmates of jenmi mansions demanding and receiving such 'high feeding' and 'presents' that their hosts were heartily glad to see their backs.[71])

With such benefits to be gained it did not need kanamkar leaders and instigators before an outbreak occurred. In the August 1852 outbreak three young 'day labourers' tried to deal with the Nair jenmi who had moved the father of one of the 'fanatics', a *verumpattomdar*, from a plot he had held for many years.[72] No kind of role appears to have been played by any kanamkar. The Moplah outbreak was a *popular* phenomenon, whatever use the substantial kanamkar made of it and whatever the degree of control they were able to exercise over it.

III. *Problems*

The above thesis may be applied to three problems which Moplah outbreaks pose: (i) where they occurred, (ii) whom they involved, and (iii) when they happened.

(i) The fact that, with few exceptions, outbreaks occurred in interior south Malabar rather than in north Malabar or the coastal zone (the two other areas where a rural Moplah population existed) was probably because the type of agriculture in the latter regions made the tenant far less vulnerable to the adverse exercise of jenmi power than in the former. While

---

[69] H.V. Conolly to Secretary, Judicial, 29 November 1851, *Moplah Outrages Correspondence*, *op. cit.*, I, p. 227.

[70] Winterbotham Report, p. 66.

[71] *Ibid.*, p. 53.

[72] Notes by T.L. Strange on a letter from H.V. Conolly, 30 September 1852, and T.L. Strange to Secretary, Judicial, 16 October 1852, MJ 716, 6 November 1852, pp. 4694, 4579.

interior south Malabar was primarily a paddy-growing area, the north (and the coastal strip) was one in which garden cultivation of crops such as pepper and coconut dominated. The Special Settlement Officer for Malabar gave figures in 1904 showing that the ratio of wet cultivation (paddy) to garden land in the four northern taluks was 59:100 whilst for Ernad and Walluvanad it was 203:100.[73]

As Special Commissioner Logan pointed out, it was 'notoriously the grain-crop cultivators who [were] worst off'. The fact that the yield of gardens was extremely sensitive to the quality of husbandry had 'taught many of the Janmis that they cannot rack-rent gardens as they can grain lands'.[74] The attempt of the jenmi to maximize his gains at the expense of his garden tenant merely resulted in the latter taking as much as he could from the garden and thus ruining it. An agriculturalist in north or coastal Malabar was far less likely to have to resort to violence to curb jenmi power than the tenant in the outbreak zone. Indeed, social divisions appear to have been sharper in the latter where the jenmi was more likely to be a 'big man' clearly distinguishable from the rest of the population. Logan in 1882 gave figures for 'principal' jenmis (those holding a 100 or more pieces of land in any one amsom), showing that while Ernad and Walluvanad had 15.6 per cent of the total number of jenmis in Malabar they had 35.5 per cent of 'principal' jenmis.[75] It was always claimed that in north Malabar land was more widely distributed and that the same person was often both a tenant and a landlord.[76] The outbreak zone was that part of the area of Malabar inhabited by Moplahs in which the jenmi stood out most clearly as an oppressor.

(ii) As already indicated, for the lower Hindu social orders there was no congregational form of religious organization to compensate for the many obstacles to combination existing in rural Malabar. Far from presenting them with opportunities to resist jenmi power, the religion of these groups bound them the more securely to their high-caste landlords. For the Hindu there could be no possibility of sanction and leadership from his religious

[73] N. MacMichael to Revenue Settlement, Land Records and Agriculture Department, 10 June 1904, Board of Revenue (Revenue Settlement, Land Revenue and Agriculture) Proceedings, 93, 10 April 1905, p. 28.

[74] Logan Report, I, pp. xxi, xxxv.

[75] Ibid., I, p. lvii.

[76] District Munsif, Badagara (North Malabar) to District Magistrate, 28 December 1907, in an inquiry concerning the working of the Tenant's Compensation Act, MJ 308, 25 February 1910, p. 19.

leaders, on the contrary he who incurred the displeasure of his Nambudri Brahmin jenmi was liable to an excommunication which was religious, social, and economic. The *desa virodham* (enmity of all the residents of the *desam* or 'village') and *svajana virodham* (enmity of one's own caste people) which were brought into play denied the excommunicant access to necessary religious facilities, wells, and all kinds of village services. The 'smallest show of independence' was 'resented as a personal affront' and though the jenmi was liable to prosecution in such cases `it was found impossible to get the people to come forward to complain for fear of the utter consequences—*eviction and ruin of families*'.[77]

Of course it was open to any Hindu to mitigate this formidable array of sanctions he was subject to at the hands of the jenmi by becoming a Muslim. In fact British administrators recorded that considerable numbers of low caste Hindus exploited this opportunity[78] while census returns indicate that the proportion of Muslims in Malabar rose from 25.7 per cent in 1871 to 31.6 per cent in 1911.[79] The material advantages from conversion for the Cheruman in particular were clear. In 1844 it was pointed out that by the custom of the country the pay of a Cheruman was less than that allowed to free labourers (like Moplahs),[80] while a low-caste convert also experienced a rise in the social scale.[81] Logan stated in 1887 that in the event of a Cheruman convert being 'bullied or beaten the influence of the whole Muhammadan community comes to his aid' and that 'with fanaticism still rampant the most powerful of landlords dares not to disregard the possible consequences of making a martyr of his slave'.[82]

By becoming a Moplah a Cheruman labourer or Tiyyan verumpattomdar was joining a body which functioned as a self-defence organization of the rural subordinate whose ultimate weapon in this period was the outbreak. Conversion certainly curbed the field of the power of the Nambudri and Nair landlords and this and the undoubted benefits the convert derived from membership of his new community must have been important elements in the striking zeal for proselytizing for which the

[77] Logan Report, I, p. xxvi.

[78] See for example Fawcett Report, p.97, and Strange Report, p. 4593.

[79] *Census of Madras, 1871* I, pp.9, 12; *Census of India, 1911,* XII, Part 2, pp.2, 24.

[80] H.V. Conolly to Secretary, Judicial, 24 November 1843, MJ 277, 20 April 1844, p. 1339.

[81] Minute by J.D. Sim cited in footnote 1.

[82] W. Logan, *Malabar Manual*, I, p. 148.

Moplah was renowned.[83] In most methods of resistance to the local power structure adopted by Moplahs conversion tended to play a conspicuous part. In several outbreaks the pattern of action for participants was to murder jenmis and 'convert', forcibly or otherwise, any lowly Hindu who fell into their hands, these proselytes occasionally becoming members of the gang.[84] One Moplah gang in 1843 'went out' specifically to deal with a Nair jenmi who had angrily forced one of his female 'outdoor menials' to apostatize, after her conversion had given her the temerity to dispense with the deference she had previously been obliged to render him.[85] Conversion seems to have been an important weapon in the struggle against the jenmi.

No doubt for this reason, a number of outbreaks (such as those in 1884 and May 1885) were directed against lowly Hindus, for these were apostates who had reneged on their previous commitment to Islam. Perhaps the outbreak directed against the low-caste apostate was a particularly brutal, and no doubt therefore a somewhat effective, means of ensuring solidarity, of helping strengthen the 'horizontal' links between Moplah and Moplah at the expense of the 'vertical' bonds between the jenmi and his Hindu dependants.

This is not to argue that Islam in Ernad was 'nothing but' a kind of proto-trade union or peasant league. It *does* seem, however, to have been the medium through which functions, which in other circumstances are discharged by such bodies, were performed. But precisely because in south Malabar Islam was not, like a peasant league, specifically designed to discharge these functions, the instrument proved to be clumsy and less than satisfactory. The frenzied Moplah 'fanatic' was not always carefully discriminating in his choice of victim, and the lives of a number of Hindus were attempted in outbreaks almost certainly merely because they were Hindus.[86] In other cases the religious form of the instrument may have permitted some Moplahs to murder Hindus for personal reasons having nothing to do with jenmi oppression, and to attempt to claim social and

[83] See for example report of Special Commissioner Graeme, in *ibid.*, p. 197.

[84] Report of H. Bradley, Acting District Magistrate, Malabar, 16 May 1894, MJ 2186-92, 8 September 1894, pp. 93-94, 99.

[85] H.V. Conolly to Secretary, Judicial, 4 November 1843, MJ 69, 27 January 1844, p. 198.

[86] See for example the account of the murder of a Hindu goldsmith in the report of J.T. Gillespie cited in footnote 7, p.22.

religious sanction for their crime by subsequently 'going out' as shahid.[87] None of this however should obscure the undoubted fact that the outbreak was overwhelmingly directed against the power network the Ernad Moplah felt was oppressing him. Moreover, incidents such as that in 1919 when immediately after a murderous assault on an oppressive Nambudri jenmi's wedding feast a Moplah gang released unharmed one Nambudri on the grounds that he was said to be a merely poor guest going for meals,[88] indicate that even the so-called 'Moplah religious fanatic' was capable of a nice discrimination.

Most importantly, the outbreak was a weapon which could be of use only against *infidel* oppressors. Whatever sanctions a Moplah could use against *Muslim* oppressors, they could hardly include those derived from a religion which taught the sinfulness of the murder of a fellow believer. On the contrary, there were several cases in which the enmity of a Moplah for one of his own religion who had oppressed him found outlet in outrage against Hindus. Thus in 1849 one indigent young Moplah was led to 'go out' after his uncle had enriched himself at the expense of the participant's mother when the family property was divided.[89] Again, in 1915 three Moplahs with grievances about what they regarded as the swindling behaviour of a Moplah 'lessor' assaulted him and burnt his house, and then made as if to commit an outbreak, looting the houses of three Nairs who had given information to the authorities earlier in the year about an outbreak two of the three Moplahs were allegedly preparing.[90] Under these circumstances the outbreak was of no use to the Moplah poor against that section of the Muslim rural community which was most likely to oppress them, the substantial Moplah kanamkar intermediaries who often stood between the jenmis and the verumpattomdars, and who could rack-rent just as well as the jenmi.[91]

(iii) As has been indicated, the outbreak persisted because the mass of Ernad Moplahs wanted it to. However, British policy from 1854 of fining amsoms which were involved in outbreaks and deporting many breadwinners as suspects diminished the value of the outbreak to the community.

[87] The abortive outbreak of 1877 seems to have been such a case. See report of W. Logan, Malabar Magistrate, 25 April 1877, MJ 1134, 5 May 1877, pp. 586-598.

[88] J.F. Hall, Malabar Magistrate to Secretary, Judicial, 25 April 1919 IOP & JDP 4582/19, 9 June 1919, p.5.

[89] Robinson Report, pp. 4981-2.

[90] Report of F.B. Evans, Malabar Magistrate, 22 November 1915, MJ (Confidential) 3008, 6 December 1915, p.6.

[91] Innes Note, p.34.

And yet, since the British were unwilling to tackle the root cause of the problem, the powers of the local high-caste hierarchy, and since there was little other means of resistance, outbreaks persisted, though on a lesser scale.

That Moplahs hoped the government would be influenced to react favourably to their grievances seems indicated by the number of occasions on which Moplah 'writings', often left behind by outbreak participants, complained of jenmi oppression and threatened worse outbreaks should the government continue to be supine. As Athan Gurikal put it in his last testament, referring to the machinations of certain 'landlords and Rajahs', if these were 'not put a stop to by the Sirkar, there will be... no other alternative, but to cut off the heads of these Hindoos'.[92] The outbreak must be seen partly as a peremptory demand for government intervention. In view of this, it seems significant that, after the first very modest gesture by the government to curb evictions through the Tenant's Improvements Act, support for outbreaks dwindled until they were so dramatically ended for a seventeen year period in 1898.

No doubt those who would be most in favour of alternative forms of action which did not bring down fines on the heads of the whole community would be those tenants best able to make the kind of improvements the Act provided for, the richer kanamkar.[93] This group may well have had sufficient influence to curb those in favour of the continuation of the traditional methods. In the evidence collected for the 1896 report there are signs that some Moplahs were arguing for the alternative of relying on redress for grievances through approach to the collector.[94] Certainly during the 1898-1915 period there were signs that, perhaps under the influence of 'moderates' who were prepared to respond to the concession of the 1887 Act, the Ernad Moplahs were prepared to rely on more constitutional methods of pressing their claims. In 1902, during his tour of Malabar, the Governor was presented with a petition from the *Hidayut-ul-Muslimin Sabha* about the difficulties experienced by Moplahs in persuading jenmis to sell land for mosque sites, an issue which had previously led to outbreaks.[95]

[92] Statement cited in note 15, p. 4788.

[93] The limited extent to which the Act assisted 'the lowest Moplahs' was noted by R.B. Wood, Acting Malabar Magistrate, to Inspector-General of Police, 9 October 1910, MJ 87, 19 January 1911, p.16.

[94] Statement of Puzhutini Kunyayu, 14 March 1896, IOP & JDP 996/96, 10 April 1896, p.12.

[95] Government Order 407, 7 March 1902, p. 22, MJ.

In 1915 however the period of abstention from violence which the British appear to have won by the 1887 Act came to an end with the outbreak of that year, followed by another in 1919. It had been clear for some time that even the limited protection conferred on the tenant by the Improvements Act had diminished as jenmis discovered ways of circumventing its provisions.[96] Moreover, there is evidence that the effect of the passing of a tenancy Act in 1914 in Cochin State, purporting to confer fixity of tenure on certain categories of kanamkar, was to create apprehension among the jenmis of Malabar that similar measures were to follow there. (In fact the government of Madras rejected Innes's proposals of 1915.[97]) The result seems to have been a movement to convert kanam holdings to inferior tenancies-at-will[98] and a consequent increase in discontent which ended the previous constitutional period.

Conclusion

As a form of resistance to jenmi power the Moplah outbreak was crude and unsophisticated. It was subject to no fine control, leadership was elementary or non-existent, demands were formulated in a haphazard way, organization was of the most rudimentary and ad hoc kind.

Moreover, it was useful only as a rough and ready way of limiting the *exercise* of the jenmi's power, not of permanently reducing or eliminating that power, which derived from the rights bestowed on the jenmi by the government. Only if British rule in Malabar could be ended or a massive change effected in the administration's policy of cultivating the jenmi as a 'political force on the side of Government' could the latter be achieved. But it was evident that in the last analysis the administration would act decisively against the jenmis only if it were clear that support for them was more likely to weaken than strengthen British rule. Any such massive change in British policy could be brought about only by posing a threat to that rule.

That the materials for the creation of such a threat existed in Malabar, officials well acquainted with the district were ready to acknowledge. In 1844 Collector Conolly spoke of the 'considerable number of needy and lawless men, Moplahs in especial' in the outbreak zone ready to rise 'at any

---

[96] Innes Note, p.21.

[97] MJ (Confidential) 3021, 26 September 1917, p.6.

[98] Survey of Kothachira desam, Ponnani, by A. Krishna Wariyar, in G. Slater (ed.), *Some South Indian Villages* (Oxford, 1918), pp. 168-9.

time' against the government 'if a sufficient prospect of plunder and impunity were held out to them'.[99] But though an outbreak of necessity ended in a physical clash with the forces of authority, the objective was not to eliminate government power but self-immolation at the hands of authority. In 1849 with the whole country at their feet after the defeat of the sepoys sent against them, Athan Gurikal's gang made no attempt to seize control, but merely awaited more troops so as to die in 'fair fight with the Cirkar', as Gurikal put it.[100] As a challenge to British rule the Moplah outbreak was mere ritual.

As long as British power in India was unchallenged, any defiance of it in Malabar had to be a formality. As one district official said in 1880: 'Our safety lies in the want of leaders, in the want of organization and in the knowledge of the Moplahs themselves, that any attempt at rebellion must end in failure.'[101] When, in 1921, there appeared to exist not only leaders and organization, but also sufficient challenge to British rule to make Moplahs believe that attempt at rebellion might succeed, the stage was set for defiance of government beyond formality. 1921 was the year of the Malabar Rebellion.

ABBREVIATIONS

Innes Note	C.A. Innes, Strictly Confidential Note, MR 3021, 26 September 1917
IOP & JDP	India Office Public and Judicial Departmental Papers
Logan Report	W. Logan, 'Report on Malabar Land Tenures', in *Report of the Malabar Special Commission*, 3 volumes (Madras, 1882), I
MJ	Government of Madras, Judicial Proceedings
MLT	Government of Madras, *Malabar Land Tenures (Madras, 1885)*
MLTCR	Government of Madras, *Malabar Land Tenures Committee Report* (Madras, 1887)
MR	Government of Madras, Revenue Proceedings

[99] H.V. Conolly to Secretary, Judicial, 6 February 1844, MJ 187, 8 March 1844, p. 752. See also T.L. Strange to Conolly, 27 May 1842, quoted in Minute of T.M. Lewin, Acting Third Judge, Court of Foujdaree Udalut, 30 June 1842, MJ October 1842, pp. 4941-42.

[100] Robinson Report, p. 4937.

[101] H. Wigram, Officiating District Judge, South Malabar, to Chief Secretary, 8 November 1880, *MLT*, p. 17.

Robinson Report   Report of W. Robinson, Head Assistant Magistrate, Malabar 1849, MJ 794, I December 1849

Strange Report   Report of T.L. Strange, 25 September 1852, MJ 483, 23 August 1853

Winterbotham Report Final Report of H.M. Winterbotham, 5 May 1896, IOP & JDP 2060/96, 30 September 1896.

Chapter Four

The Deccan Riots of 1875

RAVINDER KUMAR

The Deccan Riots of 1875 highlight the social transformations brought about in rural Maharashtra in western India during the first five decades of British rule. The riots are of special interest to the social historian since they hinged upon relations between two important and well defined rural social groups, namely, the cultivators and the moneylenders. This paper will focus on the social changes which precipitated this conflict. I shall also attempt to link these changes with the social ideals and the political objectives which inspired the new rulers of Maharashtra and determined their administrative policy.

The Deccan districts acquired by the British in 1818 possessed a marked social consensus. The belief of elite castes like the Chitpavan Brahmins in the philosophy of *advaita*, and the influence of advaita upon the middle and lower castes through the devotional religious orders, meant that all the caste groups of Maharashtra bore allegiance to common religious values. The resulting consensus was reinforced by the structure of rural institutions and by the manner in which social power was distributed in rural society.

At the hub of the rural world lay the village 'community.' And at the head of this community stood the village headman or *patel*, the leading member of the caste of cultivators or *kunbis* who dominated the village. The dominance of the kunbis determined the conditions for a harmonious relationship between the cultivators and other social groups like the *vanis* or the *bullotedars*, who played such an important role in making the village self-sufficient. Yet the dominance of the kunbis found expression primarily in social styles. The village community managed the internal affairs of the village and shaped its relations with the political authorities at Poona, the capital of Maharashtra, through the intermediate and counterposed institutions of an hereditary landowning aristocracy and a bureaucracy created for the collection of land-revenue.

For ten years after the British conquest of 1818, the Deccan territories were governed by Mountstuart Elphinstone, an administrator in the Burkean tradition, who looked for continuity in social change. But when Elphinstone left Bombay in 1828, the British administration in the Deccan became influenced by Utilitarians like Robert Keith Pringle, the man who laid the foundations of land-revenue policy in western India. Because he was a Utilitarian, Pringle had an atomistic approach to social phenomena. He concerned himself with the behaviour of conceptually isolated individuals and inferred the properties of larger social groups through the insights which he thus gained into social processes.

To his atomistic view of society Pringle added a belief in the effectiveness of rational action. His *ryotwari* system of land-revenue embraced both these key elements of Utilitarian dogma. It reflected social atomism in the creation of a contract between the individual peasant and the State to the exclusion of intermediaries; and it embodied rational action in fixing the land-revenue according to the Ricardian law of rent. The institution of a rational bureaucracy and the creation of Courts of Law which were guided by *laissez faire* principles completed the Utilitarian programme of reform for Maharashtra.

The realignment of power in rural society which stemmed from Utilitarian measures of reform led to a growth of social tensions within the villages of the Deccan that erupted in the disturbances of 1875. While all the social groups in rural society were affected by the patterns of authority which took shape after 1818, the tensions which resulted from administrative changes found their clearest expression in the relationship between the kunbi and the vani. Even before the British conquest, rural indebtedness was widespread in Maharashtra, and an inquiry into the state of a village like Lony in 1820 illustrates its extent.[1] But under native rule the kunbis dominated the village despite a burden of debt because they enjoyed a numerical preponderance and because the vani was isolated from his caste-fellows in other villages. The officers of the Poona Government remained unconcerned about the vani's fate so long as he kept the rural economy on

---

[1] T. Coats, 'Account of the Present State of the Township of Lony,' *Transactions of the Literary Society of Bombay*, III (London, 1823), 183-250. Of the 84 peasant households in Lony, 79 were indebted to the six *vanis* who lived in the village. While the total owed by these 79 households was Rs. 14,532, it was divided into small loans ranging from Rs. 50 to 200.

the move. The only judicial institution to which the vani could appeal for the recovery of his debts was the *panchayat*. Since the panchayat was dominated by the patel and other influential kunbis in the village, it was hardly an institution that gave fair consideration to the vani's claims. Because the kunbis controlled the judicial institutions, the vanis were prevented from exercising social dominance over the village.

The changes brought about in the Deccan after the British conquest effected a redistribution of social power within the villages. The most crucial of these changes was the introduction of the ryotwari system of land-revenue. The ryotwari system weakened the cohesion of the village by abolishing the collective responsibility which the kunbis had formerly borne for the village rental. It was also responsible for re-organizing rural credit along novel lines. Under native rule the role of the village vani had been sharply differentiated from that of the urban *sowcar*. The vani was a member of the village community and subject to its judicial and executive authority. He was a shopkeeper as well as a moneylender, and his meagre capital resources were tied up either in small monetary advances to the kunbis or in grain loans. Due to his isolated position, and because he was dependent upon the kunbis for the safety of his life and property, the vani never presented the village with any threat.

The urban sowcar, on the other hand, was a somewhat different person, not only because of the scale on which he conducted his financial transactions, but also because of his position vis-a-vis the village community. Instead of dealing with the cultivators directly, he advanced loans to each village as a community in order to enable it to fulfil its revenue obligations. The sowcar, therefore, very often controlled all the surplus produce of the village. However, he did not desire to establish a more intimate control over the village economy since to do so would have been to contravene the social style of his caste. Thus when in 1827 the Bombay Government tried to free the village communities of their burden of debt by compensating the sowcars with land-grants, the sowcars revealed their unwillingness to participate directly in agricultural production by refusing these land-grants.[2]

The introduction of the ryotwari system, however, changed the role of the sowcar in the supply of credit to the village. Since the new system emphasized individual responsibility for the payment of land-tax, credit

[2] R.K. Arbuthnot to Bombay Government, 23 September, 1826, *Bombay State Archives* (hereinafter cited as BA), Revenue Department (hereinafter cited as RD), 1827, Vol. 43/196.

was now required by the peasant and not by the village community. The sowcar therefore no longer had any dealings with the patel as the head of the village community. Instead, he dealt with each peasant. Since it was difficult for the sowcar to conduct business directly with the peasant, he preferred to work through his caste-fellow, the vani, in the village. An immediate result was that the vani was now united with his caste-fellows outside the village to a far greater extent than ever before; the sowcar supported him in every possible way to fulfil the new role which had become open to him. All this not only strengthened the position of the vani vis-a-vis the kunbis, but it also increased the occasions for conflict in rural society. Indeed, in the decades following the British conquest, the antagonism between the kunbi and the vani became the most significant cleavage in the villages of the Deccan.

Nothing did more to sharpen this antagonism than the operations of the new Courts of Law which were instituted by the British. These Courts not only wrested judicial authority from the kunbis, but their operations were guided by concepts of social equity and contractual responsibility which favoured the vani rather than the kunbi. The consequences of the new Courts of Law were brought to the notice of the British authorities in a petition addressed by the *ryots* of Thana to the Bombay Government in 1840. The petition opened with an enumeration of the blessings which British rule had conferred on Maharashtra. Life and property, the ryots pointed out, were secure as they had never been before; arbitrary taxes had been abolished, and the territorial aristocrats who had formerly oppressed the cultivators were firmly under control. But, the ryots continued, 'though we live under such protection and prosecute our labour free of any apprehension of oppression, yet our families are reduced to a miserable condition, so much so that their ordinary wants cannot even be supplied.' The reason behind this state of affairs was simple. To carry out their agricultural operations, the ryots were forced to borrow from the vani. Formerly the moneylenders had levied interest on money loans at rates varying from 25 to 50 per cent and on grain loans at rates varying from 30 to 60 per cent. In the case of *usmani sultani* (natural or artificial calamities), however, the vanis had recovered their advances with moderation, since the Maratha Government

never allowed its ryots to be oppressed by usurious demands, and consequently the Sahookar did not carry any complaint to the Government. Considering the Sahookar as our parent and that he would save our lives at a critical moment, we settled our claims according to our circumstances. Thus both the ryot and the Sahookar were able to sustain their situations.[3]

But the harmonious relationship between the ryot and the vani had been transformed, through the institution of the Courts of Law and the promulgation of the Regulations, into one of acute antagonism. The vani now inveigled the ryot into ruinous agreements, and if the ryot failed to fulfil legal obligations, the vani instituted a civil suit against him. As a result, the 'whole of his [i.e. the ryot's] property is disposed of; and he is reduced to such a condition as never to regain his footing in society.' Because the ryots were ignorant of the Regulations they could be enticed into agreements, unaware of their significance. In contrasting their wretched condition under British rule to their state under the Peshwas, the ryots pointed to the obvious solution of the problem:

> Under the late Government we suffered great oppression, but no one could sell our immovable property or lands etc., and therefore we were able to endure the oppression both of government and of the Sahookars... Under the present government, by the sale of our immovable property we are reduced to a starving condition in the same manner, as a tree when its roots are pulled out, dies.

> We are neither Shroffs, nor traders, and we are not acquainted with the Regulations of the Courts. The Vakeels whom we employ extort money from us in the first instance under various pretences, and when the cause is lost, advise us to make an appeal. Let government therefore consider whether the cultivator is able to litigate with the Sahookar... [We] beg that our cases may be referred to the panchayats, who should decide on the claims and liabilities of the parties with reference to the circumstances of each, agreeably to the ancient custom.[4]

The Thana appeal pointed to a social malaise of alarming intensity. It is true that, in representations to the government, the peasants as a matter of habit presented their misery in heightened colours; but the accuracy of the kunbis' indictment was confirmed by an official inquiry into the condition of the Deccan districts. Because rural indebtedness had been widespread even before 1818, the magnitude of the problem caused little surprise to the authorities, even though a District Officer pointed out that in the *talukas* of Khair and Mamul under his charge there was scarcely a village in which it was possible to find 'three persons, ryots of zamindars,

---

[3] Petition signed by 7,215 ryots of Thana to the Bombay Government, dated 27 July, 1840, *BA*, RD (1840), Vol. 110/1194.

[4] Petition of 27 July, 1840.

not in debt for sums above rupees one hundred.'[5] But what perturbed the Bombay Government was the predominance which the moneylender was gradually acquiring. By the 1840's, this predominance had become so characteristic a feature of the villages of the Deccan that not a single British administrator questioned its existence, even though there were differences of opinion on its implications for the future of rural society and the administrative policies of the Government. The majority of executive officials linked the problem with the rapacity for which the vani was notorious and with the usurious character of his financial dealings.[6] But administrators like Pringle held that while the vani was no less selfish than other men, it was wrong to assert that the rates of interest he charged were exorbitant. Looking to the condition of rural society in Maharashtra, Pringle characterized the vanis as a class which formed the only connecting link between 'civilisation and barbarism.' He also pointed out that although the ryots were 'loud in their outcry against the creditors, yet I imagine they would be the first to suffer by, and not less ready to complain against, any restrictions which would deprive them of the aid of so powerful a class.'[7]

In focusing attention on the important part played by the vanis in the rural economy, Pringle was emphasizing a point on which there was no difference of opinion. For the administrators who looked upon the vanis as a rapacious class recognized only too well the significance of their social role. The ryots rarely possessed any reserves of capital, and they seldom had access to ready money. It was the vanis who helped them pay their instalments of land-revenue and who provided them with money for the expenses of caste ceremonials and religious festivities. The vani was, consequently, a valuable member of the village community, without whose assistance the cultivators could be greatly distressed for want of money. What the 'anti-vani' administrators deplored, however, was the lack of consideration with which the moneylenders conducted their business. In the districts around Poona, for instance, an experienced administrator like Bartle Frere regarded a return of ten per cent as reasonable in a region where there was so little employment for capital. Yet

---

[5] B. Frere, Assistant Collector of Poona, to P. Stewart, Collector of Poona, 2 July, 1840, *BA*, RD (1840), Vol. 1664.

[6] M. Rose, Assistant Collector of Poona, to P. Stewart, Collector of Poona, 15 July, 1840, *BA*, RD (1844), Vol. 1664.

[7] R.K. Pringle, Collector of Kandesh, to Bombay Government, 17 July, 1840, *BA*, RD (1844), Vol. 1664.

interest on loans in the Poona countryside ranged from 25 to 60 per cent. With respect to the cost of social functions, which the vani helped the cultivators meet, Frere described him as 'one of the greatest obstacles upon the prosperity of society.'[8]

The relations between the kunbi and the vani, barely two decades after the British conquest, had assumed a pattern which boded ill for the future. The vani was becoming more and more of a power in the rural world. His progress towards social dominance was facilitated by the Courts of Law. The increase in civil suits instituted against the ryots proves that the moneylender recognized in the new judicial institutions an instrument for self-aggrandizement. In Ahmednagar district, for instance, the number of cases involving ryots increased from 2,900 to 5,900 between the years 1835 and 1839. This increase is convincing evidence of the 'knowledge that the Marwarree has acquired of the Regulations, and of the powerful regime they afford him for exacting the fulfilment of the most usurious contracts which ever disgraced any country...'[9] A considerable portion of these suits led to the transfer of holdings from the kunbis to the vanis. But since the caste style of the moneylenders prevented them from cultivating the land, they were mostly content to let the former proprietors labour in their fields and to appropriate all the profits of their labour, apart from what was necessary for their subsistence.

The social values of the vanis constituted the most frustrating feature of the agrarian scene, for they stood in the way of a social transformation through which the small peasant proprietor would have yielded to a class of capitalist farmers owning large landed estates, and possessing the resources necessary for efficient agriculture. The vani, Frere pointed out, did not

> by a liberal expenditure of his part of his gains make up for the poverty of the ryots... Seldom do you see them [the vanis] improving any property that may have come in hand, or in embarking on any speculation such as sugar plantations, cotton or the cultivation of silk. Their thoughts and speculations are confined to their ledger and money transactions, and in no instance have I ever found a banian step forward.... to aid in any work of public activity.[10]

[8] B. Frere, Assistant Collector of Poona, to P. Stewart, Collector of Poona, 2 July, 1840, *BA*, RD (1844), Vol. 1664.

[9] Letter from the Collector of Ahmednagar, dated 30 October, 1840, *BA*, RD (1844), Vol. 1664.

[10] B. Frere, Assistant Collector of Poona, to P. Stewart, Collector of Poona, 2 July, 1840, *BA*, RD (1844), Vol. 1664.

The transfer of economic dominance over the village from the kunbis to the vanis proceeded at a rapid pace after the completion of the Goldsmid Settlement in the 1840's, which put British land-revenue policy in the Deccan on a firm basis. Barely three decades after the completion of this Settlement, the Commissioners investigating the disturbances of 1875 discovered that in one village after another the landowning families had been gradually dispossesed of their holdings by the moneylenders. In each village a basic pattern of change was in evidence. The patel and the principal cultivators, who had formerly guided the affairs of the village, were reduced to the status of tenants tilling the fields of the moneylenders. Their position as the most privileged and dominant group in rural society was a thing of the past. It was the vanis who now dominated the rural world. Yet caste prejudices and conservative habits of thought prevented the vanis from assuming that active leadership over the village which had formerly been exercised by those whom they had disposessed.

The process through which a nouveau riche caste of vanis rose to positions of dominance in rural Maharashtra can be reconstructed through looking closely into the affairs of a village like Parner in Ahmednagar Collectorate.[11] Parner was a rather large village, the headquarters of a taluka, and a *mamlatdar's* station. In it resided 50 moneylenders, mostly *marwaris* by caste, whose financial dealings were not confined to the village but extended over the neighbouring countryside. The patelship of Parner was held by the Kowrey family. At the time of the British conquest it was a coveted office, with the patel owning 160 acres of fertile land. In 1840 the Kowrey estate was partitioned into two shares of 80 acres each. A further subdivision took place soon afterwards, with Rowji Kowrey and Babaji Kowrey, two grandsons of the patel at the time of the British conquest, receiving shares of 40 acres each. In 1863 Rowji Kowrey borrowed a sum of Rs. 200 from Rajmull Marwari to buy a standing crop. He paid Rs. 150 on the original bond by the sale of his own standing crop and signed a second bond of Rs. 100 for the balance. He then paid Rs. 24 yearly for three years, and in 1866 the bond was renewed for Rs. 175. Rajmull subsequently sued him for a sum of Rs. 388 in court and obtained a decree on the strength of which he was able to appropriate Rowji's share of the Kowrey estate. Being dispossessed of his holdings, Rowji Kowrey drifted to his wife's village, where he eked out a miserable existence as a labourer on daily wages. Babaji Kowrey's fate was no better than that of

[11] *Report of the Deccan Riots Commission* (Bombay, 1876), II, Appendix C, 66-69.

his brother, since his share of the family estate was in possession of Vittoo
Marwari of Parner. 'There is not now one yoke of bullock or acre of land
in Parner village held by the Kowreys,' the Deccan Commissioners stated,
'though some of the family are still cultivating land in the hamlets.'[12]

The decline of the cultivators and the rise of the vanis can be illustrated
from a comparison of the eclipse of the Kowreys with the emergence of a
Marwari family like the Karamchands of Parner.

The first immigrant of this family was Karamchand, who came to Babulwari in
Parner, about 60 years ago. Karamchand had four sons. Tukaram, the eldest,
came to Parner about 39 years ago as his father's agent; served him in that
capacity for two years. Then his father lent him Rs. 150 at annas 12 per cent,
per mensem, and he set up on his own account. Now his *khata* in Parner and
Nagar talukes is Rs. 664 for government assessed land... (which) represents an
annual produce of Rs. 3,600. How much land is mortgaged to him, and what
may be the account of his annual dealings, it is impossible to say with any
accuracy.... (The) *kulkarni* states that Tukaram was assessed in 1871-72 at Rs.
2,000 per annum.[13]

The fate of the Kowreys and the Karamchands is not an isolated
instance of the decay of landowning families or the rise of the moneylenders. The *deshmukhs* of Parner, for instance, held 500 acres in 1818, but in
the intervening period all their land passed into the hands of local
marwaris. Typical members of the vani caste whose rise paralleled the
eclipse of the landowning families were individuals like Chandrabhan
Bhuban, who started from humble beginnings in 1840 and acquired land
assessed at Rs. 2,000 by 1875.

The extent of Marwari dominance over rural society in 1875 is also
reflected in the affairs of a village like Oorli, located in the taluka of Haveli
in Poona district.[14] Oorli had a population of 1,264 souls, an annual
assessment of Rs. 3,735, and 2,158 acres of cultivable land. It was also the
happy hunting ground of five marwaris, whose claims amounted to Rs.
16,000. Tukaram Marwari was the most substantial of these. His biggest
debtor was Buggaji Panduji who owed him a sum of Rs. 1,300, and who
paid him in return Rs. 75 worth of produce and cash yearly. Of the other
moneylenders only two, Muniram Marwari and Govinda vani, were of any
substance. Muniram had claims amounting to Rs. 2,500 in all, while
Govinda had loaned Rs. 1,000 to the various cultivators in the village.

[12] *Report.*
[13] *Ibid.*
[14] *Ibid.*

Between them the vanis of Oorli controlled all the ryots living within the village.

While the extent to which the cultivators were being dispossessed of their holdings is revealing in itself, the full repercussions of the growth of vani predominance can be appreciated only if account is taken of the social values which shaped the cultivators' style of life, and of the wider economic setting in which this transformation was being consummated. The agrarian changes described above were not accompanied by the growth of urban centres of industrial activity. Without an outlet to the urban world, the dispossessed peasant was forced to live as a landless labourer, often on those very fields which he had formerly cultivated as an independent proprietor. However, even if an expanding urban economy had provided an alternative source of employment for the kunbis, it is doubtful whether caste prejudices would have permitted them to adapt themselves readily to a new style of life.

The increasing bitterness between the cultivators and the moneylenders of the Deccan threatened to provoke a major agrarian crisis. But while the dispossession of the ordinary ryot was serious enough in itself, the simultaneous decline of important landed families, like the Kowreys, or the deshmukhs of Parner, generated intense social frustration bound to result in an upheaval. What was implicit in the social climate became explicit through two factors which heightened antagonism between the kunbis and the vanis and produced unrest in the country districts. The revision of the Goldsmid Settlement in the 1860's combined with the economic dislocation, caused in the Deccan by the American Civil War, to transform the dormant antagonism between the kunbi and the vani into open conflict.

Revision of the land settlement was an important undertaking in an agrarian society like Maharashtra, where the entire population subsisted on the land. The importance of the revision was enhanced by the Ricardian principles which informed British agrarian policy. According to these principles the State was the supreme landlord and could alter at will, and according to the dictates of political expediency, the share of the profits of agriculture awarded the cultivators. A resurvey involved, too, an evaluation of changes in economic conditions as they affected the profits of agriculture and of the portion of these profits which the State could claim as its own. A revision of the Settlement was therefore bound to stimulate excitement and could easily arouse discontent among the cultivators.

Establishment of the *Pax Britannica* had transformed the Deccan in a new way. Peace, political stability and the rule of law were at work in the new order. However, there were other equally important forces reshaping Maharashtra. Communications, for instance, had been altered beyond recognition. Since it surrounded a city of the size and importance of Poona, Haveli [taluka] was particularly fortunate in this respect. But what happened in Haveli was also characteristic of the rest of the Deccan. In contrast to the primitive conditions prevailing under the Marathas, a railway line now ran through the taluka, having four stations within its limits. There were in addition roads of good quality linking Poona with places like Bombay, Nasik, Satara and Sholapur. Since all the roads converged on the former capital, they afforded easy access to the vast quantities of supplies required for so large a city. Increasing facilities for the movement of agricultural surpluses, and expansion of markets for their disposal, stimulated a growth in population and extended cultivation in the Deccan. The population of 81 villages in Haveli (for which figures are available) rose from 37,695 in 1840 to 58,829 in 1870. (While statistics are not available for the preceding decades, it is probable that the population was static before the British take-over.) Increased population forced an increase in cultivated land, and inferior soils inevitably were pressed into cultivation. In the Haveli villages already referred to, the area under cultivation rose from 176,974 to 204,135 acres during the period of the Goldsmid Settlement. Already the Deccan appeared to be poised on the edge of a Malthusian abyss.[15]

These were the altered conditions which confronted J. Francis and W. Waddington, Superintendents of 'The Revenue Survey and Assessment of the Deccan,' when they began the revision of the Goldsmid Settlement in 1867. Like their predecessors, it was to the taluka of Indapur that Francis and Waddington first turned in the hope of devising a scale of assessment which could be extended with local variations to other districts. The spirit in which they entered upon the revision of the survey was divorced from the doctrinaire predilections which had made the Pringle Survey of 1830 so disastrous. But they did subscribe to the Ricardian view that the extension of cultivation which had taken place under the Goldsmid Settlement indicated a rise in the profits of agriculture and therefore justified an increase in the rents levied by the State. With this assumption George Wingate, who along with Goldsmid had laid the foundations of the 'Bombay System of Survey and Assessment,' was in complete agreement:

[15] J. Francis, 'Report on the taluka of Indapur,' dated 12 February, 1867, *Selections from the Records of the Bombay Government*, New Series, No. 151, (hereinafter cited as Francis' Report).

The land assessment... is not a tax at all [stated G. Wingate in a memorandum approving of Francis' Report on Indapur], but a share of the rent which the land yields to its possessors. This share of the land rent has from the dawn of history formed the great fund from which expenses of government in India have been defrayed, and in an agricultural country like India the land must ever remain the great source of taxation from which the expenses of government will have to be supplied. The government right to increase the land assessment is the property of the public, and forms a sacred trust, which, in my humble opinion, the government is bound to transmit... to its successors unimpaired...[16]

The Ricardian argument for the revision of the Goldsmid Settlement was reinforced by the errors discovered by Francis in the measurement of holdings by the former surveyors. Not all these errors were fortuitous. The Goldsmid Survey had concerned itself exclusively with measurement of the better land, and did not take into account inferior soils. However, apart from the land deliberately left unmeasured, holdings had increased in area because field boundaries had not been erected by the Survey Department until a decade after the first Settlement; and during this interval the ryots had begun to cultivate unclaimed land surrounding their fields. Thus in the villages of Baura, Kalas, Shetphal and Nagunda the ryots had appropriated between 1,500 and 2,000 acres of land of which no account existed in the official survey. Surreptitious extension of their holdings by the ryots obliged Francis to devise a new classification of fields. The need for reclassification became obvious to Francis when remeasurement revealed a particular field to be 40 acres instead of 30, as under the old survey. If the 30 acres formerly taken into account were of medium quality, and the ten acres added by the ryot poor, the entire field had to be reclassified if its owner was to escape over-assessment.

However, since the revenue system was based on Ricardian principles, the need for a revised scale of assessment was independent of inaccuracies in measurement. There were no intermediate proprietors in the Deccan between the State and the ryots, whose increased rentals could provide Francis with an index of the rise in profits of agriculture over the period of the Goldsmid Settlement. He was therefore obliged to use a less reliable criterion in calculating the increase in the net surplus of different soils during the thirty years of the lease. The rise in the price of *jowri*, the staple crop of the region, during the Goldsmid Settlement constituted the best index on which a revised assessment could be based.

[16] Memorandum dated 12 February, 1869, *Selections from the Records of the Bombay Government*, New Series, No. 151.

During the first decade of the Goldsmid Settlement the price of jowri fluctuated from 72 seers per rupee in 1843-44 to 36 seers in 1845-46; and the average over the ten years was 56 1/2 seers. The following decade began with a poor season, when near famine conditions raised the price of food grains above the previous average, and jowri was selling at 48 seers per rupee. However, in the very next year jowri plummeted to 72 seers per rupee. A steady increase in price then set in, and in 1856-57 the grain was selling at 32 seers, the average for the entire decade being 45 3/4 seers per rupee. The beginning price of 32 seers in the last decade of the lease did not change significantly in the first few years and until 1861-62 remained 30 seers.

The outbreak of the American Civil War caused a sudden dislocation in the economy of the Deccan. The export of raw cotton from the United States to England ceased abruptly, and India was called upon to supply the English market. Indapur had not been a cotton-growing district, but as soon as reports of the profits from cotton production reached the ryots, they immediately planted this crop. By 1867-68, 30,000 acres of land had been planted in cotton. The diversion of such a large area to the cultivation of cotton boosted the price of food crops, and jowri sky-rocketed to twice its normal price, raising the average to 26 seers per rupee in the last decade of the Settlement.[17]

The exceptional circumstances arising from the American Civil War, Francis reasoned, could hardly form the basis of a new scale of rates. During the thirty years of the Goldsmid Settlement, therefore, the price of jowri in Indapur in effect rose from 56 to 32 seers per rupee. But before this rise could be related to increase in rents, Francis had to answer two questions: (1) at what stage during the Settlement had the ryots accumulated the capital necessary for the efficient cultivation of their fields, and (2) what was the price of jowri at this stage.

To answer these questions Francis turned to an appraisal of economic conditions in Indapur during the thirty years of the Goldsmid Settlement. Introduction of the Goldsmid rates lowered the rental of the taluka from Rs. 203,000 to Rs. 89,000, encouraging cultivation considerably. In the five years following 1836, cultivated area increased by 60,000 acres, and there was a proportionate increase in land revenue. But it soon became obvious that low rentals had tempted the ryots to spread their resources too thinly in order to cultivate as large an area as they could secure for themselves. When the inevitable reaction set in, the ryots were not only forced to yield

[17] Francis' Report.

a considerable portion of the land they had taken, but they were also compelled to seek remissions in land revenue. It was only late in the 1840's that conditions stabilized and there was need neither for large scale remissions in revenue nor reduction in cultivated land. This favourable trend continued throughout the 1850's, indicating that the condition of the cultivators had improved to an extent where they could pay their rents in poor as well as in good seasons, and also indicating that they had built up a substantial reserve of capital and could cultivate their land with reasonable efficiency. From this turning point in the state of the Deccan districts in the 1850's, Francis drew the following conclusion:

> I think, therefore, we may fairly assume that towards the latter end of the second decennial period (1846-47 to 1855-56) the cultivators had acquired that amount of capital and that well to do position which we could assign to them in the name of profit to be left to them after payment of the government assessment. I am consequently of opinion that we may take the average price of grains during the latter half of the second decennial as the index by which we may estimate from prices what our present assessment ought to be. In other words, the percentage increase which has taken place during the last ten years will represent generally the percentage addition to the present assessment which may now be made.[18]

With the increase in price of grain during the second decennial preceding the Civil War as the index on which the increase in assessment was to be based, Francis' task was relatively simple. The average price of jowri during the last five years of the second decennial of the Goldsmid Settlement stood at 42 seers per rupee; the corresponding figure for the decennial ending 1865-66 was 26 seers. 'We may assume, therefore,' Francis stated, 'that between 50 and 60 per cent is the addition (calculated solely with reference to the price of grain) which may be made to the present assessment.'[19] The rate at which Goldsmid assessed the best soil in Indapur was 12 annas per acre. Francis raised it to one rupee per acre, an increase of 33 per cent. The total assessment, however, was raised by 50 per cent, because of the undisclosed expansion of cultivated area discovered by Francis, and all told, the revision of the survey raised the rental of Indapur from Rs. 89,000 to Rs. 124,000.

When confronted with the high rates imposed by the Pringle Survey in 1830, the ryots of Indapur forced the Survey Department to reassess the taluka by migrating in very large numbers to surrounding districts. The events of the 1830's implied that even though British Survey Officers did

---

[18] *Ibid.*
[19] *Ibid.*

not take the peasants into their confidence in making a revenue settlement, the peasants could exercise a decisive influence on the land-tax. The success of the new rates recommended by Francis, therefore, hinged on their acceptance by the kunbis. However, the timing of a new Settlement was just as crucial an issue as the pitch of the assessment, and the Bombay Government could not have chosen a worse moment for launching a revised settlement. By 1870, the stimulus given the rural economy by the American Civil War had yielded to an acute depression which left the peasants impoverished and discontented.

An additional tax burden imposed on a peasantry whose economic life had been dislocated by a temporary demand for a cash crop could hardly evoke a favourable reaction, particularly in view of the effect of the new rates on individual ryots and specific villages. While the total rental of Indapur had only been raised by 50 per cent, the increase in rent was as high as 200 per cent in certain villages.[20] What Francis never bothered to investigate, however, was how individual ryots were affected by the changes he had recommended. The fate of the ryots of the village of Kullum was, in this respect, typical of the consequences of the new survey. It is clear from the alterations proposed in the rent roll of Kullum that a kunbi like Madoo Mitoo, whose assessment had been increased from Rs. 6 to Rs. 24 because of the ten acres of land he had surreptitiously added to his original holding of 20 acres, was bound to become hopelessly indebted under the new Settlement. A similar fate awaited ryots like Madhoo Bawanee and Naroo Rowji, whose assessment had been raised from Rs. 7 to Rs. 19, and from Rs. 12 to Rs. 21, respectively.[21]

Yet the mass migration to which the peasants had resorted in 1830 was no longer possible in 1870 because of the increase in the population of the Deccan in the intervening period. Apart from population pressures, the creation of property rights and the generation of an acquisitive spirit through the low rates of the Goldsmid Settlement had tied the ryots to their villages in a way which made it impossible for them to abandon their fields and wander about the countryside. However, the abandonment of their villages was not the only form of protest traditionally adopted by the peasants against arbitrary bureaucratic action. Under the former rulers, the deshmukhs and the territorial families had in comparable situations communicated the grievances of the ryots to the authorities at Poona.

[20] *Ibid.*

[21] Petition from the ryots of Kullum to the Bombay Government, dated 5 October, 1874, *BA*, RD (1874), Vol. 96.

While an increase in the power of the bureaucracy under British aegis had virtually shorn the deshmukhs of their status as patrons of rural society, the traditional leaders still exercised a certain hold over the imagination of the peasants. It was, consequently, not surprising that they turned to a traditional leader like Gopal Narsingh Deshmukh, who adopted the new title of 'Agent of the Ryots of Indapur,' to communicate their grievances over the new rates to the Bombay Government.

The petition presented by Gopal Narsingh Deshmukh on behalf of the ryots of Indapur to the Bombay Government was a remarkable document, both for the values it expressed and for the use it made of a traditional and supposedly defunct institution for communication between the peasants and the State. The petition originated at a meeting held in Indapur in July of 1873, which was attended by the more influential cultivators from the villages of the taluka, and it was circulated until 2,694 ryots had signed it. The peasant origin of the appeal is evident from the old style sentiments and the traditional values which inspired it: a romantic vision of Maharashtra under the Peshwas; a traditional belief that the ruler would protect the peasantry from the acts of an unjust bureaucracy; and an impression that the mere expression of discontent would oblige the Government to yield.[22]

The principal theme hammered home by the ryots of Indapur was the tragic contrast between their affluence under the liberal Goldsmid Settlement and the abject destitution to which the rates introduced by Francis and Waddington were reducing them. Pringle's attempt to settle Indapur in 1830 had left the taluka desolate, and when the 'popularly beloved Mr. Goldsmid' was appointed to resurvey the taluka, he had realized that the peasants were so poor that only a very light assessment could induce them to recultivate their fields.

A moddel [sic] survey of the taluka was aimed at [the ryots pointed out], and such were the settled rates, that after defraying all expenses of cultivation, etc., including the assessment, they [i.e., the peasants] received no less than 1/8 of the produce as a reward to the cultivators. Moreover, an ample provision had been made by these truly circumspect officers for our cattle, etc., in excluding all and every sort of waste land, amounting to about 43,000 acres, which had been used by us as grazing lands, but are since the late survey about all assessed along with the cultivable lands proper.[23]

[22] Petition from the ryots of Indapur to the Bombay Government, dated 15 February, 1875, *BA*, RD (1875), Vol. 96.

[23] Petition of 29 July, 1873.

To the ryots of Indapur it appeared as though Francis and Waddington were inspired by motives completely opposed to the generous principles which had guided their predecessors. The new rates of taxation, they observed, were not only too high in themselves, but the inclusion of the former waste land under assessable categories made them positively ruinous. Yet the enhanced scale had been imposed by Francis 'without reflecting for a moment what calamities he was about to bring on the helpless poor...' The effect of an oppressive scale of rates was heightened by the harshness with which the revenue officials forced payment of the land-tax. Because of inadequate rain in 1871 and 1872, the yield of agricultural produce had been only 50 to 75 per cent of the normal figure; but when this had been brought to the notice of the Collectors, 'not the slightest notice had been taken of all our cries for exemption from this heavy tax.' Did the Collectors (the ryots posed the rhetorical question) pause to consider the damage they were inflicting by selling landed property worth thousands of rupees for the non-payment of a few beggarly instalments? Faced with the cruel demands of the revenue officers, the ryots were obliged to turn to the moneylenders for assistance. This ensured for them a fate worse than death for they were reduced to being the bonded slaves of the vanis. 'The monstrosity of our subordinate rulers has been so great that words cannot express them,' the ryots stated in conclusion:

The usurpation of our rights by both the moneylenders and the government together, really brings to our recollections the jolly old times when our fathers swayed the sceptre in prosperity. It really cuts one to the core to reflect on the past and present conditions; i.e. the freedom and affluence in which our fathers lived and died, and the serfdom in which we are doomed to live and die!

From past history it is evident that during the 2,000 years back a great many rulers have been in possession of India; but notwithstanding their great abilities and power, no sooner they manifested a desire for worthless gain... the Almighty God supplanted them by others more human; for God will hear the cries of the afflicted and punish the wicked.[24]

The significance of the Indapur petition lies in the light it throws on the hold of traditional institutions over rural society and on the ryots' reaction to the new Settlement. Whether the petition was able to prove the existence of high rates of assessment, which it did not, is irrelevant to either of these considerations. For the ryots' problems stemmed basically from the economic dislocation caused by the American Civil War and a succession

[24] Petition of 29 July, 1873.

of bad harvests, which had combined to cut into their slender accumulations of capital and to render their economic condition precarious.[25]

It took, however, civilians like Auckland Colvin, a revenue official who served on the Deccan Riots Commission, and W.H. Havelock, the Commissioner of the Northern Division, to reveal through a rational analysis what a deshmukh playing his traditional role could only express in emotive terms. Colvin questioned from the outset the assumption of increasing rural prosperity which lay behind the rates proposed by Francis. In support of his view that a high assessment had helped 'disturb [for the worse] the relation of creditor and debtor in the Poona district,' he recalled the Malthusian nightmare which had already cast its shadow over the Deccan. Colvin, observing that the increase in population in the four talukas of Haveli, Pabul, Supa and Bhimthari during the thirty years of the Goldsmid Settlement had been accompanied by an increase in economic resources, found it 'very significant that the growth in population is out of all proportion to the growth of plough cattle or homes.... [It seems] that individual property in stock has declined....'[26] The problem became more intense in view of the manner in which population pressure had forced the ryots to cultivate poor soils. The extent to which the waste land was being pressed into cultivation expressed the ryots' dilemma clearly and precisely. For here was evidence beyond refutation that the increase in the area of cultivation had kept pace with the increase in population 'until, generally speaking, the whole available area had been occupied. For further rise of population there is no further margin of waste.'[27]

The most devastating indictment of the revised rates came from Havelock, who was requested by the Bombay Government to look into a complaint by some villagers from Sholapur that the new rates weighed more heavily on inferior soils than on superior soils. To ascertain the accuracy of this charge, Havelock analyzed the effect of the revised rates on a number of villages in Sholapur, of which Alipur can be taken as typical. On doing so, he discovered that the assessment on soils of the most inferior category in Alipur had been raised by 67 per cent, while medium and superior soils were paying only 18 per cent more than they had formerly paid. The inquiry he had instituted convinced Havelock that

[25] Petition from the ryots of Indapur to the Bombay Government, dated 15 February, 1875, *BA*, RD (1875), Vol. 106.

[26] A. Colvin, Memorandum dated 8 November, 1875, *Report of the Deccan Riots Commission*, Vol. I.

[27] A. Colvin, Memorandum.

there has been a judicious increase [of rates] in the highest class of lands; but that, notwithstanding a most salutary and well designed reduction in the two lowest classes of the... [Goldsmid] scale, and a slight reduction in the 7th class, the new scale has not suited the special circumstances of the region; that the application of the revised classification at too high a rate on much of the former unculturable land, and on the lower and medium lands, has raised the assessment to be too high on such lands.[28]

To what could the imbalance in the scale of assessment be attributed? Havelock traced the high pitch of the new rates to Francis' belief that Goldsmid and Wingate had left a standard proportion of the net surplus of the land to the rich as well as the poor peasant. This belief, he pointed out, was founded on a misapprehension because Goldsmid and Wingate had not made any rigorous use of the Ricardian law of rent in devising a scale of assessment for the Deccan. Their rates were based on the pragmatic principle that the poor cultivators deserved a larger share of the net surplus than the more prosperous cultivators. In revising Pringle's rates, which were based on the law of rent, Goldsmid and Wingate had reduced the rent on superior soils by 40 per cent; but the rent on soils of medium and inferior quality had been reduced by 60 per cent and 70 per cent respectively, 'with the entire success which had been recognised at all hands.' Francis had repeated Pringle's error and had tried to ensure a profit to the cultivator bearing the same proportion of the net produce irrespective of the quality of the soil he cultivated. But a successful scale of rates, Havelock concluded, had to be based on a progressive decrease in the pitch of assessment as it proceeded from the best to the worst soils.

The reasons why the ryots of Indapur opposed the introduction of the Francis rates were pinpointed by Colvin and Havelock with a cogency that completely eluded a traditional leader like Gopal Narsingh Deshmukh. In his failure to build a convincing case against the Francis Survey, the Deshmukh revealed the inability of the traditional leaders of Maharashtra to direct rural discontent in any meaningful way in the new circumstances. The influence of the traditional landed families had not disappeared by the 1870's. But the gulf in values which lay between them and the new rulers restricted their effectiveness. Contrast, in this context, the inadequacy of the Indapur petition with the powerful case for a revision of the Francis Settlement made by Colvin and Havelock. How much more effective would have been the arguments of a Colvin or a Havelock if they had been put forth by the 'Agent of the Ryots of Indapur'? To raise this question is

[28] W.H. Havelock, Memorandum dated 20 April, 1874, *BA*, RD (1874), Vol. 97.

to indicate the need for new and more effective leadership in the community.

In response to this challenge a number of Poona Brahmins founded in 1867 an organization called the Poona Association, the dual object of which was to educate the public on the crucial issues of the day and communicate the views of the people to the British Government. In 1870 the Association was reorganized along more ambitious lines as the Poona Sarvajanik Sabha. The Sabha not only claimed the sympathy of traditional aristocrats like the Pant Pratinidhi of Aundh and the Maharaja of Kolhapur, but also counted among its members Brahmin intellectuals like Mahadev Govind Ranade, the most sophisticated political thinker of his time in western India, and Ganesh Vasudev Joshi, a liberal and a leading figure in Poona politics. While the association of the traditional aristocrats gave legitimacy to the Sabha, it was the Westernized elite represented by Ranade and Joshi which shaped the style of its political activity.

The popular foundation of the Sabha consisted of 95 representatives from all over Maharashtra who attended the inaugural meeting on 2 April 1870. These representatives had been 'elected' by 'over 6,000 persons, representing all castes, creeds, and interests...'[29] A number of affiliated bodies were founded concurrently with the Poona Sabha in the principal towns of Maharashtra such as Satara, Wai, Solapur, Nasik, etc. The purpose in creating these Sabhas was set out clearly in the preamble to the constitution of the Poona branch.

> Whereas it has been deemed expedient that there should exist between the government and people some institution in the shape of a mediating body which may afford to the latter facilities for knowing the real intentions of Government, as also adequate means of securing their rights by making timely representations to government of the real circumstances in which they are placed, an association has been formed and organised with the appellation of Poona Sarvajanik Sabha.[30]

The Sarvajanik Sabha has often been regarded as a body which concerned itself merely with presenting cautiously-phrased petitions to the British Government. But its leading members like Ranade and Joshi did not conceive of their role in a purely passive light. For they were

[29] Memorandum on the founding of the Poona Sarvajanik Sabha, unpubl. collections of *Bombay State Committee for a History of the Freedom Movement in India.*

[30] V.M. Potdar, an article on the history of the Poona Sarvajanik Sabha, unpubl. collections of *Bombay State Committee for a History of Freedom Movement.*

anxious to establish themselves in the affections of the peasants and to build a social base for the Sabha in the rural areas. The Sabha resembled a caste organization to the extent that it provided an institutional framework for the political activities of the elite castes (Chitpavan Brahmins, etc.) in the urban areas. But it simultaneously tried to assume the traditional role of the rural aristocracy as a bridge between the peasants and the State. The new elites of the Sabha were not only *advaitic* intellectuals who fashioned their politics in the style of John Stuart Mill; they also stood as deshmukhs defending the interests of their rural clients through the use of Ricardian and Utilitarian values.

The resettlement of the Deccan by Francis and the opposition of the kunbis to the increased rates of assessment afforded the Sabha a unique opportunity to champion the cause of the peasants. Immediately after the introduction of Francis' rates, Ranade and Joshi tried to express the peasants' opposition to the new rates and to secure a diminution in the land-tax, in conscious imitation of the traditional role of the deshmukhs. Their attempt to do so was expressed in political activity at two levels: the drawing up of petitions which communicated the grievances of the peasants to the government; and the despatch of trained cadres to the villages in order to acquaint the peasants with reasons for their miserable plight. An elite which combined traditional techniques of agitation with a commitment to Western political ideals presented a serious threat to British authority, and the significance of this development was not lost on the British Government.

The 'Report of the Sub-Committee of the Poona Sarvajanik Sabha' on the Francis Settlement, which was presented to the British Government in 1873, concerned itself with themes identical to those elaborated in the Indapur Petition. But in contrast to the 'romantic' idiom of the Petition, the Report was couched in the rational language of political economy, and a wide gulf separated its arguments and the traditional principles on which Gopal Narsingh Deshmukh had based his indictment. Though written contemporaneously, the Petition and the Report stood worlds apart in intellectual attitudes. They displayed the cleavage in Maharashtra between those who subscribed to traditional styles of politics and the emerging Westernized elite which was attempting to secure positions of leadership within the community. There can be no doubt of the greater impact the agitation of the Sabha had on Government as well as on the rural community. For the leaders of the Sabha understood the political beliefs of the new rulers and could use techniques of agitation which the kunbis understood.[31]

The rational terms in which the Report spelled out the charge of the impoverishment of the kunbis put the Bombay Government on the defensive, and a searching inquiry was instituted into the condition of the Deccan districts of which the Sabha had presented so gloomy a picture. District Officers who were intimately associated with the administration of the rural areas did not reject the Report as a polemical tract devoid of all objectivity. They emphasized, however, that the changed quality of social experience in the villages of the Deccan had to be given the same prominence as the statistical realities of rural life if any balanced assessment of the agrarian situation was to be achieved. Civilians like A. Wingate, the Collector of Satara, argued that unfavourable reports on rural prosperity mostly stemmed from changes in expectations and in social values resulting from British rule. Under the former rulers the material requirements of the ordinary peasant were extremely modest. '[The]... wants of man scarcely exceeded those of animal... [A] few bits of rags, a hut, and a cooking pot or two constituted the family accumulations.' All this was related to the fact that in a community where the means of transport were primitive and markets undeveloped, surplus agricultural produce could easily be appropriated by the State and dominant sections of rural society. Such conditions in turn produced a peasantry which was passive and apathetic, and reconciled to its wretched existence. The situation had changed dramatically under British rule. A generation of peace and stability and the introduction of a rational revenue policy had opened the eyes of the kunbi to a new range of material wants and shown him how he could satisfy these wants, making him thoroughly discontented with his lot.

Western energy [Wingate pointed out] is introducing the element of labour to Eastern apathy and with the desire to accumulate comes the necessity to work. People are no longer content with what satisfied them 50 years ago; their own and their neighbours' estimate of the fitness of things has changed... For example, one mamlatdar states that during the decade 1820-30 the ryots'

[31] The author has been unable to trace the report presented by the Poona Sarvajanik Sabha to the Bombay Government. This inadequate precis of the report is taken from extensive quotations from it in a despatch to the Secretary of State for India by the Bombay Govt, 27 December 1875, *BA*, RD (1875), Vol. 44A.

condition was all that could be desired. The crops ripened well, grain was plentiful, and the instalments in kind were easily given. Great men in the State supported numerous retainers, and these in turn supported their families, so that labour was cheap, the necessity for buying little, and so long as the rains fell seasonally, everybody got enough to eat... It is true that there were no shops, no roads, no trade, and little encouragement for labour; no one could afford either to leave his village or to purchase his clothes or ornaments or fair stock. But then... [these things] were not wanted...[32]

The *Pax* had thus created an undercurrent of discontent among the[TU] peasants by generating a desire for a higher standard of living. It had also stimulated the rise of a class of kunbis with new standards for consumption and accumulation of capital, who were 'in every way so superior to what the same people were 40 years ago as it is possible to conceive.'[33] In contrast, however, to the prosperity of these rich kunbis stood the poverty of the great majority of the cultivators. This was reflected in the widening gap between the wages of agricultural labourers and the prices of the major food crops, on the one hand, and the decreasing margins of profit on the inferior soils which were being taken under cultivation, on the other. In Ahmednagar district, for instance, the wage of an agricultural worker rose from Rs. 28 per annum to Rs. 60 per annum during the Goldsmid Settlement. Yet this rise did not represent any real gain since the price of basic food grains like jowri and *bajra* had risen by 250 and 185 per cent respectively over a corresponding period.[34]

The inquiries instituted in response to the Sabha's Report demolished the picture of increasing rural prosperity which formed the background to Francis' revision of the Goldsmid Settlement. But while the Sabha had attributed the deterioration in economic conditions to revenue policy in general and to the Francis survey in particular, the Bombay Government ascribed it to the population upsurge which followed the British take-over and to the lack of prudence and thrift among the peasants. In the Government's view, an uncontrolled rise in population in a country which did not have a diversified and an expanding economy could produce only one result: The rising pressure on the land would oblige the cultivators to cultivate inferior soils and would in the course of time lower the profits of

---

[32] A. Wingate, Collector of Satara, to Bombay Government, 3 October, 1874, *BA*, RD (1875), Vol. 44B, Part 2.

[33] Wingate to Bombay Government, 3 October, 1874.

[34] H.B. Boswell, Collector of Ahmednagar, to Bombay Government, 28 August, 1874, *BA*, Rd (1875), Vol. 44B.

agriculture. 'Some experienced officers under us are already of opinion that the land is less productive now than formerly,' the Bombay Government pointed out in a despatch to the Secretary of State for India, which rebutted the Sabha's accusations. But there were obvious limitations to what the Government could do in the circumstances. All 'civilised' communities exercised a voluntary restraint on the increase of population. In India the limit was set by famine and epidemics. Marriage was enforced by religious sanctions, and the possession of heirs was the precondition of salvation. Persistence of such social beliefs was leading to disaster, for while

> a government can do much to foster the development of the country, it can do little or nothing to enrich such of its subjects as are wanting in thrift or enterprise ... [Progress] depends more on the desire of the people to learn than on the capacity of the government to teach, and after all the principal object of government is to afford protection to life and property, and the accumulation of wealth must be left to individual action. It cannot be denied that as regards security of life and property, opportunity of education, etc., there is no comparison between present and former ideas... But still the people are worse off then they were ten or fifteen years ago...[35]

The Sarvajanik Sabha's ability to put the Bombay Government on the defensive indicates how effective the new elite was becoming. The Sabha's success also illustrates the impotence of a traditional leader like Gopal Narsingh Deshmukh. But to play the role of a deshmukh to the full, it was essential for the Sabha to secure participation by the peasants in its campaign against the Francis Settlement, not least because the elites who comprised its backbone were urban-based and urban-oriented and lacked that intimate connection with the kunbis which lay behind the power and influence of the natural leaders of rural society. In pursuit of this tactic, the Poona Sabha and its branches sent agitators to the villages to whip up the kunbis' opposition to rates imposed under the new survey and to explain the reasons for their poverty. The Sabha's success in arousing the peasants against the new Settlement was dramatic, and became immediately apparent to the authorities. One Superintendent of the Revenue Survey, for instance, ran into organized opposition the moment he tried to introduce the new rates to the ryots of Barsi taluka.

> I had scarcely concluded my explanatory remarks [Waddington wrote to Francis] when most of the assembled ryots stood up, refusing to have their

[35] Despatch to Secretary of State for India, dated 27 December, 1875, *BA*, RD (1875), Vol. 44B, Part 2.

'khatas' examined, and declaring their intention to pay no more assessment than they had hitherto been in the habit of paying. I endeavoured to reason with them and to point out the grounds on which it was but equitable that their payments should be raised; but without success, and so excited and disrespectful was their demeanour that I felt myself bound to report the matter to you for further instructions.... I have little doubt that the opposition has been fostered by the Sarvajanik Sabha of Sholapur.[36]

Waddington's experience was not an isolated example of the Sabha's ability to arouse opposition to the new scale of rates. The revised assessment had been introduced without any trouble in the talukas of Madhe and Sholapur in 1872. But while the ryots had quietly accepted the new rates in the first instance, they refused to pay the land-tax the moment agitators connected with the Sabha appeared in their midst, and told them that in raising the rental the Survey Department had encroached upon their rights. The Collector of Satara believed that these agitators had urged the cultivators 'to refuse to pay the assessment, trusting to them, the Sarvajanik Sabha, to make it all right for them to do so.'[37]

The disaffection aroused by the Sabha's agitation thoroughly alarmed the Bombay Government, particularly because the depth of this disaffection was obvious from the alarming reports, which poured in from one taluka after another, of the increasing resistance which the ryots were offering. It was true that most revenue officers felt that the ryots' intransigence stemmed from the poverty which had overtaken them as a result of a succession of bad harvests. But an inability to pay rent due to the State was not the only reason behind the kunbis' truculence, as Waddington had discovered to his astonishment. The high pitch of the Francis rates had led to an alliance between the traditional leaders of rural society and the new elites of the Sabha, and this alliance was an important factor behind the peasants' hostility to the new Settlement. Pointing to the active role which a few leading families of the taluka of Bhimthari had played in the Sabha's 'no-tax' campaign, the Collector of Poona observed, 'It may seem strange to connect a general failure in the collection of a whole taluka long under our rule, with the discontent of a few families... But it must be remembered that these families had considerable influence under the former Government, which still exists to some degree...'[38]

[36] W. Waddington, Letter dated 18 April, 1874, *BA*, RD (1874), Vol. 104.

[37] Letter from the Collector of Sholapur, dated 3 December, 1873, *BA*, RD (1874), Vol. 97.

[38] Letter from the Collector of Poona to Bombay Government, dated 14 December, 1873, *BA*, RD (1874), Vol. 97.

The situation took so serious a turn that in April, 1874, Havelock, the Revenue Commissioner, called a meeting of the revenue officials of the affected districts in order to chalk out a course of action that would relieve the prevailing tension. In a letter written to the Collector of Poona on the eve of the Conference, Havelock outlined his views on the problem and the means he considered expedient to solve it. He was prepared to view with sympathy appeals for reductions in tax from ryots who were unable to pay their rents, but he would not yield an inch to the spirit of opposition fostered by the Sabha and a few disgruntled territorial families.[39] However, the discussions which took place at the Conference revealed how difficult it was to arrive at any definite conclusions regarding the peasants' ability to pay their dues. Rao Saheb Balwant Sitaram, the mamlatdar of Bhimthari, felt that the peasants were genuinely unable to pay the new rates because the opposition 'was as great in the revised villages that had good crops, as in those that had bad crops.' But the mamlatdar was opposed by the Collector of Poona, who held that the opposition sedulously propagated by the Sabha in alliance with the disaffected territorial families had alienated the ryots to such an extent that they were thirsting for a collision with the authorities. The solution suggested by the Collector was of a piece with his prognosis. Once the spirit of the leading families had been broken by the confiscation of their estates, he pointed out, the peasantry would no longer dare to oppose the revised rates.[40]

But the Bombay Government refused this course of action because economic deterioration had been disclosed by the inquiries provoked by the Sabha's charges and because the revised rates had hit the poor cultivators harder than the prosperous ryots.[41] Instead of accepting the suggestions of the Collector of Poona, the Government extended a series of concessions to the ryots in order to reduce their hostility. It resolved that in no case should the assessment of a taluka or a group of villages be raised by more than 50 per cent, or that of a single village by more than 75 per cent. It also decided that if a cultivator were unable to pay his tax, the Revenue Department would in the first instance attach his movable property, the

[39] H.D. Havelock, Letter to the Collector of Poona, dated 26 December, 1873, *BA*, RD (1874), Vol. 97.

[40] Letter from the Collector of Poona, dated 13 May, 1874, *BA*, RD (1874), Vol. 97. See also Minutes of a departmental conference held in Poona on April 14, 1874, *BA*, RD (1874), Vol. 97.

[41] Resolution by the Bombay Government, dated 29 October, 1874, *BA*, RD (1874), Vol. 97.

ryot's holdings being auctioned only when his movable property proved insufficient to cover the full amount of his tax.[42]

The imposition of a ceiling on the assessment mitigated the ryots' hostility to the authorities and to the Francis Settlement. But the decision to attach the ryots' movable property if they failed to pay the land-tax held implications which suddenly created a tense situation. It is important to bear in mind that despite the opposition provoked by the Francis Settlement, the disquiet prevailing in the Deccan stemmed basically from the loss of their lands by the kunbis and the dominance which the vanis had consequently come to acquire. No doubt the American Civil War and a series of bad harvests provoked a degree of rural ferment which was unusual. But the most important cleavage in the rural world still came from the antagonism between the peasants and the moneylenders. Because of the dependence of the ryots on the vanis for the payment of the land-revenue, the decision to impose a preferential alienability on movable property was a concession which was immediately exploited by the vanis. The vani had hitherto advanced the assessment to the ryot in the knowledge that if the latter defaulted, his land had to be attached by the Revenue Department. Since the Government now declared its intention to alienate the ryot's movable property in the first instance, the vani did not see any threat to the ryot's land, and refused to advance him the land-revenue.

The latent antagonism between the moneylenders and the cultivators consequently burst into open warfare, and the pressures which had been building up within rural society erupted into violent conflict between the kunbis and the vanis. It would be misleading to look upon this upheaval as stemming from a single cause. The dislocation of the rural economy by the American Civil War; an ill-conceived Settlement by Francis; the agitation begun by the Poona Sarvajanik Sabha and backed by the dominant rural families; and finally, the simmering hostility between the kunbis and the vanis, combined to create the tension and social frustration which resulted in the riots of 1875. The social complexion of the disturbances was determined by the Government's decision to make the ryots' movable property alienable, which put an abrupt end to loan transactions in rural society. It was also influenced by the fact that the tie between the kunbis and the vanis was the weakest link in the chain which bound different classes and castes in rural society.

In contrast to the opposition to the Francis Survey, the riots of 1875 were spontaneous and did not bear any evidence of the organization that

---

[42] Resolution of 29 October, 1874.

had gone into the Sabha's 'no tax' campaign. Moreover, while the anti-Settlement agitation had revealed the extent to which the old landed families still influenced rural society, the upsurge against the moneylenders brought into sharp focus the fragmentation of castes and groups in the village whose close interdependence was formerly a distinguishing feature of rural life. Thus the patel had become so closely associated with the administration that he was now incapable of leading the cultivators and was looked upon by them as the member of a hostile institution. Similarly, the partial introduction of the cash nexus and the spread of acquisitiveness had driven a wedge between the cultivators and the bullotedars which had been absent earlier.

In the initial stages of the upsurge of 1875, the cultivators only imposed social sanctions against the vanis in an attempt to browbeat them into acquiescence. Their moderation is brought out in a *sama* patra (bond of agreement) executed by the ryots of Kallas in Indapur, which illustrates at the same time the fragmentation along caste and functional lines in the village. The main ire of the sama patra was directed against the moneylenders, since it pointed out that any kunbi cultivating the field of a vani

> will neither be allowed to come to caste dinners nor intermarry amongst his own society. Such person will be considered outcaste. He will not be allowed to join the community without their unanimous consent, and will have to pay the fine which the community may inflict on him, and further to give one meal to the community.[43]

But there were many indications of the lack of co-operation, if not actual friction, between the kunbis, who were the driving force behind the anti-vani agitation, and the village officers and the bullotedars. The latter were threatened with the termination of their customary dues if they did not join the rest of the village in the boycott of the vanis; and the patel, instead of leading the kunbis to whom he was related by bonds of caste and class, had to be warned that 'if he joins the moneylending people, his hereditary rights will be discontinued...'

Coercion of the vanis along the lines set out by the ryots of Kallas soon spread extensively to the villages of Poona, Ahmednagar, Sholapur and Satara. Lacking any centralized control, it is difficult to trace the lines along which the movement gained momentum. But a letter written by the ryots of Kallas to the neighbouring village community of Akola indicates the mechanism used to spread the anti-vani agitation. Rebuking the ryots

---

[43] Substance of a *sama patra* by the villagers of Kallas in the taluka of Indapur, dated 7 May, 1875, *Report of the Deccan Riots Commission*, Vol. II.

of Akola for their apparent refusal to cooperate with their caste fellows in other villages, the cultivators of Kallas asserted in their letter:

> It is very wrong of your people to keep communication with the marwaris whom we excluded from the community of this village. Unanimity is very important at this time. You will perhaps know this if you give the subject mature consideration, and we therefore refrain from making further remarks about the matter.[44]

They appealed to Akola in the name of the traditional ties binding the two villages ('We have always regarded Kallas and Akola one'); they asked them to send two responsible elders to thrash out the issue at a meeting; and they concluded with a renewed plea for unity at so critical a period: 'We shall be helpless,' they argued, 'should you take no measure to [prevent cooperation with the moneylenders].... For the good of all of us it is necessary that we should cooperate with each other.' The widespread adoption of social sanctions against the moneylenders created so tense a situation in the Deccan that a violent clash between the kunbis and the vanis became inevitable. The first outbreak occurred at Supa, a rather large village in Bhimthari taluka, on 12 May, 1875. The victims of the ryots were the vanis, of whom there were a goodly number in Supa. Their houses and shops were stripped of everything the rioters could find, and then burnt, but no violence to any person was committed. Within 24 hours of the outbreak at Supa, the leading marwari of Khairgaon, a village 14 miles away, had his residence burnt. In the following days riots occurred in four other villages of Bhimthari and threatened in 17 more. The contagion then spread to the talukas of Indapur and Purandhar. Outside Poona district the disturbances were concentrated in the talukas of Parner, Shrigonda, Nagar and Karjat in the Collectorate of Ahmednagar. The riot at Supa was alone in the wholesale destruction of property, and that at Damareh in the murderous assault on a vani. In a few other cases the ryots threatened the vanis with violence. But on the whole there was little violent crime. The object of the rioting cultivators was to obtain and destroy the bonds and decrees possessed by the moneylenders. Where these were given up without resistance, no bodily harm was done; but if the vanis refused to yield the legal documents in their possession, violence followed. The 'non-violent' character of their proceedings impressed itself forcibly on the Commission investigating the disturbances.

---

[44] Letter from the villagers of Kallas to the villagers of Akola, dated 15 May, 1875, *Report of the Deccan Riots Commission*, Vol. II.

In reviewing the character of the disturbances generally [ran the report of the Commission] the most remarkable feature presented is the small amount of serious crime. A movement which has a direct appeal to physical force was over a large area usually restrained within the limits of a mere demonstration; the moderation is in some measure to be attributed to the nature of the movement itself. It was not so much a rebellion against the oppressor, as an attempt to accomplish a very definite and practical object, namely, the disarming of the enemy by taking his weapons [bonds and accounts], and for that purpose mere demonstration of force was usually enough.[45]

The unique features of the Deccan Riots of 1875 were a reflection of the tensions generated within rural society through the legal and administrative reforms carried out by the British Government. Before 1818, rural society in the Deccan was characterized by an intimate interdependence between the different caste and functional groups within each village. The dominance of the ryots, a feature which found expression in the power and influence of the patel, created the consensus which held these groups together. It was also responsible for generating a social climate in which collective action rather than individual endeavour formed the basis of the social order. All this was transformed under the impact of Utilitarian policies which tried to mould rural life along individualistic and acquisitive lines. Changes in Maharashtran society were keynoted by the increasing dominance of the vanis and the growing antagonism between them and the ryots.

Tension between the kunbi and the vani created maximum antagonism in rural society. The climate of individualism and acquisitiveness encouraged by Utilitarian measures of reform undermined the collective structure of the village and split it into discrete functional groups; alienating the patel from the kunbis by stripping him of his traditional authority; setting the bullotedars against the ryots by introducing them to the notion of a cash nexus; and transforming the harmonious relationship between the vani and the kunbi into one of conflict by the introduction of new legal principles and the creation of new judicial institutions. To an extent these changes were calculatedly fostered by the advocates of reform in the fond hope that the breakdown of the village community would set in motion a revolution in rural society. But the caste styles which fashioned the pattern of rural life stood in the way of the transformations which the Utilitarians regarded as politically expedient and socially desirable. The caste values of the vanis, for instance, prevented them from becoming capitalist farmers even after

---

[45] *Report of the Deccan Riots Commission*, Vol. I, para. 15.

they had acquired large holdings of land. Instead, they preferred to lease their estates at rack rents to the former proprietors. The marwaris thereby generated a social climate of conflict and strife which had been absent earlier, and which found expression in agrarian upsurges like the Deccan Riots of 1875.

GLOSSARY

Advaita a monistic philosophy propounded by Shankaracharya, an eighth century religious teacher, which has left a profound impact on the Hindu religion.

Bullotedars the marathi term for village artisans like carpenters, potters, blacksmiths, etc.

Deshmukh a landed aristocrat.

Jowri barley.

Khata an account book.

Kulkarni a village accountant.

Kunbi the Marathi term for a peasant.

Mamlatdar a native revenue officer who presides over a *taluka*.

Sowcar a financier in an urban setting.

Taluka a territorial unit, several of which comprise a district, presided over by a native revenue officer.

Vani a village moneylender. (Also *Bania*, or *Marwari*.)

Chapter Five

Agrarian Disturbances in Nineteenth-Century India*

I.J. CATANACH

I

This paper deals with some aspects of the history of popular disturbances in nineteenth-century rural India, partly in the light of the literature on popular disturbances in European history.[1]

I first became aware of the potential interest of my subject when I came across the official report on a series of agrarian riots which took place in the districts of Poona and Ahmednagar in 1875, the so-called 'Deccan Riots'.[2] At first sight, there appeared to be a strong similarity between the development of the Deccan Riots and the development of the *Grande Peur* in the early days of the French Revolution. In both cases, it seemed, a countryside, swept with rumour, rose up largely spontaneously and with some rapidity. Here, I thought, was a field for further research in India: what Indian peasant history required, I concluded, was the application to it of some of the techniques and the theories of Professor Rudé, and of a scholar amongst whose disciples Professor Rudé must surely be included, Georges Lefebvre.

In this paper, while the Deccan Riots are my point of departure, I have also tried to say a little, for purposes of comparison and analogy, about

---

* Paper presented to the International Conference of Economic History, Munich, August 1965.

[1] George Rudé, 'The Study of Popular Disturbances in the "Pre-Industrial" Age', *Historical Studies, Australia and New Zealand*, Vol. X, No. 40, May 1963; *The Crowd in History: A Study of Popular Disturbances in France and England*, 1730-1848, New York, 1964.

[2] *Report of the Committee on the Riots in Poona and Ahmednagar*, 1875, 2 vols., Government of Bombay, 1876.

some other agrarian disturbances in nineteenth century India, and even elsewhere. I am concerned as much with the course of these disturbances as with their causes. And I have concentrated my attention in this paper on disturbances of a reasonably extensive nature rather than on minor skirmishes. In the nineteenth century, as in fact today, commotions in single villages over such matters as the use of a well, or the route to be taken by a religious procession, were not uncommon; the result, more often than not, was a few bruised heads, though occasionally men died. In the earlier years of the nineteenth century, too, minor skirmishes between villages often occurred; the disappearance of such means of excitement with the coming of what is generally known as 'British law and order' is apparently still mourned by some of the inhabitants of the villages.[3] Such matters are interesting in themselves, but I do not intend to say anything further about them here. And I intend virtually to exclude from my discussion the phenomenon known as *dacoity*—sporadic raids on villages, and especially on their wealthier inhabitants, by armed bands, based generally in wild, comparatively uninhabited, areas.[4] One could perhaps note, however, that occasionally the activities of *dacoits* in the nineteenth century take on an air of what Dr. Hobsbawm calls 'social banditry': the rich are robbed for the sake of the poor.[5] India certainly has its Robin Hoods, in history and in legend. Folk songs in Malabar tell us of the activities of one Tachcholi, specifically called by Logan, the great authority on the area, 'the Robin Hood of North Malabar'.[6] The bitter crusade conducted in 1874 by 'Honya's Gang' in the foothills of the Western Ghats received the active or passive support of virtually the whole of the semi-aboriginal Koli community. 'Honya is the Coeur de Lion with no Saladin to oppose him', lamented one of the more imaginative of the younger men at the Secretariat at that time.[7]

[3] See Alan R. Beals, 'Change in the Leadership of a Mysore Village', in M.N. Srinivas (ed.), *India's Villages*, 2nd edn., Bombay, 1960, p. 155 (1st edn., Calcutta, 1951.).

[4] See George A. Floris, 'A Note on Dacoits in India', *Comparative Studies in Society and History*, Vol. IV, No. 4, July 1962.

[5] See E. J. Hobsbawm, *Primitive Rebels*, Manchester, 1959. See too J.C. Holt, 'The Origins and Audience of the Ballads of Robin Hood', *Past and Present*, No. 18, November 1960; Maurice Keen, 'Robin Hood—Peasant or Gentleman?', *Past and Present*, No. 19, April 1961.

[6] William Logan, *Malabar*, 3rd edn., Government of Madras, 1951, Vol. I, p. 96 (1st edn. 1887).

[7] J. Nugent, Under Secretary, on W.H. Havelock, Commissioner, S[outhern]

II

Honya's activities in the Western Ghats may to some extent have set an example to the caste Hindus of the Deccan plateau, further to the east, who turned against their money-lenders in the following year.

In very broad terms the background to the Deccan Riots is this. The British may be said to have brought with them the system of mortgage. At the time of prosperity in the sixties, which resulted from the high prices that were paid for Indian cotton during the American Civil War, many Deccan peasants borrowed on mortgage far more extensively than they had ever done before. The lack of an indigenous money-lending caste in the Deccan made it necessary for them normally to borrow from members of the immigrant Marwadi and Gujerati communities. With the sharp contraction of prices after the restoration of normal conditions, a large number of peasants found that their land was slipping from them, through foreclosure, into the hands of non-agriculturist money-lending groups, 'outsiders' in the Deccan situation. The Deccan Riots were undoubtedly a protest about this state of affairs.

The pattern followed in most of the riots seems to have been that men began congregating in the afternoon, sometimes within a village, often just outside it. When dark came there was a noisy rush on the money-lenders' houses. Often a hundred or more men took part; the money-lenders of one village affected by rioting claimed in a petition that eight hundred participated, though this may be an example of the hyperbole of which the historian must be continually wary when dealing with petitions, whether in India or in Europe.[8] The moneylenders were generally warned of what was to come; many of them seem to have complied with the rioters' demands, which were normally limited to the surrender by the money-lenders of their bonds and other documents—the written evidence of indebtedness—and the destruction of those documents. Little violence followed. One group of villagers was even reported to have paid the railway fare of their money-lender to a safe haven.[9] Here we may recall that Professor Rudé notes the 'discriminating purposefulness' of the rioters in

---

D[ivision] to Secretary, J[udicial] D[epartment], No. 673 of 9 April 1874. Bombay Secretariat Record Office, Government of Maharashtra [hereafter 'B.R.O.'], J.D. Vol. 52 of 1874.

[8] Petition of the Supa Sowkars to Governor of Bombay, 23 September 1875. B.R.O., J.D. Vol. 82 of 1875.

[9] Shivram Hari Sathe, Hon. Secretary, Poona Sarvajanik Sabha, to Chief Secretary, Government of Bombay, 28 June 1875. B.R.O., J.D. Vol. 82 of 1875.

many of the disturbances which he has studied.[10] In the Deccan, if the money-lenders did not comply with the rioters' demands, or had left the village earlier, the rioters would use their lighted torches to set fire to the money-lenders' haystacks (here again there are European parallels), ladders would be brought so that the rioters could enter the money-lenders' comparatively opulent two-storied houses, doors and shutters would be prised open with scythes and iron bars, and the houses ransacked for the bonds. There would be an opportunity for general looting and then the money-lenders' houses would be set on fire. Even in these cases, however, there was little real violence: according to government records only one money-lender was seriously injured.

The first outbreak of rioting occurred in Supa, a large village—more of a market town, in fact—about forty miles west of Poona. Twenty-four hours later there was a similar disturbance in another village, fourteen miles away. Police, and army units in which the dreaded Pathans seem to have predominated,[11] arrived in time to avert a good many of the outbreaks which threatened in the fortnight following the first disturbance. In all, only about thirty villages in the two districts of Poona and Ahmednagar specifically figure in surviving government records, although there were certainly minor disturbances elsewhere, probably in another thirty villages. For a time, it was a case of governing 'whole districts by dragonnade'.[12] Nine hundred and fifty-one men—a large number in those pre-Gandhian days—were arrested on various charges. Only five hundred and one of these were convicted, however. Few villagers were willing to give evidence against their fellows. The Collector of Poona, writing of those who appeared before him, had to admit that 'the evidence against most of them is very incomplete'; it is perhaps worth noting that this passage was later omitted when the document from which it came was quoted in the official report on the disturbances.[13] Summary justice was the most usual procedure—with the result that most of the records of the cases kept in the

[10] Rudé, *The Crowd in History*, p. 253.

[11] Most of the units of the Poona Horse which patrolled the affected areas came from the cantonment of Sirur, the population of which was predominantly Pathan. See *Census of the Bombay Presidency taken on the 21st February* 1872, Part III, Government of Bombay, 1875, p. 350.

[12] W.F. Sinclair, Asstt. Collector, Ahmednagar, to Collector, Ahmednagar, No. 342 of 6 September 1875, quoted *Deccan Riots Report*, Appendix A, p. 297.

[13] Compare G. Norman, Collector, Poona, to Commissioner, S.D., No. 736 of 20 May 1875, B.R.O., J.D. Vol. 82 of 1875, with *Deccan Riots Report*, Appendix C, p. 2.

Collectors' offices have been destroyed. If we may judge from the seventy-four 'Depositions of Convicted Rioters' published in an appendix to the Deccan Riots Report, quite a number of those convicted were poorer villagers—shoemakers, sweepers, basket-makers—who had had the misfortune to be found with looted property in their possession, but who may not, in fact, have participated in the original assaults on the money-lenders' houses.[14]

So much by way of a preliminary narrative of events. But the historian will not be satisfied with that. To turn first to causes. Why was it that the Deccan Riots occurred in May and June 1875 and not in, say, May 1874, or December 1876? The time of year is reasonably easily explained. The months of May and June are the hot 'lean' months before the monsoon breaks, months in which there is not a great deal for the Deccan peasant to do, months in which tempers become frayed.[15] The solitary outbreak of rioting in the Deccan after June 1875 came towards the end of October, when the rainy season, which is also the planting season, had virtually finished. But still, we must ask, why 1875, of all years? I have fairly diligently scoured the available published statistics of prices and productivity for data that might lead one to give a strictly 'economic' answer to this question. The published statistics are of doubtful reliability and normally in this period do not refer to areas smaller than a 'district'; a jeep or two and an American-style 'team' approach—or the patience of a Lefebvre—might yield some material from the *taluka* (sub-district) offices, and from the villages themselves, though its reliability would still be doubtful. All that I can say at the moment is that, in general, the areas affected by the riots were not amongst the wealthiest in the Deccan. They had never been able to grow much cotton in the boom years; in the Ahmednagar talukas, especially, the rainfall was too low. Yet to some extent the areas in which riots occurred in 1875 had shared in the prosperity and easy credit of the boom years, and now they shared in the fall of prices, even for basic millets, which accompanied the decline in cotton prices. Perhaps the general decline affected them more than it did the cotton-producing areas, since their inhabitants were already living closer to subsistence level. But, for all that, it must be said that, for the areas affected by the riots, 1875 was not a particularly bad year as compared with the

[14] See *Deccan Riots Report*, Appendix B. These depositions must be used with care, however; they were taken from prisoners already convicted, and almost all plead complete innocence.

[15] Cf. Beals, *op. cit.*, pp. 155-6.

immediately preceding years, although it was not a particularly good one.[16]

A most important precipitating factor in the riots, but by no means the only one, seems to have been an order which the Government made on 5th February 1875,[17] with the laudable intention of preventing to a considerable extent the sale of land in order to pay arrears of land revenue. In future, movable property was to be attached first by the Government in cases of non-payment of land revenue; land was to be attached only as a last resort. But the Government reckoned without the long-standing relationship between the money-lenders and the peasants. Many peasants had normally handed over all their surplus produce to the money-lenders in payment of 'interest' on their debts; the money-lenders had then paid the land revenue on the peasants' behalf. The money-lenders had paid it because if they did not the peasants' chief security, their land, would disappear, and that would mean the end of a set of money-lender-peasant relationships. But in the weeks following 5th February 1875, because the money-lenders no longer considered the land to be in immediate danger of government attachment, some of them refused to pay the second instalment of the land revenue demand, even though the peasants' surplus had already been handed over to them in the belief that this instalment would be paid. It seems fairly obvious, then, that it was the money-lenders acting in concert who finally provoked the peasants to act in concert. But this point cannot be over-emphasized, since the Government order of February 1875 applied to areas in which riots did not occur, as well as to areas that were riot-stricken in May and June.

There remains the assertion, common in the writings of Poona's early

[16] See *General Reports on the Administration of the Bombay Presidency*, 1869-76, *Administrative Reports of the Cotton Department, Bombay*, 1869-70 to 1875-76. For some taluka figures see the 'Settlement Report' for 'Old Indapur', Bhimthari, Pabal and Haveli talukas, Poona District (*Selections from the Records of the Government of Bombay*, New Series, No. *CLI*, 1877), and especially, Government of Bombay, R[evenue] D[epartment], Resolution No. 5739 of 29 October 1874, and attached 'Leading Points Regarding Revision Settlements', by W.G. Pedder, National Archives of India, Revenue, Agriculture and Commerce (Land Revenue and Settlement) 'A' Proceedings, March 1875, Proceeding No. 18. Statistics are given on a taluka-wise basis in the published *Report of the [Land] Revenue Settlement of the Collectorates of the Southern Division, Bombay Presidency*, 1874-75. Statistics are given only on a district-wise basis in the S.D. Revenue Report for 1871-72. I have not been able to see the Revenue Reports for the vital years 1872-73 and 1873-74.

[17] Government of Bombay R.D., Resolution No. 726 at 5 February 1875.

'nationalists', that the riots were a protest against the imposition by the Government 'of new and higher revenue settlements'.[18] In most parts of the Bombay Presidency the amount of land revenue to be demanded annually was not permanently fixed; a 'revision settlement'—in effect a type of revaluation—was made about every thirty years. In 1875 revisions had recently been made, or resurveys undertaken, in many parts of Poona district; it would be not unnatural for a protest by the peasants against a new settlement to be deflected so that it became an agitation against the money-lenders, who, after all, were frequently intimately connected with the collection of the revenue. But the fact remains that the riots spread from Poona district into talukas of Ahmednagar district where no revision or resurvey had taken place. Not even the Brahman intelligentsia of the Poona Sarvajanik Sabha, rapidly becoming expert in the art of needling the British, could explain the riots in Ahmednagar district in terms of peasant opposition to new revenue settlements.[19]

III

It may well be that in an effort to find the most important immediate causes of the Deccan Riots I shall eventually be forced back to a re-examination of the published statistics, and the collection of new economic data. In the meantime, I have become interested in other aspects of the Deccan Riots; some of these aspects are perhaps not so very important in a discussion of

[18] See e.g. S.H. Sathe to Chief Secretary, letter cited; *Dnyan Prakash*, 28 October 1875, quoted *Report on the Native Papers of the Bombay Presidency* for the week ending 30 October 1875.

[19] See S.H. Sathe to Chief Secretary, letter cited. Dr. Ravinder Kumar has recently made some interesting discoveries, which may be of considerable significance, regarding agitation by members of the Poona Sarvajanik Sabha against the new settlements in the rural areas in the years 1873 and 1874. (Ravinder Kumar, 'State and Society in Maharashtra in the Nineteenth Century', Ph.D. thesis, Australian National University, 1964). Here we seem to have the first ventures of the new, westernized (though still very consciously Brahman) political elite in Western India. Unfortunately, when in January 1964 I tried to see the archive material in the Bombay Record Office which deals with this agitation and which Dr. Kumar had used a few months before, it could not be found. It must be said, however, that Dr. Kumar does not see the Deccan Riots as being primarily the result of the Poona Sarvajanik Sabha's agitation; in his thesis he does not, in fact, offer any proof that the Poona Sarvajanik Sabha agitated in the areas specifically affected by the riots of 1875.

the Deccan Riots *per se*, but they are interesting because of the theoretical problems which they raise, and the excursions into other territory to which they lead.

I have become interested, for example, in the curious notion, that appears to have been held by many of the rioters in the Deccan, that the government was somehow on their side, sometimes that the great white queen herself—the lady whose image appeared on the silver rupee— would intervene in their favour. According to one Indian newspaper, the story circulated that:

> a Marwadi creditor brought an attachment on the hat of a European gentleman and on the gown of a European lady. The gentleman was greatly enraged at the insolent proceeding of the Marwadi, and reported the same to Government, which has issued a circular order to plunder Marwadis wherever they be found.[20]

Some villagers were reported to have temporarily restrained themselves from riot against the money-lenders while they waited for the mythical government order to be received.[21]

A belief that the government approved of their actions was not uncommon amongst participants in peasant disturbances in nineteenth century India. The Santals, a semi-aboriginal people of a remote corner of Bihar, who revolted against their money-lenders in 1855, had a rather similar belief,[22] as did the Bengal peasants who took part in the Indigo Disturbances of 1859-60.[23] Notions of a like nature, of course, were found amongst the French peasantry at the time of the *Grande Peur*,[24] and amongst the peasant followers of Pugachev in Russia in 1773.[25] A general explanation of such phenomena is probably to be found in the fact that in all these situations the government was distant and mysterious, and yet surrounded to some extent by an aura of paternalism. In India the

[20] *Dnyanodaya*, 10 June 1875, quoted *Report on the Native Papers of the Bombay Presidency* for the week ending 19 June 1875.

[21] H.B. Boswell, Collector, Ahmednagar, to Commissioner, S.D., No. 2093 of 16 July 1875. B.R.O., J.D. Vol. 82 of 1875.

[22] *Deccan Riots Report*, p. 106.

[23] *Report of the Indigo Commission appointed under Act XI of 1860*, Mins. of Ev. Q. 3203, 3201, 3294-5, 3564. G.B.P.P., 1861 (72.I) xliv.

[24] Georges Lefebvre, *The Coming of the French Revolution*, trans. R.R. Palmer, Vintage Books, edn., New York, 1957, p. 125.

[25] Jerome Blum, *Lord and Peasant in Russia from the Ninth to the Nineteenth Century*, Princeton, 1961, p. 555.

government was *Ma-Bap*, mother-father; in Russia the Czar was the 'Little Father' of his people.[26]

Yet in the case of the Deccan Riots I think that we can attempt to be a little more specific in our explanation. It was suggested at the time that the rumours that flew in 1875 could have originated in the fact that not long before the Riots, in 1874, various officials had been called upon to make extensive enquiries into the economic conditions of the peasants under their charge for the purpose of the compilation of the first Bombay *Gazetteers*.[27] The fact that government officers were making these enquiries, which appear to have included a close investigation of the activities of money-lenders,[28] could well have encouraged the peasants to believe that the Government was planning a campaign against the money-lenders. This hypothesis is given added weight so far as the Ahmednagar district is concerned by the fact that in that district the enquiries were directed by Captain Harry Daniell, District Superintendent of Police, and by all

[26] The notion of the divinity of kingship has been present in popular Indian thought since earliest times. See the discussion of the notion of the *cakravartin*, the Universal Emperor, in A.L. Basham, *The Wonder that was India*, London, 1954, p. 83, and also in Robert Heine-Geldern, 'Conceptions of State and Kingship in South-east Asia', *Far Eastern Quarterly*, Vol. II, No. 1, November 1942. For a modern Buddhist cult in Burma which draws on Hindu ideas of the *cakravartin*, and incorporates Queen Elizabeth II and possibly Queen Victoria into its system, see E. Michael Mendelson, 'A Messianic Buddhist Association in Upper Burma', *Bulletin of the School of Oriental and African Studies* (University of London), Vol. XXIV, pt. 3, 1961.

[27] *Deccan Riots Report*, p. 106. However the letter of Boswell, Collector, Ahmednagar, No. 2899 of 5 October 1875, quoted in *Deccan Riots Report*, Appendix A, p. 253, leads one to suspect that government enquiries made in 1874, as a result of allegations made by the Poona Sarvajanik Sabha, may also have had some connection with the Riots. Here again one is handicapped by the loss of the relevant archive material. One is further handicapped by the fact that Bombay Gazetteers cannot be found; almost certainly it has been destroyed. Thus an extraordinarily valuable source of economic and social history has been lost.

[28] Some blank copies of the detailed questionnaires for use in gathering Gazetteer material in Gujarat have survived in J.M. Campbell, Gazetteer Compiler, to Chief Secretary, No. 20 of 11 September 1874, B.R.O. G[eneral] D[epartment] Vol. 7 of 1874, and G.D. Memorandum No. 61 of 12 January 1875, B.R.O., J.D. Vol. 36 of 1875. I have not found a copy of the Deccan questionnaire, but I assume that it asked the same sort of detailed questions about money-lenders as were asked in the Gujarat questionnaire.

accounts a crack policeman.[29] The dangers of conducting any sort of government social investigation or enumeration in the rural areas, especially through the agency of the police, had already been illustrated in both Bombay and Bengal at the time of the first all-India census in 1872. 'All sorts of rumours were spread abroad', declared a report on the Bengal Census. 'One man hid his babies, on the ground that they were too young to be taxed. In Murshidabad it was believed that the authorities intended to blow the surplus population away from the guns; in other places it was reported that it was to be drafted to the hills, where coolies are wanted'.[30] And a Collector in the Bombay Presidency wrote, 'Among the mass of the community a superstitious notion exists that enumeration is the precursor of an epidemic, or that its object is the levy hereafter of a tax of some sort or other.'[31]

In the Deccan, the imposition of a new tax in fact followed, in some cases, the 'enumerations' conducted for Census and Gazetteer purposes. So, too, it is worth noting, had an epidemic. The Deccan Riots took place at a time when a cholera epidemic of quite serious proportions was raging. Of recent years a number of historians have become interested in the connections between the fears generated by the occurrence of cholera and social disturbance in nineteenth-century Europe.[32] But the possible connections between cholera and social disturbance in the home of cholera, India, have as yet hardly been investigated. In nineteenth-century India,

[29] J.M. Campbell, Gazetteer Compiler, to Secretary, G.D., No. 952 of 17 December 1875, enclosed Government Resolution, G.D., No. 101 of 10 January 1876, B.R.O., J.D. Vol. 24 of 1876. Daniell was described by Sir Richard Temple as 'the ablest Police officer in the Presidency'. (Temple to Lord Lytton, 3 July 1879, Temple Papers, India Office Library, quoted G.R.G. Hambly, 'Mahratta Nationalism before Tilak: Two Unpublished Letters of Sir Richard Temple.....', *Journal of the Royal Central Asian Society*, Vol. XLIX, pt. 2 April 1962).

[30] *Statement exhibiting the Moral and Material Progress and Condition of India during the Year* 1872-73, p. 121. G.B.P.P. 1874 (Cmd. 196) xlix.

[31] L. Reid, Collector, Dharwar, quoted *Census of the Bombay Presidency... 1872 Pt. 1, General Report on the Organization, Method, Agency and c., employed for Enumeration and Compilation*, Government of Bombay, 1875, p. 12, see also p. 53 and 213.

[32] See Louis Chevalier, ed., *Le cholera: La premiere epidemie du xixe siecle* (Bibliotheque de la Revolution de 1848, Tome xx), La Roche-sur-Lyon, 1958; Chevalier, *Classes laborieuses et classes dangereuses a Paris pendant la premiere moitie du xixe siecle*, Paris, 1958; Asa Briggs, 'Cholera and Society in the Nineteenth Century', *Past and Present*, No. 19, April 1961.

cholera was undoubtedly the disease most feared by the populace. A cholera goddess came into being; a number of rituals were invented so that the disease might be induced to bypass an area during an epidemic.[33] One of the these rituals seems to have included the circulation of wheaten cakes, *chapatis*, from village to village; the mysterious chapatis that circulated immediately before the Indian Mutiny of 1857 were very probably connected with the cholera epidemic that had attacked northern India at that time.[34] I have not been able, from the statistical material available in the Bombay Record Office,[35] to work out any significant correlation between the areas actually affected by cholera and the areas where riots occurred. Yet I remain convinced that there is some significance in the fact that in one of the villages in which rioting took place the villagers would not allow a money-lender's carts, containing grain, to be moved from the village, 'alleging that they were afraid of cholera.'[36] It may be significant, too, that in the solitary case of rioting after the end of June—in a village in Satara district—the participants 'all assembled about nine o'clock at night at the temple of Muryama (cholera) on the outskirts of the village, whence they proceeded to the Goozar's houses.'[37]

IV

I turn now to the question of the ways in which rumour and actual rioting spread in 1875.

Perhaps the first point that is noticeable is that the riots did not spread with the lightning speed that one sometimes associates with the *Grande Peur*. However, Lefebvre himself points out that the *Grande Peur* often did not move at more than walking pace—about four kilometres an hour.[38] The Deccan Riots spread at a much slower rate: they took a fortnight to

[33] William Crooke, *Religion and Folklore of Northern India*, Oxford, 1926, pp. 125-7.

[34] Surendranath Sen, *Eighteen Fifty-seven*, Delhi, 1957, pp. 398-400; Pratul Chandra Gupta, *Nana Sahib and the Rising at Cawnpore*, Oxford, 1963, p. 36.

[35] Chiefly to be found in G.D. Vols. 13 and 15 of 1875. Perhaps I could note here the fact that an enormous amount of work has yet to be done by historians on the subject of epidemics and public health generally in India.

[36] Norman, Collector, Poona, to Commissioner, S.D., No. 1051 of 13 July 1875. B.R.O. J.D. Vol. 82 of 1875.

[37] Memorandum on the Riot at Kukrur, *Deccan Riots Report*, Appendix C, p. 11.

[38] Lefebvre, *La Grande Peur de 1789*, Paris, 1932, p. 182.

spread over an area of only about forty miles from north to south and sixty from east to west. One of the biggest and quickest leaps taken in the Deccan seems to have been the first—fourteen miles in a day. In these circumstances, then, there was plenty of time for messages to be passed from village to village by the normal means—by the use of runners, frequently *Mhars*, scavengers by caste.[39]

Yet rumours did not spread only in this traditional but rather haphazard way. The occurrence of rioting appears to be fairly closely related to the occurrence of market days. Many of the villages in which the riots took place seem to have been of some importance, with a population of 2,000 or more, and holding weekly markets.[40] The market-place was a natural centre for the passing on of rumour, and, of course, it was also the headquarters of the men against whom the Deccan Riots were directed. The crowd often moved a short distance outside the village before returning at night-fall, but it was to the market area that they returned.

The importance of markets as meeting places and centres for the distribution of rumour is a common enough characteristic of peasant disturbances the world over. The Bengal peasants who participated in the Indigo Disturbances of 1859-60 quite obviously gained many of their notions in the market-place,[41] and Lefebvre has noted the role of markets in the spread of rumour in revolutionary France.[42] It may be possible, however, to probe even further into the question of the connections between markets and the Deccan Riots. Markets in the Deccan were held in different places on different days of the week. I suspect, but I have not yet been able to prove, that the timing of the Deccan Riots was sometimes tied to the sequence of market days which is a mark of each of the marketing 'circles' into which the Deccan, and indeed much of the rest of the peasant world, appears to be divided.[43] The existence of a weekly

[39] See 'Substance of a Letter addressed to the Mokadam Patel [headman] and the Village Community of Akola by four persons of the Village of Kalas on behalf of the whole Community', *Deccan Riots Report*, Appendix C, p. 210.

[40] This assertion is based mainly on the details in *Bombay Census Report*, 1872, pt. iii, pp. 268-274, and, so far as some Poona villages are concerned, on *Gazetteer of the Bombay Presidency*, Vol. XVIII, Poona, pt. 3, Government of Bombay, 1885, *passim*.

[41] *Indigo Commission, Mins. of Ev.*, Q. 3210.

[42] Lefebvre, 'Foules revolutionnaires', in *Etudes sur la Revolution Française*, Paris, 1954, p. 273.

[43] Cf. G. William Skinner, 'Marketing and Social Structure in Rural China', pt. i, *Journal of Asian Studies*, Vol. XXIV, No. 1, November 1964; Eric R. Wolf,

sequence of markets could well mean that a man who had witnessed or heard of a riot at one market could spread the news a few days later at another market some miles from his home. For, even in 1875, the Deccan peasants probably did not always patronise the one market, especially in the slack season, when there was time for long journeyings. One market might be 'good' for the sale or purchase of one commodity, another market for the sale or purchase of another commodity.[44] Certainly the 'Depositions of Convicted Rioters' show that some peasants had dealings with money-lenders in more than one centre, though such proceedings must have made them thoroughly unpopular with the money-lenders.[45]

Yet in India the links between villages are not, I think, primarily economic ties. This is one way, perhaps, in which Indian peasant culture differs from Chinese peasant culture.[46] In India there are bonds of caste between villages, and, in particular, very strong bonds of marriage. And yet, as a distinguished Indian anthropologist, Professor M.N. Srinivas, pointed out over ten years ago—and the observation is still true today—we have hardly begun to examine the social and more particularly the 'political' implications of Indian marriage networks.[47] When, during the Deccan Riots, a group from one village wrote to a group in a neighbouring village 'we consider Kallas and Akola as one village',[48] they were almost certainly not thinking solely of economic bonds. But because, so far as marriage networks are concerned, the Deccan is something of a transitional area between the distinctly different systems of North and South India, it is almost impossible to say definitely, without investigating on the spot, whether or not marriage ties between the two villages are referred to

---

'Closed Corporate Peasant Communities in Mesoamerica and Central Java', *Southwestern Journal of Anthropology*, Vol. XIII, No. 1, Spring 1957; Sidney W. Mintz, 'Peasant Markets', *Scientific American*, Vol. CCIII, No. 2, August 1960.

[44] Cf. articles cited above, and Walter C. Neale, 'Kurali Market: A Report on the Economic Geography of Marketing in Northern Punjab', *Economic Development and Cultural Change*, Vol. XIII, No. 2, January 1965.

[45] Here, probably, is the reason why in many cases those convicted seem to be those who were picked on as scapegoats by the money-lenders when the police asked for reports on the disturbances.

[46] Skinner, *op. cit.* p 32, claims that 'Insofar as the Chinese peasant can be said to live in a self-contained world, that world is not the village but the standard marketing community'.

[47] Srinivas, *India's Villages*, p. 12.

[48] Letter cited *Deccan Riots Report*, Appendix C, p. 210.

here. The historian at this point must become an anthropologist, and do some fieldwork.

V

To return from such speculative realms: what can we say about the vital question of leadership during the Deccan Riots? A short time after the riots had occurred the Bombay Government claimed that there was 'little doubt that the riots were preconcerted and instigated by wiser heads than simple village *kunbis*'.[49] The Government probably had members of the Brahman dominated Poona Sarvajanik Sabha, of Poona city, in mind; after all, there had been plenty of sniping from this organisation over the imposition of new revenue settlements. But, after the report of the Committee which investigated the Riots had been received, the Government of Bombay had to admit that 'there is no evidence to show that the ryots were urged to their excesses by persons of position and education, though they had the sympathy of many such.'[50] This conclusion would appear to be correct.

The most striking feature of leadership within the village at the time of the riots is that the authority of the leaders rarely appears to have been 'purely local and temporary', as Professor Rudé believes was often the case in agrarian riots in France and England in his period.[51] Few 'N.C.O's.'[52] thrust themselves forward at a time of crisis from amongst the ranks of those who had never before assumed positions of leadership. There is one recorded case of a riot being led by a Mhang, a ropemaker—an Untouchable, in fact—but this is in the solitary, but comparatively well documented, October riot in Satara district, in which a number of Mhangs appear to have taken part.[53] It is conceivable, in fact, that this riot began as

[49] *General Report on the Administration of the Bombay Presidency*, 1875-76, Government of Bombay, 1876, p. 84.

'Kunbis' = (in this context) peasants, ryots.

[50] E.W. Ravenscroft, Chief Secretary, Government of Bombay, to Secetary, Government of India, Revenue, Agriculture and Commerce Department, No. 2207 of 6 April 1877. National Archives of India, Home (Judicial) 'A' Proceedings, April 1879, Proceeding No. 8.

[51] Rudé, *op. cit.*, p. 251.

[52] Professor Rudé used the term in a public lecture at the University of Canterbury in August 1962.

[53] *Deccan Riots Report*, Appendix C, pp. 10-12, 'Memorandum on the Riot at Kukrur'. See also, J.E. Oliphant, Acting Commissioner, S.D., to Chief Secretary, Government of Bombay, No. 1605 of 25 October 1875, and enclosures. B.R.O.,

a Mhang affair. In the main series of riots, leadership most frequently appears to have been assumed by the *patels*, the village headmen.[54] The position of the headman in the British system of government in the Deccan was a peculiar one: he was both village leader and the servant of government, and thereby subject to many strains.[55] The Deccan Riots showed that at a time of conflict the headman was liable to assume his traditional leadership role in the village, or at least to forget his generally newer duties as policeman and informant. And yet, as the Deccan Riots Report itself points out, very acutely, 'an assembly of villagers acting with their natural leaders for a definite object, was a less dangerous body than a mob of rioters with no responsible head would have been.'[56]

VI

It is normally not possible for a government to continue to govern 'whole districts by dragonnade' for any length of time. Sooner or later some sort of *modus vivendi* has to be found between the antagonists in a situation such as that which caused the Deccan Riots. Life has to go on; compromises have to be reached within a society by the members of that society themselves, who have no option but to continue living in that area.[57] This is true of any community, but perhaps it is especially true of the Indian

---

J.D. Vol. 81 of 1875. Extracts from the diaries of the District Superintendent of Police are included in the enclosures. I have been told that the originals of the Police Diaries for the period have been destroyed: even if they have survived it would certainly be very difficult to gain access to them. Later in the century the valuable Police Abstracts began to be compiled.

[54] See the correspondence regarding the 'dismissal' of headmen in B.R.O., J.D. Vol. 82 of 1875, especially Norman, Collector, Poona, to Commissioner, S.D., No. 1051 of 13 July 1875, and in B.R.O., R[evenue] D[epartment] Vol. 35 of 1875, especially Boswell, Collector, Ahmednagar to Commissioner, S.D., No. 2929 of 7 October 1875.

[55] See Kumar, *op. cit.*, pp. 206–209. Cf. Max Gluckmann, 'The Village Headman in British Central Africa', in Gluckmann, *Order and Rebellion in Tribal Africa: Collected Essays with an Autobiographical Introduction*, London, 1963, pp. 146–152.

[56] *Deccan Riots Report*, p. 6.

[57] Cf. Gluckmann, 'The Peace in the Feud', *Past and Present*, No. 8, November 1955.

village community.[58] The Indian village, even today, generally has certain built-in mechanisms for finding a consensus in a situation of conflict. I am referring, of course, to the institutions known as *panchayats*. There is no need to believe that in pre-Bitish times India was covered with largely self-sufficient, self-governing village communities, each ruled by permanent village councils. Panchayats, even in pre-British times, do not often appear to have been permanent; rather, they were largely *ad hoc* bodies, summoned informally for the purpose of settling a particular dispute.[59] Such panchayats continued to be called together in British times, though perhaps somewhat less frequently, and they are still very important factors in the village today.[60]

After the Deccan Riots panchayats were formed spontaneously in some of the affected villages. At first there was some hope that they would reach a rough solution to the problem. 'In places where the *soukars* (money-lenders) were in the most terror they were, and as far as I know still are, quite ready to compromise', wrote an official late in 1875. 'Several have told me that they were ready to take 4 annas in the rupee in full payment of all demands, and in Supa many bonds have been handed to the Panchayat which has been constituted there, which it was understood was

[58] The whole subject of Conflict and Consensus in India is a most interesting one. The aim of consensus is an aspect of Indian 'politics' which has deep roots in the Hindu psychology, as well; I am not one of those who believe that Hinduism is basically a matter of tolerance and non-violence. To some extent there may be a division within Hinduism between a consensus-seeking group of Vaisnavites—worshippers of God in the form of Vishnu, the Preserver, and a more violently inclined group of Saivites—workshippers of God in the form of Siva, the Destroyer. I have a strong suspicion that in the history of the Indian national movement the Saivites tend to supply the bomb-throwers and the revolutionaries. But the violent aspect of the Hindu psychology often coexists in the same person with the consensus-seeking aspect. See, for a discussion of some of these matters, Nirad C. Chaudhuri, 'Subhas Chandra Bose—His Legacy and Legend', *Pacific Affairs*, Vol. XXVI, No. 4, December 1953, and Susanne Hoeber Rudolph, 'Consensus and Conflict in Indian Politics', *World Politics*, Vol. XIII, No. 3, April 1961. I eagerly await the arrival in this part of the world of Henry Orenstein's recently published anthropological study, *Gaon: Conflict and Cohesion in an Indian Village*, Princeton, 1965, which deals with a Maharashtran village in Poona district, very close to the areas in which the Deccan Riots occurred.

[59] Kenneth Ballhatchet, *Social Policy and Social Change in Western India, 1817-1830*. London, 1957, pp. 106-107.

[60] Srinivas, 'The Study of Disputes in an Indian Village', in Srinivas, *Caste in India and Other Essays*, Bombay, 1962, p. 118.

going to award payment on a graduated scale.'[61] But there were many peasants who could not pay even four annas in the rupee, or who did not wish to do so. And no Deccan panchayat could avoid overlooking the fact that the money-lenders were generally 'outsiders' in the Deccan. Further-more, the institution of a full-scale inquiry into the Riots had an effect similar to that of previous enquiries: the peasants were encouraged to believe that the Government was about to come to their rescue, while the money-lenders tried to extract as much as they could while they were still able to do so.[62] Passions cooled to a certain extent in the years that followed, but, even so, it may be said that officials spent the next thirty years in a somewhat dilatory search for a permanent solution to the problem posed by the money-lender in the Deccan. Co-operative credit societies, it may be added, which were produced as a virtual panacea in the years following 1904, did not, on the whole, fulfil the high hopes that some held out for them.[63]

VII

The problem of the money-lender was, of course, an India-wide one in the nineteenth century.[64] To what extent did peasants in other parts of India revolt against the money-lender in that period? The topic is a very large one, and can be touched on only very briefly here.

One of the most significant results of the observance of the centenary of the Indian Mutiny in 1957 has been a new interest amongst Indian historians in civilian unrest in the rural areas at the time of the Revolt. It now appears that during the temporary withdrawal of the British presence from parts of northern India the peasants quite often took the opportunity to take their revenge on their money-lenders. This aspect of the Revolt can

[61] W.G. Pedder, Acting Under Secretary, R.D., Demi-official note [late 1875] to J.B. Richey [Chairman of the Committee investigating the Riots] B.R.O., R.D. Vol. 118 of 1875, p. 427.

[62] See Boswell, Collector, Ahmednagar, Ahmednagar Administration Report, 1875-76, No. 1952 of 20 July 1876. B.R.O., R.D. Vol. 5 of 1876.

[63] I have dealt with the co-operative movement in my unpublished London Ph.D. thesis (1960), 'The State and the Co-operative Movement in the Bombay Presidency, 1880-1930.'

[64] A convenient article on British policy on the matter in the latter half of the century, though one that is in need of certain qualifications, is Thomas R. Metcalf, 'The British and the Moneylender in Nineteenth-century India', *Journal of Modern History*, Vol. XXXIV, No.4, December 1962.

be exaggerated, as it has been by some Indian Marxist historians.[65] Nevertheless the evidence on the matter which Sulekhchandra Gupta has found in the *Narratives of Events*, compiled for each district of the North-Western Provinces[66] shortly after the Mutiny, is of considerable interest. An official at Bareilly reported: 'I would except from the imputation of the desire to change the government, those classes among the Hindus who owe all they possess to the existence of the avowedly strongest and most just government, I mean—the trading and banking classes'. At Muzaffarnagar, we are told, 'the *bunyahs* (money-lenders) were in the majority of cases the victims.' And from Sahranpur comes this account: '...money-lenders and traders were forced to give up their books of accounts and vouchers for debts... All the government records with the *mahajans*' (money-lenders') accounts, bonds, etc., were torn up and scattered over the neighbouring gardens.'[67] The Deccan Rioters were obviously not alone in the type of action that they took. Perhaps the men of 1857 tended to be a little more violent—but the times *were* violent in 1857.

In some ways the most interesting of all nineteenth-century agitations against the money-lending classes is the 1855 movement of the Santals, one of the semi-aboriginal groups of Bihar, a movement which I have already mentioned briefly. The complicating factor here is that the Santals were and are, as I have described them, 'semi-aboriginal'. It is true that some anthropologists do not see any basic distinction between the Hindus of the rural areas and the pockets of 'semi-aboriginal', or 'aboriginal', or 'tribal' peoples still found in isolated parts of India. 'In India', writes Dr. Irawati Karve, 'the primitives and the others have lived together for over two to three thousand years. The primitives have almost all the social institutions and elaborations of the non-primitives while the Hindus actively share all the beliefs of the primitives. In such a context one must use great precaution before dubbing something as primitive'.[68] There is undoubtedly much wisdom here. But the village organization of the two semi-aboriginal groups which I have briefly studied—merely as an historian, I hasten to add—seems to be significantly different from the village

[65] See P.C. Joshi (ed.), *Rebellion 1857: A Symposium*. New Delhi, 1957, especially Talmiz Khaldun, 'The Great Rebellion'.

[66] Later included in the United Provinces, now Uttar Pradesh.

[67] *Narratives of Events attending the Outbreak of Disturbances*, quoted Sulekhchandra Gupta, 'Agrarian Background and the 1857 Rebellion in the North-Western Provinces', *Enquiry*, No. 1, New Delhi, [1959].

[68] Irawati Karve, *Kinship Organization in India*, Deccan College Monograph Series No. 11, Poona, 1953, p. 21.

organization of much of the rest of India, North or South. During the Santal rebellion of 1855, and during the Birsite disturbances amongst the Mundas (another semi-aboriginal group in Bihar) from 1895 to 1900, large assemblies of up to 10,000 were brought together with little apparent difficulty.[69] Such assemblies would be extremely unusual on the 'Hindu' plains. The enabling factor in the hilly, semi-aboriginal areas in Bihar seems to have been the strong alliances of several villages under the one leader which were common amongst the inhabitants of those areas.[70]

The Santal rebellion seems to have had about it a strongly religious air. Two of the four brothers who led the revolt claimed to have had a vision: '....suddenly the Thakur (god) appeared before the astonished gaze of Seedoo and Kanhu; he was like a white man though dressed in the native style; on each hand he had ten fingers; he held a white book and wrote therein; the book and with it 20 pieces of paper, in 5 batches, four in each batch, he presented to the brothers; ascended upwards and disappeared'.[71] One is tempted to look for the influence of Christianity here, though missionaries do not appear to have begun work in the Santal areas until 1862.[72] The so-called Kharwar movement amongst the Santals in 1871 and the following years was certainly influenced by Christianity;[73] its participants wished to revert to a golden age, when they were not oppressed, and when only one god was worshipped.[74] Obviously, there are distinct overtones of millenarianism here.

The Birsa movement amongst the Mundas was directed against the incoming 'Hindu' *zamindars* (landholders) rather than against moneylenders, but it, too, is notable for its millenarian and indeed its messianic overtones, again, it would seem, largely the result of the impact of Christian missions. Birsa, a Munda youth, claimed to have been through a process resembling the Transfiguration, and to have worked miracles. He was proclaimed by his followers to be Bhagwan, or God Himself. The Mundas, he said, were to worship only one god, to wear clothes, to wear the sacred thread and to cease eating animal food (here obviously is the

---

[69] Kalinkar Datta, *The Santal Insurrection of 1855-57*, Calcutta, 1940, p. 15; Sarat Chandra Roy, *The Mundas and their Country*, Calcutta, 1912, p. 329.

[70] See *Bengal District Gazetteers*, Vol. 22 *Santal Paraganas*, by L.S.S. O'Malley, Calcutta, 1910, pp. 107-9; Vol. 20, *Singhbhum, Saraikela and Kharsawan*, by L.S.S. O'Malley, Calcutta, 1910, pp. 60-61; Roy, *op. cit.*, p. 412.

[71] *Calcutta Review*, 1856, quoted Datta, *op. cit.*, p. 15.

[72] *Santal Paraganas Gazetteer*, p. 68.

[73] *Ibid.*, p. 147.

[74] *Ibid.*, p. 150.

caste Hindu influence) and to wait for the promised day when fire and brimstone would destroy all but the elect.[75]

Much more could be said—another paper could be written, in fact—on the subject of such movements amongst India's semi-aboriginal peoples. But we are hardly dealing here with the mainstream of Indian rural history. We are certainly not dealing with the 'Hindu' mainstream. We are dealing with groups of people who are unlike most caste Hindus in that they are particularly susceptible to Christian influence. I could perhaps add, however, the final bald assertion that I do not regard a type of millenarianism and indeed a type of messianism as being altogether impossible within the mainstream of Hinduism. Hinduism, it is true, does not think in terms of a once and for all end to life as we know it. Things are continually either growing worse or growing better. But under the Vaisnavite system, anyway, when things are growing worse and worse God sometimes intervenes in the form of an *avatar*, an incarnation. And indeed what else was Gandhi in the minds of many Indian peasants by the nineteen forties but an avatar, a form of God Himself, come to bring about the 'reign of justice and happiness' that the *Bhagavadgita* speaks of?[76]

---

[75] See Roy, *op. cit.*, especially pp. 326-343, and *Singhbhum Gazetteer*, pp. 42-46. I owe the original reference to Birsa's movement to Mr. P.G. Ganguly of the Department of Anthropology, University of Otago. The study of millenarian and messianic movements is now, of course, a most fashionable pursuit amongst historians. See, e.g., Sylvia L. Thrupp, (ed.), *Millennial Dreams in Action: Essays in Comparative Study. Comparative Studies in Society and History, Supplement II,* The Hague, 1962; Vittorio Lanternari, *The Religions of the Oppressed: A Study of Modern Messianic Cults,* London, 1963.

[76] *Gita*, iv, 7-8.

Chapter Six

The Myth of the Deccan Riots of 1875

NEIL CHARLESWORTH

I

In the early summer of 1875 agrarian rioting occurred in the Bombay Deccan. The disturbances began at Supa, a market village in Bhimthari taluka of Poona District, where on 12 May an unruly peasant mob attacked the houses and shops of the local Gujarati moneylenders. ¡The combustible elements were everywhere ready'[1] and the riots spread through the poor, eastern regions of Poona and Ahmednagar Districts. The riots were directed entirely at the village *sowkars* (moneylenders), to whom most of the peasant agriculturists of the area were indebted.

> The object of the rioters was in every case to obtain and destroy the bonds, decrees, etc. in the possession of their creditors.[2]

> The riots were confined to 33 places,[3] though we are assured that disturbances were threatened in many other villages.[4]

and in September there was an isolated outbreak at Kukrur, Satara District. The riots died out quickly. Although approaching one thousand ryots (peasants) were arrested[5] a show of force sufficed to restore the disturbed tract of country to its normal state of peace and quiet in a short space of time.[6]

[1] *Parliamentary Papers* [hereafter P.P.], 1878, LVIII. *Report of the Deccan Riots Commission* [Hereafter *Deccan Riots Report*], para. 4.

[2] *Deccan Riots Report*, para. 12.

[3] These were in Poona District, 5 villages of Bhimthari taluka and 6 villages of Sirur taluka, and in Ahmednagar District 6 villages of Parner taluka, 11 of Shrigonda, 4 in Nagar and one in Karjat taluka. *Deccan Riots Report*, Paras 8-9.

[4] *Deccan Riots Report*, para. 9.

[5] 392 in Ahmednagar and 559 in Poona. *Deccan Riots Report*, para. 10.

[6] *Bombay Presidency Annual Police Report, 1875, Southern Division*, p. 25.

These disturbances have received a great deal of attention. As a major crisis in agrarian history, they provide the occasion for a study of change and conflict in rural society. Amidst the apparent tranquillity of the Western Indian countryside, the Deccan riots suggest great discontent and the riots commissioners were impressed that the violence was committed by people normally passive and law-abiding. Hence it has always been assumed that the Deccan riots must be indicative of great social and economic change. The report of the special commission aroused widespread public comment at the time. Its proposals ranged from reducing the rigidity of the Bombay revenue system to legislation against fraud and abolishing imprisonment for debt. The Bombay government included some of these provisions in the Deccan Agriculturists Relief Act in 1879. Recently, historians have also emphasized the importance of the events. Ravinder Kumar has cogently argued that the disturbances evidence a social revolution in Maharashtra, whereby the ryot had been ousted from his traditional holding of land by the sowkar.[7] I.J. Catanach has been sceptical of the extent of land transfer, but he too has seen the Deccan riots as a prime example of rural dislocation in nineteenth-century India.[8] Yet the vast amount of literature about the Deccan riots cries out for a re-examination of the disturbances and of the explanations for them.

[7] Kumar argues that the riots were 'a reflection of the tensions generated within rural society through the legal and administrative reforms carried out by the British Government'. Maharashtran society had been 'transformed under the influence of Utilitarian policies which tried to mould rural life along individualistic and acquisitive lines'. Ravinder Kumar, 'The Deccan Riots of 1875', *Journal of Asian Studies*, Vol. 24, No. 4, p. 634 [reprinted herein]. He also argues that by 1875 'it was the vanis (money-lenders) who now dominated the rural world'. Ibid., p. 618. Many ryots had lost their lands and without 'an outlet to the urban world, the dispossessed peasant was forced to live as a landless labourer, often on those very fields which he had formerly cultivated as an independent proprietor'. *Ibid.*, p. 619.

[8] His conclusion is: 'it seems that we must accept, on the combined evidence of the statistics and the official reports, the notion that land transfer from agriculturist to non-agriculturist classes was occurring, and probably occurring increasingly, in the twenty or thirty years preceding the Deccan Riots. We must accept the notion, too, that this situation was regarded seriously by the Deccan peasantry. But this is not to say that the Deccan Riots were necessarily a result mainly of increasing land transfer.' I.J. Catanach, *Rural Credit in Western India, 1875-1930* (Berkeley, Los Angeles and London, 1970), pp. 19-20.

II

Ostensibly the Deccan riots were about indebtedness. The rioters owed large sums of money to the sowkars whom they attacked. In Bhimthari taluka, the heartland of the Deccan riots,

at least 90 per cent of the cultivators are not merely in debt but hopelessly involved in it. Men paying Government Rs 10 to 20 owe the sowkars Rs. 1,000 to 2,000...[9]

However, peasant indebtedness has always been a commonplace of agrarian life in Western India. In 1832 the French traveller, Victor Jacquemont, remarked how

dans cet Eldorado de notre imagination europeene, dans l'Inde la tres-grande majorite de la population au lieu d'avoir, doit[10]

and he claimed that

Nulle part peut-etre dans I'Inde, il ne produit plus de misere que dans le Deccan.[11]

Nearly a hundred years later a Bombay enquiry revealed:

a very large proportion of the agricultural population is born in debt, lives in debt and dies in debt.[12]

For many Kunbis (the large agricultural caste of Poona and Ahmednagar) this meant the total control of their economic life by the moneylender. Normally the sowkar took all the ryot's crops, bar a bare subsistence, and in return paid his land revenue and provided seed for the following year's operation. But borrowing became such a habit that

a Kunbi will sometimes borrow a sum at heavy interest when he actually has it buried in his house.[13]

[9] Maharashtra State Archives, Bombay. Revenue Department Papers [Hereafter M.A.R.D.], Vol.7 of 1875, No. 960. C.G.W. Macpherson, 2nd Assistant Collector, to G. Norman, Collector of Poona. No. 72, 17 July 1875, para 11.

[10] Victor Jacquemont, *Voyage dans l'Inde pendant les annees 1828 a 1832* (Paris, 1841), Vol. 3, p. 558.

[11] *Ibid.*, p. 559.

[12] *Report of the Bombay Provincial Banking Enquiry Committee, 1929-30*, para. 45.

[13] India Office Library and Records, London [Hereafter I.O.L.]. Political and Secret Department Library. Papers of Sir William Lee-Warner, Deccan Ryot

Often the wealthiest cultivators, because they had the best credit, were the most in debt. In the late 1920s, for example, an enquiry into the marketing technique of cotton cultivators found that 82.4 per cent of this rural elite in Middle Gujarat and 71 per cent in Khandesh owed money.[14] In 1875 borrowing was universal throughout western India. The Bombay *Gazetteers* chronicled this situation. In Khandesh District it was estimated that

> not more than ten per cent of the agricultural population, including Bhils and others who are mere field labourers, can afford to begin the year's tillage without the moneylender's help.[15]

In a society where everyone borrowed and debts were handed down from father to son, indebtedness alone would not have provoked agrarian disturbances. Indeed it was a necessary part of the rural economy and there need be no basic conflict of interest between ryot and sowkar. In the far northern district of the Panch Mahals in the 1870s and 1880s indebtedness among the Bhil tribes which lived there was rampant. In 1881, one settlement officer reported that of every 8 rupees which the average Bhil earned, at least 6 went straight into his moneylender's pocket.[16] And yet

> the relations between these rude faith-keeping and truthful rayats and the astute savkar are most amicable.[17]

The ryot-sowkar relationship was a traditional tie and, though there were some tensions, its motto was usually 'live and let live'. Without the moneylender, the peasants had to

> admit they would neither be able to find seed to sow nor money to meet their necessary expenditure.[18]

Series, 2. W.G. Pedder, *Note on the indebtedness of the Indian agricultural classes, its causes and remedies*, 29 July 1874, p. 1. (I am grateful to Dr I.J. Catanach for his help in locating these papers.)

[14] The Indian Central Cotton Committee, *General Report on eight enquiries into the finance and marketing of cultivators' cotton, 1925-28*. Table VII.

[15] Gazetteer (hereafter B.G), XXII, Khandesh.

[16] Selections from the Records of the Bombay Government [Hereafter B.G.S.], New Series, 165. Settlement Report on Jhalod taluka, Panch Mahals District, by N.B. Beyts. No. 79, 16 February 1881, para. 42.

[17] *Ibid*.

[18] B.G. XVIII, Poona, Part I, p. 97.

Throughout the nineteenth and into the twentieth century the Bombay masses were usually indebted and usually quiescent. The central problem remains: why did indebtedness lead to rioting in a few Deccan villages in 1875?

A possible explanation lies in the fact that revision of the Presidency's land revenue assessment was under way. This had begun in Indapur taluka of Poona District in January 1867. Wingate and Goldsmid's settlements of the 1840s had been based on moderate charges to encourage the expansion of cultivation, but by 1867 the survey teams of Francis and Waddington felt that price rises and apparent improvements in agricultural techniques justified higher taxes. Francis proposed a total revenue increase for Indapur from Rs. 81,391 to Rs. 124,700[19] and officials thought that a higher assessment would have been feasible. Sir George Wingate supported the proposals on the grounds that 'an error on the side of liberality is a safe one'.[20] In 1871 Waddington increased the land revenue due from 54 villages of Bhimthari taluka by 69 per cent.[21] In the next few years four talukas of Sholapur District, Madha, Sholapur, Pandhapur and Barsi, and two more of Poona, Haveli and Pabal, were reassessed and their land revenue raised by at least 50 per cent.

The possible significance of these tax increases was not lost on contemporaries. Carpenter, a member of the riots commission, and most of the press saw high taxation as the flame that kindled the Deccan riots. Yet of the areas that rioted only Bhimthari taluka in Poona had recently been reassessed.[22] Ahmednagar District, which experienced most of the riots, faced no reassessment for seven years. On the other hand the few reassessed riot villages suffered smaller proportionate increases than the other revised talukas which stayed completely pacific in 1875. Supa village, where the disturbances first flared up, had in 1873 been assessed

[19] B.G.S. New Series, 151, Settlement Report on Indapur taluk, Poona District, by J. Francis. No. 147, 12 February 1867, para. 163.

[20] B.G.S. New Series, 151. Notes upon Colonel Francis' Report on the Resettlement of the District of Indapur by G. Wingate, para. 12.

[21] B.G.S. New Series, 151. Settlement Report on Bhimthari taluka, Poona District, by W. Waddington. No. 440A, 12 July 1871, para. 41.

[22] The areas reassessed by summer 1875 were Indapur, Bhimthari (including the Supa Petha), Haveli and Pabal talukas of Poona and the four Sholapur talukas. The Deccan riots took place in four different talukas of Ahmednagar District and in Bhimthari and Sirur talukas of Poona.

at Rs. 9,009 total land revenue, as against Rs. 6,202 in 1843.[23] But in the Sholapur villages 70 per cent assessment increases were the norm and in Haveli taluka of Poona, where the proposed revenue enhancement was from Rs. 77,352 to Rs. 1,53,517,[24] some rich villages had their land tax more than doubled. Yet the immediacy of the issue anyway had gone off the boil by 1875 and the very large increases of 1867-72 had been reduced as prices fell off.[25] The Deccan riots could hardly have been a protest against high taxation.

In other parts of India agrarian disturbances often occurred as a result of money-lending groups securing title to land and becoming the effective owners of it. Most of those who have studied the Deccan riots have come to favour this explanation. Yet had the Kunbi been ousted by his sowkar by 1875? If he had, it was clearly a recent development. For, from the British conquest to the Joint Report of 1847, land in Western India had little market value. Wingate and Goldsmid found that

sales of land have become as little known in the Deccan as in Madras.[26]

and one visitor to Bombay commented

in England the land is looked upon as the most secure investment for money, and is certainly the favourite one. But who hears of a wealthy native in Guzerat investing his money in land?[27]

Moneylenders shared this aversion for 'they never sue for possession

[23] B.G.S. New Series, 151. Settlement Report on Supa Petha, Poona District, by W. Waddington. No. 846, 5 September 1873, para. 25.

[24] B.G.S. New Series, 151. Settlement Report on Haveli taluka, Poona District, by W. Waddington. No. 840, 30 November 1872, para. 29.

[25] In Poona District, for example, the enhancements were lowered from 55 per cent to 39 per cent in Indapur taluka, from 63 per cent to 37 per cent in Bhimthari, from 65 per cent to 41 per cent in Haveli and from 51 per cent to 38 per cent in Pabal. Minute on the Connection of the Recent Enhancement of Land Revenue with the Riots by Mr Carpenter, *Deccan Riots Report*.

[26] *P.P.* 1852-53, LXXV. *Official Correspondence on the System of Revenue Survey and Assessment in the Bombay Presidency*. H.E. Goldsmid and G. Wingate to John Vibart, Revenue Commissioner, Poona, 17 October 1840, para. 28.

[27] Alexander Mackay, *Western India. Reports addressed to the Chambers of Commerce of Manchester, Liverpool, Blackburn and Glasgow* (London, 1853), p. 134.

of land'.[28] After 1850, as more land was taken up, agricultural prices rose and land assessment became lighter, land acquired some worth and transfers of occupancies did begin to take place. Even so, numbers of voluntary sales remained small. For example, in Rahu, one of the riot villages, there were only five deeds of sale between 1867 and 1874 and in Ambi Khoord, another scene of disturbances, only two holdings were sold during the same period.[29] Such sales were mainly between traditional agriculturist families, as in Bhimthari taluka where Waddington remarked:

> of the 13 instances of sale of land quoted by me, in 10 cases the buyers were patels or ryots.[30]

The moneylender therefore had no opportunity to work a social revolution through the market. But perhaps the main charge against him remains valid: that he won his debtor's land after its confiscation and compulsory sale by court decrees.

There was a perceptible growth in litigation in Bombay in the years before the Deccan riots. In Poona the number of suits for debt shot up from an average of about 6,000 per year in the mid 1860s to over 13,000 in 1872 and 1873.[31] These were usually cases over extremely petty issues. In 1873 nearly three-quarters of Bombay civil cases were suits for sums of Rs.20 and under.[32] The defendant very rarely won. Out of over 120,000 suits adjudicated in 1861 all but 10,096 of them were decided for the plaintiff,[33] and many defendants did not even bother to put up a fight. The *ex-parte* decree, which by-passed a legal conflict, was used for over half the Presidency's total cases.[34] The pattern was clear to all. Creditors were simply using the courts to collect their debts and to obtain weapons to

[28] *P.P.* 1852-53, LXXV. *Official Correspondence on the System of Revenue Survey and Assessment in the Bombay Presidency.* Memorandum by H.E. Goldsmid and G. Wingate, dated 21 December 1850, para. 32.

[29] *Deccan Riots Report*, Appendix C, Return on pp. 317-23.

[30] B.G.S. New Series, 151. Settlement Report on Bhimthari taluka, Poona District, by W. Waddington. No. 440A, 12 July 1871, para. 12.

[31] *Deccan Riots Report*, para. 71.

[32] *Deccan Riots Report*, para. 91.

[33] *Bombay Presidency Annual Administration Report*, 1861-62, Appendix A, No. 13, p. 120.

[34] In 1873, 57 per cent of Bombay's civil cases were decreed *ex-parte*, compared with only 25 per cent in Madras, the second highest proportion in British India, *Bombay Presidency Annual Administration Report. 1874-75*, p. 89.

coerce their debtors. The verdict on Bombay's civil litigation was delivered by an official report:

> in this Presidency nearly all plaintiffs are moneylenders and they have evidently little trouble in getting a case decided in their favour.[35]

This situation was bemoaned in the riots commission report. But the new litigiousness—a Presidency-wide phenomenon—was partly the result of the Limitation Act of 1859, which had restricted the validity of bonds and suits for money and debts to a period of three years. It also represented the sowkar's reaction to the possibility of ryots breaking out of the subsistence cycle following the cotton boom days of the 1860s. The peasant's traditional dependence on his moneylender gave the sowkar control over the crops and hence usually a monopoly of local trade. As late as 1883 one official wrote of

> the trade monopoly of the Banias, a monopoly which now amounts to a despotism in many parts of the country.[36]

But the continuance of this monopoly depended on the subservience of the debtor and continual court decrees were the surest way of ensuring maintenance of a *status quo* very lucrative for the sowkar.

However, the evidence that moneylenders went further and assumed ownership of debtor's lands is scanty. Some might acquire holdings in the large market and head-quarter villages, which were particularly rich in Marwari and Gujarati traders. Parner village, where the sowkar Tularam Karamchand acquired around 1,000 acres of land in the 20 years before 1875,[37] was the supreme example.[38] But away from such centres moneylenders' landholdings were few. Thus, of the villages in Indapur taluka examined by the riots commission, Marwari moneylenders owned 2,189 acres out of a cultivated area of 19,443 acres in the big market village of

---

[35] *Ibid.*

[36] M.A.R.D. Vol. 29 of 1883, No. 1934. A.B. Fforde, Assistant Superintendent, Revenue Survey, to T.H. Stewart, Survey and Settlement Commissioner, No. 40, 13 November 1883, para. 2.

[37] *Deccan Riots Report*, para. 76. The case of Tularam is also quoted by Ravinder Kumar, See 'The Deccan Riots of 1875', *Journal of Asian Studies*, Vol. 24, No. 4, August 1965, p. 619. [Rpt. herein.]

[38] See *Deccan Riots Report*, Appendix C, Mr Sinclair's notes on the sowkars of Parner. Kumar makes much of the considerable landholding changes in Parner. But, as Sinclair indicates, Parner, with its large number of notorious moneylenders, was untypical.

Indapur; they owned 1,438 acres out of 18,503 acres in another large village, Bhowri; yet they held just 17 out of 8,252 acres in Nimgaon Khedki, 294 out of 12,396 acres in Kullus, 584 out of 8,643 acres in Lasume and 467 out of 7,417 acres in Palasdeo.[39]

Most of the riots did not occur in large villages like Parner or Indapur, nor were they directed against the capitalist banker-sowkars like Tularam Karamchand. But even in those areas where moneylenders had moved into landholding, sowkars owned only a small proportion of the Deccan's cultivated area in 1875. In Parner taluka, 308 Marwari moneylenders held 660 fields in their own names with a total area of 17,885 acres.[40] Yet the cultivated area of Parner amounted to over 350,000 acres when it was revised less than a decade later:[41] so the Marwaris held only around 5 per cent of the fields in occupation. This was still more than they owned in Akola taluka, where, with a cultivated area a little smaller than Parner's, Marwaris held 7,771 acres made up of 976 fields. Marwari sowkars owned a larger proportion of Sangamner taluka. Here, with a cultivated area of around a quarter million acres,[42] they held 702 fields totalling 13,004 acres in area. But this was well under 10 per cent of the total. Overall, alien sowkars seem to have owned about 5 per cent of the land of the Deccan under the plough in 1875, and most of the traditional land holders were not threatened by them.[43]

In fact the very idea of ryot proprietors being dispossessed conflicts

[39] *Deccan Riots Report*, Appendix C, Statement on pp. 200–1.

[40] M.A.R.D. Vol. 3 of 1875, No. 971, A.F. Woodburn, Assistant Collector, to H.B. Boswell, Collector of Ahmednagar. No. 271, 5 July 1875, para. 11. The following statistics for Akola and Sangamner talukas are also from this source.

[41] B.G.S. New Series, 177. Settlement Report on Parner taluka, Ahmednagar District, by G.A. Laughton. No. 352, 12 March 1884, para. 4.

[42] See B.G.S. New Series, 440. Settlement Report on Sangamner taluka, Ahmednagar District, by J. Ghosal. No. 1293, 15 September 1903, para. 6.

[43] The evidence of years after the Deccan riots appears to confirm this. Bombay's reply to Elgin's great land transfer enquiry of the late 1890s provided information for 289 villages in Poona District. Here, out of a total occupied area of over 720,000 acres, non-agriculturist money-lenders owned 108,437 acres and other non-agriculturists 68,085 acres. But these figures must have included some traditional holdings by non-agriculturists and the classification of 'non-agriculturist' may have included village servant castes like Mahrs and Mangs as well as aliens like Marwaris. Clearly a minority of Poona's land had changed hands by 1899. See I.O.L. Bombay Confidential Revenue Proceedings, Vol. 5777, October 1899.

with the way most moneylenders operated. The Marwari does not seem to have been very interested in his debtor's land. He wanted a new debt bond to bind the ryot to him and the formal transfer of a ryot's land was usually only a last resort. As Auckland Colvin noted:

the Marwaris do not, as a rule, desire to possess themselves, or to record themselves as proprietors, though in some districts they may be found willing to do so. It is only, as I understand it, when a Kunbi has resort to a second moneylender, or has in his khata exceptionally good lands that the Marwari steps in and causes a transfer of the proprietary title to his own name.[44]

It was good tactics for the sowkar to threaten but not use the ultimate sanction of dispossession. When the moneylender became pressing, the standard procedure was for the ryot to sign a bond for an increased debt. Both parties regarded this as more satisfactory than any arrangement of land transfer.

This was true throughout Western India in the years before the Deccan riots. In Gujarat in the 1850s

money is advanced upon mortgage, but it is rare indeed to find a mortgage foreclosed for default... They hold the bolt suspended over the head of the borrower, well knowing that he will make every possible effort to prevent it being hurled against him.[45]

The relationship between sowkar and ryot was regulated by traditional, unwritten rules which both usually observed. Since 1860 the moneylender had become tougher, resorting to the courts more often to assert his control over debtors. Yet his interest still lay in controlling the ryot's agricultural

---

Settlement reports of the Great War era give a more detailed picture. When Parner taluka, at the very centre of the 1875 troubles, was revised in 1914–15, all non-agriculturists owned only 47,538 acres of land compared with 314,238 acres held by agriculturists. Only 13.1 per cent of Parner was thus the property of interlopers, a 'quite satisfactory' situation. B.G.S. New Series, 567. Settlement Report on Parner taluka, Ahmednagar District, by J.H. Garrett. No. S. 393, 22 May 1916, para. 57. In Sirur taluka, which had also seen disturbances in 1875, in 1915 non-agriculturist money-lenders owned 15 1/2 per cent of the total occupied land as against 78 per cent held by traditional agriculturists. B.G.S. New Series, 558. Settlement Report on Sirur taluka, Poona District, by R.D. Bell. No. S.I, 25 September 1915, para. 20.

[44] *Deccan Riots Report*, Memorandum by Mr Auckland Colvin, para. 22.

[45] Alexander Mackay, *Western India. Reports addressed to the Chambers of Commerce of Manchester, Liverpool, Blackburn and Glasgow*, p. 134.

production and directing the tight-knit web of local commerce. Land ownership was a mere encumbrance to this and thus

> the sowkar finds it to his advantage to keep the ryot in nominal possession of the land and to take the difference between what it produces and what is absolutely required for the support of the cultivator.[46]

Indeed it seems that indigenous high-caste moneylenders were much more interested in land than were the Marwaris. One official, describing classes of moneylenders in the Deccan, wrote of

> the rich bankers or traders of large towns, among whom, besides the races mentioned, are found a good many Brahmins chiefly of the Yajur-vedi Deshast caste.[47]

He remarked of this group:

> they deal much less in advances of grain than the traders of lower caste and have a much greater taste for getting land into their own hands and names than the immigrant traders.[48]

And yet the Deccan riots were almost exclusively directed against immigrant moneylenders. Wealthy Brahmin sowkars were not usually threatened at all. Like the other customary explanations for these disturbances, a social revolution on the land seems inadequate as an explanation in itself.

III

One particular feature of the Deccan riots, however, here claims our attention: they were directed at a single homogeneous group and not aimed indiscriminately at all moneylenders.

> The Marwari and Gujur sowkars were almost exclusively the victims of the riots, and in villages where sowkars of the Brahmin and other castes shared the money lending business with Marwaris it was usual to find that the latter only were molested.[49]

---

[46] M.A.R.D. Vol. 7 of 1875, No. 960. C.G.W. Macpherson, 2nd Assistant Collector, to G. Norman, Collector of Poona. No. 72, 17 July 1875, para. 11.

[47] Notes by Mr Sinclair, Assistant Collector of Ahmednagar. Quoted in *Deccan Riots Report*, para. 39.

[48] *Ibid.*

[49] *Deccan Riots Report*, para. 12.

There were clear reasons for the extensiveness of immigrant sowkars in the Deccan. Throughout the Bombay Presidency the onset of British rule had produced a struggle for power in the localities, as traditional elites fought a rearguard action against the centralizing forces of the raj. The new legal and revenue arrangements were naturally suspected by the old leaders of society, who sought to maintain their authority by denying local knowledge to the British. The Joint Report of 1847, a detailed guide-book for the organization of the Bombay revenue system, had laid down a *ryotwari* ideal which produced challenges from groups who felt that their status was above such arrangements. Hence there were clashes and campaigns against the revenue survey in Khandesh in 1852,[50] in Karwar, North Kanara District in 1870-71 and in Ratnagiri District throughout the 1860s.

In Bombay, as throughout India, the European rulers were often forced to compromise their idealism in order to secure revenue. So the ryotwari system was tempered to suit several village elite groups. The supreme example was in Ratnagiri and Thana Districts of the Konkan, where village renters called *khots* claimed proprietary rights. In fact these claims were bogus. Wingate noted in 1851 that khoti rights

> appear to me to be nothing more than might be anticipated to have grown out of their original position as farmers of the Government revenue.[51]

In 1886 an exhaustive government enquiry confirmed Wingate's impression. Out of the 141 khoti sanads examined only seven mentioned proprietary rights and no khot could produce documentary evidence of such rights from pre-British times.[52] Yet because the khots had persistently resisted ryotwari revenue surveys—by Dunlop in 1820-24, by Dowell in 1827-30, by Wingate in 1851 and by Francis and Waddington in 1866-69—with non-cooperation and no-rent campaigns, the government had had to come to terms with them. A Khot Act in 1880 sacrificed the principles of the Joint Report by granting the khots the right to settle with Government; with the registration of customary tenants as a sop of official consciences. This was not an isolated case. Indigent *talukdars* in Gujarat

[50] See J.F.M. Jhirad, 'The Khandesh Survey Riots of 1852', *Journal of the Royal Asiatic Society*, 1968, Parts 3 and 4, pp. 151-65.

[51] B.G.S. No. 2. Report by Captain Wingate on introducing a survey and settlement in the Ratnagiri Collectorate. No. 44, 30 January 1851, para. 35.

[52] B.G.S. New Series, 446. Papers regarding the proprietary rights of khots in the Ratnagiri District, 1907.

were shored up, and *inamdars* and *jagirdars* bought off. Bombay could boast a ryotwari system, but not a ryotwari monopoly: in 1880 over 8 million acres in the Presidency were held on some overlord or revenue-free tenure against just over 28 million acres of ryotwari villages.[53]

The survival of the old official elites, like the Ahmedabad talukdars and the Ratnagiri khots, gave Gujarat and the Konkan a political stability they might otherwise have lacked. It provided those areas with a traditional social elite to act as village officers, moneylenders and traders. But in the Deccan the customary officials, notably the Patel, the village headman, had fared much less successfully.

The Deccan had always felt the firmest smack of governmental authority. Poona, the Peshwa's capital, was located there and the Mahratta government, like its British successor, had since Shivaji's day striven to limit the autonomy of the local official.[54] When the British took over, the village officers in some parts of the Deccan no longer possessed *huks*[55] on land and were remunerated entirely by Government, a clear sign of loss of independent power. This was true of Newasseh and Parnèr talukas, Ahmednagar, where Naroo Babajee and his dependants had cut back the Patel's privileges.[56]

The Deccan Patel was traditionally one of the most authoritative and influential of village officers. Sykes wrote

> of such importance and so profitable was the office, or in such estimation was the dignity of the Patel anciently, that princes of the Maharatta empire established themselves wholly or in part in the office in various towns and villages; Holkur, for instance, at Munchur; Seendeh [Sindiah] at Jamgaon.[57]

This social prestige was based on economic advantage. The Patel's emoluments included free land, grain, fees and presents on investitures and marriages, and

[53] B.H. Baden-Powell, *The Land Systems of British India* (Oxford, 1892), Vol. 3, p. 251.

[54] A.R. Kulkarni, 'Village life in the Deccan in the seventeenth century', *The Indian Economic and Social History Review*, Vol. 4, No. 1, March 1967, p. 50.

[55] 'Huks' were the emoluments village officers received from their villagers as rewards for carrying out their duties.

[56] B.G.S. No. 4. Report on the village communities of the Deccan by R.N. Gooddine [Hereafter Gooddine's Report], para. 15.

[57] W.H. Sykes, 'Special Report on the statistics of the four Collectorates of Deccan under the British Government', *Report of the Seventh Meeting of the British Association for the Advancement of Science* (London, 1838), p. 288.

annual presents from the shoemaker of shoes, from the potmaker of pots, from the shopkeepers of coconuts, etc.[58]

The grain-share the Patel received from the village was normally enough to feed his family. At Karjat, Ahmednagar District, the Patel was entitled to 128 seers of grain for every 120 beegas of land cultivated. In 1827, 8,491 beegas were under cultivation and so the Patel received 9,057 seers, enough, according to Sykes, to feed 25 people for a year.[59]

Yet this village leader, usually a Maratha by caste,[60] had to face not only the snipings of government at his local autonomy, but sometimes the obstruction of his villagers who had an obvious financial interest in keeping his huks at a minimum. In the central Deccan this tacit alliance had cut heavily into the Patel's perquisites, making the headman largely a government-paid civil servant by the 1840s. In 1843-44, for example, the Patels of 33 villages in the Patoda taluka of Ahmednagar received a total remuneration of Rs. 4,207, but Rs. 3,588 of this was government pay and independent huks totalled only just over Rs. 600.[61]

Then came the ryotwari revenue settlement, fixing land taxation with each individual Deccan ryot. Now

the principal onus is removed from the Patel, and he is deprived of uncontrolled action, and shorn of a great portion of his dignity.[62]

The system reserved the right to settle the Patel's huks, and valuable rights, like the power of appropriating or selling the lands of extinct families and waste lands, were taken from him. Sykes was soon talking of 'his degradation to the level of other cultivators'.[63] The decline, too, in the financial value of the Patel's office was intensified by its division. It became perfectly normal, by the 1850s, for a dozen men to share a village's Patelship. The Rs. 4,207 which the Patels of the 33 villages in Patoda taluka received in 1843-44 seems a reasonable reward, at over Rs. 100 per

---

[58] *Ibid.*

[59] *Ibid.*, p. 289.

[60] Sykes wrote in the 1830s: 'originally the Patels were Mahrattas only; but sale, gift, or other causes, have extended the right to many other castes. A very great majority of Patels, however, are still Mahrattas.' *P.P.* 1866, LII. W.H. Sykes, *Report on the land tenures of the Dekkan*, p. 11.

[61] Gooddine's Report, para. 36.

[62] Gooddine's Report, para. 21.

[63] *P.P.* 1866, LII. W.H. Sykes, *Report on the land tenures of the Dekkan*, p. 29.

office. Yet there were 284 sharers in these Patelships, so they each received well under Rs. 20.[64]

In this way, though there had been no dispossession of traditional landholders in the Deccan by 1875, there had been a decline in the social standing and financial position of the old elite. This had one vital result. In Poona and Ahmednagar there was no wealthy and powerful elite to conduct the money-lending. In contrast, in Ratnagiri District:

> when his circumstances allow, the khot secures the monopoly of the village moneylending and grain-dealing business. His position gives him a great advantage over professional usurers such as Marvadis who, as a consequence, have little inducement to settle in the district.[65]

Yet Marwaris had every inducement to settle in the Deccan where they ran most of the village money-lending businesses. The years before the Deccan riots saw a steady immigration of petty moneylenders and traders, increasingly with a 'get rich quick' attitude. The riots commission noted how

> the smaller class of sowkars, who are also the most unscrupulous, have increased very considerably during the last 10 years.[66]

In the tight-knit caste structure of the central Deccan the foreigners were highly conspicuous. Ahmednagar and Poona were overwhelmingly Kunbi districts. In 1881 of Ahmednagar's 751,228 population, 304,818 were Maratha Kunbis and well over a third of Poona's people belonged to the same caste:[67] many others were from associated agriculturist castes. In contrast there were at most only 15,565 Marwaris in the whole Presidency, just over half of whom lived in the Deccan. Gujarati Vanias, their fellow moneylenders, too, boasted only 32,693 persons in all Bombay.[68]

The Deccan moneylender was therefore very much an outsider. As Sir Erskine Perry put it:

> every large Deccan village under the Peshwah no doubt had its moneylender, but he belonged to the village and to its quasi corporation, ...Now into almost every Maratha village four or five moneylenders from Marwar have penetrated,

[64] Gooddine's Report, para. 36.

[65] *B.G.* X, Ratnagiri, p. 138.

[66] *Deccan Riots Report*, para. 63.

[67] *Imperial Census of 1881. Operations in the Bombay Presidency.* Vol. 2, Appendix A, Tables 1 and 8.

[68] *Ibid.*, Vol. 1, p. 125.

having no sympathy and no connection whatever with the villagers in language, race, or religion...[69]

The Deccan Riots were directed against such immigrant sowkars, and the 'foreignness' of the Marwari moneylender, the product of the decline of the Patel, was the clear occasion of the riots.

However, we must still look very carefully at the nature of the disturbances. After all, the agrarian riot, particularly at times of economic hardship, was a standard mode of political expression in a rural community. These 33 disturbances were no more intense than the isolated grain riots of the famine years of 1876-77 and 1899-1901, or indeed than similar outbursts by Russian peasants or by Scottish crofters and Welsh hill-farmers in the Great Depression of the 1880s.

Poona and Ahmednagar, particularly the eastern regions where the riots occurred, were very poor areas: millet-growing districts with holdings typically of around five acres in size.[70] All the riot areas were at the heart of the severe famine-risk region. In Bhimthari taluka 20 inches of rainfall meant a reasonably good year. So here any agricultural slump was liable to drive the cultivators over the threshold of subsistence. In fact, despite the bland assurances of settlement officers that all was well, there is clear evidence of an agricultural crisis in the central Deccan in 1875.

This was the inevitable reaction to an unreal boom. In the early 1860s the expansion of cotton cultivation, together with a big railway extension and public works programme, sent prices and wages soaring and poured money into the mofussil. The wages of unskilled labourers on public works shot up from an average of Rs. 7 3/4 per month in 1860-62 to Rs. 13 1/2 in 1863.[71] Cotton prices rose by over four times between 1860 and 1864.[72] But prices began to turn in 1866 and then declined for 10 years, returning to the 1860 *status quo*.[73] This was an all-India phenomenon, but in Western

[69] M.A.R.D. Vol. 3 of 1878, No. 140. Note on the Deccan Riots Commission Report by Sir Erskine Perry, 1 December 1877.

[70] In Poona District in 1882-83 there were 86,193 holdings of 1 to 5 acres in size, 43,898 of 6-10 acres, 45,359 of 11-20 acres, 30,677 of 21-30 acres and 21,744 over 30 acres in size. *B.G.* XVIII, Poona, Part 2, p.6.

[71] *Deccan Riots Report*, para. 66.

[72] Average prices of cotton per candy of 784 lbs in Bombay: 1860: Rs. 128. 1861: Rs. 157. 1862: Rs. 215. 1863: Rs. 314. 1864: Rs. 539. M.A.R.D. Vol. 25 of 1875, No. 1855. Table of cotton prices.

[73] Average prices of cotton per candy of 784 lbs in Bombay: 1865: Rs. 627. 1866: Rs. 382. 1867: Rs. 343. 1868: Rs. 274. 1869: Rs. 235. 1870: Rs. 294. 1871: Rs. 284. 1872: Rs. 225. 1873: Rs. 195. 1874: Rs. 184. *Ibid.*

India the price fall in the primary foodgrains seemed most perceptible. In Poona in 1863-64 a rupee bought 14 seers of jowari. By 1873-74 the price had crashed to 26 seers per rupee and the following year to 30 seers.[74] At the same time declining wages affected those who supplemented their income by labouring. The Poona skilled labourer, whose average wage in 1873-74 was 12 annas per day, received only 10 annas for a day's work in 1874-75.[75]

The effects of price and wage fluctuations on an agrarian community would vary greatly, according to whether the individual peasant was more frequently a buyer or seller. But if these developments might bring variable results, the persistent rise in the population was a consistently gloomy factor. By 1875 population had been rising steadily in the Deccan for over thirty years, pressing on the already scanty resources of the region. In the 54 villages of Bhimthari taluka settled by Waddington in 1871, the number of people had increased by over 8,000 since 1840-41—40 per cent more mouths to feed.[76] The population rise could be absorbed while there was land—in the 1840s there had been 43,000 acres waste in Bhimthari[77]—but by 1870 the margin of cultivation had been reached. Indeed in the few years before the Deccan Riots some land even began to fall out of cultivation. From Ahmednagar District the collector reported that 92 acres of hitherto cultivated land were abandoned in 1871-72, a further 12,002 acres in 1872-73, 6,793 acres in 1873-74 and another 16,266 acres in 1874-75.[78] This helped to create James Caird's impression that

> a great part of Madras and some part of Bombay seemed to me to have reached the margin of production under the present agricultural system.[79]

In such an atmosphere it only needed some minor catalyst to provoke

[74] B.G.S. New Series, 577. Settlement Report on Haveli taluka, Poona District, by R.D. Bell. No. S.3, 7 March 1916, Appendix N.

[75] M.A.R.D. Vol. 7 of 1875, No. 960. G. Norman, Collector of Poona, to J.E. Oliphant, Acting Revenue Commissioner, Southern Division. No. 1625, 9 August 1875, para. 9.

[76] B.G.S. New Series, 151. Settlement Report on Bhimthari taluka, Poona District, by W. Waddington, No. 440A, 12 July 1871, para. 6.

[77] Deccan Riots Report, Memorandum by Mr Auckland Colvin, para. 31.

[78] M.A.R.D. Vol. 3 of 1875, No. 971. H.B. Boswell, Collector of Ahmednagar, to the Revenue Commissioner, Southern Division. No. 2132, 20 July 1875, para. 14.

[79] I.O.L. Home Miscellaneous Series, No. 796. Letters and Papers of Sir James Caird, relating to India. Sir James Caird to Lord Northbrook, 19 September 1879.

a demonstration of discontent. This was provided, as Catanach has pointed out,[80] when the sowkars started to drag their feet over paying their debtors' revenue assessments, hoping to persuade government to grant remissions.

Accordingly the ryots pressed the Sowkars to pay the instalments and the refusal of the latter led to the disturbances which commenced in Bhimthadee and extended all around.[81] The riots which broke out in the Poona and Ahmednagar districts in 1875 were thus no more than minor grain riots, a simple outburst born of deprivation, rather than an articulate protest against a social revolution. Even so, they were an extremely feeble gesture. The lack of violence amazed the riots commissioners. In one village a Marwari, trapped inside his burning house with a broken leg, was rescued by rioters.[82] The Deccan Kunbi reacted to harder times by making an outburst against his creditor, who was an alien and an obvious target for discontent. This was hardly peculiar. In the Panch Mahals, for example, in September 1899 when grain first rose to famine prices the Jhalod Bhils rose in large numbers and attempted to sack Jhalod, Limbdi and other places.[83]

Thus, though the riots incidentally provide evidence on the decline of the Deccan Patel, the essential insignificance of the disturbances themselves must be stressed. Here were just 33 hardly very violent demonstrations.[84] The central problem of the Deccan riots is really not what caused them but why so momentous a non-event has been considered so important.

[80] I.J. Catanach, 'Agrarian Disturbances in Nineteenth Century India', *The Indian Economic and Social History Review*, Vol. 3, No. 1, March 1966, p. 69. The cholera outbreak, which Catanach also mentions, may have been another contributory factor in the outbreak of these disturbances.

[81] M.A.R.D. Vol. 7 of 1875, No. 960. C.G.W. Macpherson, 2nd Assistant Collector, to G. Norman, Collector of Poona. No. 72, 17 July 1875, para. 11.

[82] *Deccan Riots Report*, para. 6.

[83] B.G.S. New Series, 481. Settlement Report on Jhalod taluka, Panch Mahals District, by C.V. Vernon. No. S. 80, 3 March 1904, para. 10.

[84] Also, as Catanach has well said, the riots took a fortnight to spread over a relatively small area. I.J. Catanach, 'Agrarian Disturbances in Nineteenth Century India', *The Indian Economic and Social History Review*, Vol. 3, No. 1, March 1966, p. 72. This hardly suggests they were a spontaneous and determined outbreak against universally-felt, deep-seated grievances. Rather, the Deccan riots were a falteringly connected series of individual small riots.

IV

In fact the prominence of the Deccan riots owes much to contemporary officials. The disturbances came at a crucial time when a great debate was taking place within the government about the sort of society British rule was creating in Western India. The ryotwari ideal of the Joint Report had been strongly *laissez-faire* in tone. It envisaged rich peasants eagerly grasping the opportunity provided by a low field assessment, while their weaker and less efficient brethren were driven out of land-ownership.

Yet no sooner was the system born, than worries and disillusion spread about its beneficiaries. Already in 1852 Wingate, the protagonist of the Bombay system, was predicting

> the destruction of our peasant proprietors and the transference of land to the monied classes, who will sublet it to the former proprietors at rents which will probably leave them only a bare subsistence. We shall have a richer community on the whole but the peasants will probably be reduced to the condition of those in Bengal or of the Irish colliers.[85]

This, Wingate wrote,

> I do not contemplate... with any degree of satisfaction.[86]

Already the concept of government interference to protect the peasantry was mooted. Wingate advocated exempting immovable property from attachment or sale for debt, enacting insolvency legislation and granting the courts powers to fix interest rates.[87] This, in 1852, was an astonishingly accurate precursor of the Agriculturists' Relief Act of 1879.

After 1857 'mutiny-spotting' added a new urgency to fears about the fate of the Bombay peasant. It now became an orthodoxy that the landholders of India—in Bombay's case the ryot smallholders—had to be preserved from economic ruin to keep them politically quiescent. The first fruits of this attitude in Western India were the abandonment of the Inam Commission's detailed enquiries into rights to alienated land and the enactment of the Ahmedabad Talukdars Act in 1862. In such a mood, with

[85] Papers of Sir George Wingate, Sudan Archive, School of Oriental Studies, Durham University [Hereafter 'Wingate Papers'], Box 293/8. G. Wingate to H. Green, 27 August 1852. (I am grateful to the Librarian of the School for permission to consult these papers.)

[86] *Ibid.*

[87] Wingate Papers, Box 118. G. Wingate to the Registrar of the Court of Suddur Dewanee Adawlut, Bombay. No. 319, 24 September 1852.

the administration nervous and jumpy, it was easy for officials to talk themselves into exaggerating the dangers of social change.

Raymond West, a high court judge and a paternalist of the old school, gave voice to the prognostications of the pessimists in his pamphlet *The Law and the Land in India*, published in 1872. West was above all a moralist and he believed the indebtedness of the peasantry was producing a grave degeneration in Maratha society:

> the Rajput or Maratha yeomen disappearing or sinking into indigence and despair are replaced, as the Roman freeholders were, by a class of mere cultivators living from hand to mouth without strong local attachments or the sturdiness which independence gives.[88]

But West's fears for the state of the peasantry were shared by men of very different ilk. Among them was William Wedderburn, an Ahmednagar judge at the time of the Deccan riots. Wedderburn felt that the whole trend of British administration in Western India was dangerously subversive of old well-tried institutions. On revenue matters, for example, he strongly attacked the rigidity of the Bombay system and its demand for cash payments:

> no sliding scale of charge and of remission could be simpler or more effective than the Native system of 'Batai' or payment in kind which gave the Sirkar a proportional share of the gross produce.[89]

On the indebtedness issue, Wedderburn advocated an insolvency law for extreme cases and the constitution of village panchayats to conciliate between ryot and sowkar.[90]

For a government which had no channels of communication with the masses, fears such as those in Bombay about the state of the peasantry were apt to assume nightmarish proportions. Calmer heads had no evidence to which they could point. By 1875 a large and powerful party existed within the Bombay secretariat who believed that some ameliorative action for ryot indebtedness was vital. This group was informal in constitution and it was neither consistent nor united in its programme of reform. West, for

[88] Raymond West, *The Law and the Land in India* (Bombay, 1872), para. 24. (A copy is No. 1 of Selections from Papers on Indebtedness and Land Transfer, I.O.L. Government of India Revenue Proceedings, October 1895.)

[89] W. Wedderburn, *A Permanent Settlement for the Deccan* (Bombay, 1880), p. 9. (A copy is in the Indian papers of Sir James Caird.)

[90] *Deccan Riots Report*, Appendix B, W. Wedderburn, Report on the Indebtedness of the Ryot, 7 December 1876. para. 5.

example, was highly sceptical of the advantages of legislation and later became the fiercest opponent of the Agriculturists' Relief Act. Wedderburn argued that

> the indebtedness of the ryot to the village sowkar instead of being an evil may be made the foundation of general prosperityand that
>
> the more capital that flows towards the land the better.[91]

Wingate, however, had believed:

> what is wanted is to restrict credit. To make the moneylender more chary of his advances. To make the people look to their own industry rather than to loans...[92]

The latter argument had, by 1875, considerable support in the revenue department. At any rate, all could unite around some proposals, such as that for an insolvency law to allow ryots to declare themselves bankrupt.

The pressure for official action on indebtedness was steadily building up in the 1870s. Sir Philip Wodehouse, the governor, informed the secretary of state:

> this Government is in the constant receipt of representations of increasing poverty (which I do not believe to be quite correct) and is urged to go to the rescue.[93]

But Wodehouse himself had been won over to some support for action. In November 1873 he had presented a scheme whereby rural land in Bombay would be prohibited from sale for debt and any bond containing an encumbrance on land would be registered by a European officer.[94] However, reform on the indebtedness issue received short shrift from the Duke of Argyll, then secretary of state. He denounced any such ideas as

> a sort of sentimental favouritism for Peasant Proprietorship and involving the most violent interference with existing rights.[95]

Wodehouse had tried again in 1874, when Lord Salisbury succeeded Argyll at the India Office. In 1875 legislation to help indebted peasants was

---

[91] *Ibid.*

[92] Wingate Papers, Box 118. G. Wingate to the Registrar of the Court of Suddur Dewanee Adawlut, Bombay, No. 319, 24 September 1852. para. 12.

[93] I.O.L. Papers of Sir Philip Wodehouse, Vol. 13. Sir Philip Wodehouse to the Marquis of Salisbury, 20 October 1874.

[94] *Ibid.*, Vol. 12. Sir Philip Wodehouse to the Duke of Argyll, 27 November 1873.

[95] *Ibid.*, Vol. 11. Duke of Argyll to Sir Philip Wodehouse, 29 December 1873.

a live issue in Bombay, already occupying much thought and official paper.

In this situation the impact of the Deccan riots was electric. Here was a magnificent opportunity for those who wanted action: forceful evidence of discontent and possible grave political trouble from the Maratha masses. It paid such men to make as much of the disturbances as they could. J.B. Richey, who was on the spot, wrote that the Kunbi

is likely to become dangerous especially in his present mood.[96]

Richey was quick to press a scheme whereby the government would pay off some of the sowkars' claims against their debtors and hence effect compromises between ryot and moneylender. W.G. Pedder, a prominent revenue department activist on the indebtedness issue and Acting Secretary to the Bombay Government, strongly supported the scheme and hailed it as

nothing else but the establishment of agricultural banks for a special object, under very special circumstances, and for a limited period.[97]

Though these proposals were rejected, the pressure on government continued. A riots commission was appointed, whose members included Richey and Auckland Colvin from the United Provinces.[98] This process brought widespread publicity to the problem of peasant indebtedness in the Deccan and activists like Pedder eagerly seized the opportunity to present their case to the British public.[99] Soon the minor disturbances which had so decisively added fuel to the debate were lost sight of.

[96] M.A.R.D. Vol. 118 of 1875, No. 1867. Report by J.B. Richey, 30 August 1875, para. 2. Of course, it was also in the interests of the Poona Sarvajanik Sabha and the Press to stress the importance of the 1875 riots. They had, for some years, been claiming that the harshness of Bombay's land revenue system was hitting the Deccan ryot hard.

[97] M.A.R.D. Vol. 118 of 1875, No. 1867. Note by W.G. Pedder, 31 August 1875.

[98] Ever since Wingate and Thornton's lengthy argument over the relative merits of each of their revenue systems, Bombay and the United Provinces had waged a guerrilla war of intrigue and counter-intrigue against one another's revenue and administrative ideas. Any U.P. man might therefore be expected to support reforms which acknowledged the shortcomings of the principles by which Western India had been governed.

[99] In the *Nineteenth Century* for September 1877 Pedder published an article, largely about the Deccan riots situation, entitled 'Famine and Debt in India'

Richard Temple succeeded Wodehouse as Governor and his influence steered through the Deccan Agriculturists' Relief Act in 1879. And yet, apart from the feature of the Patel's decline which the disturbances illustrated, the Deccan riots, as any sober assessment would have acknowledged, showed no symptoms of revolutionary social or economic change. Here was one of the clearest instances of British rule in India confusing the illusion of social change with the reality of agrarian life.

Yet misconceptions are not easy to dispel. After 1879 the debate moved on to further stages and the revolutionary importance attached to the Deccan riots was never challenged. Discussions on the Agriculturists' Relief Act and on its degree of success in the 1880s and 1890s were always conducted in these terms. All shades of opinion accepted that the Deccan riots had indicated a social revolution whereby the Maratha Kunbi had been largely ousted from his land by Marwari moneylenders. In 1893 one defendant of the Relief Act in a discussion whose participants included West and Wedderburn remarked of the situation in 1875:

> the land in the possession of the peasantry was found passing away from their hands, not by private sale, but by the compulsory process of judicial sale, into the hands of foreign usurers, who, not being able to cultivate it themselves, used to retain their former proprietors as their rack-rented tenants or bond-slaves.[100]

The myths of the Deccan riots were established very early.

Readers of this journal were certainly given their fill of Bombay matters. In 1878 Florence Nightingale got in one the act, describing the situation in Western India as 'the utter demoralization of two races—the race that borrows and the race that lends'. Florence Nightingale, 'The People of India', *Nineteenth Century*, Vol. 4, 1878, p. 211.

[100] Comments by Dr A.D. Pollen on Sir Raymond West's paper 'Agrarian Legislation for the Deccan and its results', *Journal of the Society of Arts*, Vol. 41, 1892-93, p. 729.

Chapter Seven

The Punjab Disturbances of 1907: The response of the British Government in India to agrarian unrest[*]

N. GERALD BARRIER

Two major factors determined the growth of Indian nationalism: developments within Indian society and indigenous political organizations and the British response to agitation and the demands of Indian politicians and interest groups. Current studies of Indian nationalism generally focus upon the former, while the relation of British rule to political development is either ignored or misunderstood. Frequently the British are portrayed as inept bureaucrats, or the government as a monolithic structure run on inflexible bureaucratic principles and bent upon a policy of repression.[1] While the shoe often fits, this caricature ignores the basic fact that, despite mounting pressure from 1904 onwards, the British rulers managed to keep order and to choose their hour of departure. They were able to continue governing primarily because they were adroit in handling agitation, minimizing serious grievances, and retreating from policies or measures which threatened to inflame the subcontinent. Professor Anthony Low of the University of Sussex has shown graphically how the Government of India manipulated Gandhi's first and second civil disobedience campaigns to its own political advantage.[2] This paper explores the background and the

[*] Research and preparation of the study was made possible by a United States Government Title VI Travelling Fellowship and a Duke University grant. The author is also indebted to Professor William B. Hamilton of Duke University for criticism of an earlier draft.

[1] E.g., R.C. Majumdar, *History of the Freedom Movement in India*, Calcutta, 1963, II, pp. 1-6, 251-64; III, pp. 67-225. For the general historiographical problem, Robert I. Crane, 'Problems of Writing Indian History', in *Problems of Historical Writing in India*, New Delhi, 1963, pp. 41-5.

[2] 'The Government of India and the First Non-Cooperation Movement—1920-22', in *Journal of Asian Studies*, 25, 1966, pp. 241-59; 'Sapru and the First Round

British response to an earlier but equally volatile situation, the 1907 rural disturbances in the Punjab.

The Punjab Government had passed several paternal measures which were calculated to improve the economic position and the standard of living among Punjabi agriculturists. In 1907 its further attempts to help the agriculturists unexpectedly resulted in the alienation of the political support of the agrarian population in the central Punjab. This area was the recruiting ground for more than a third of the Indian army. There were indications in May 1907 that the Punjabi agriculturists, as individuals and as soldiers, had moved towards rebellion. The Government of India met this challenge by putting an end to the local government's paternal programme. By extending too far the principle of helping the rural classes, the Punjab Government aroused agitation which in turn led to a retreat from rural planning and intervention into the lives of the people. The decision-making process engendered by the unrest demonstrates that the Governor-General and his Executive Council could take into account 'public opinion' when that opinion directly affected the security of the British raj. It also shows that the government was not an impersonal monolith, but an arena in which personalities, conflicting principles, and faulty channels of communication determined official policy and action.

The Issues: The Punjab Colonization of Land Bill and the Bari Doab Canal Rates

Since the annexation of the Punjab in 1849, the provincial government had prided itself on its paternal protection of the cultivating land-owners who comprised the bulk of the population. The first British rulers and their successors believed that if a paternal district officer ruled his 'flock' with an iron hand and protected them from outside threats—whether money-lender or political agitator—the zamindars (landowning cultivators) would loyally support the British government.[3] For three decades the local

---

Table Conference', unpublished paper presented at the Working Conference on Indian Politics and Political History, University of Sussex, 1965.

[3] This attitude or administrative tradition is discussed in the following: R.N. Cust, *Memoirs of Past Years of a Septuagenarian*, London, 1899, pp. 29-30; J. Douie minute, 24 December 1900, Punjab Government (hereinafter PG) Judicial File, February 1901, 14-38A. Although a few of the PG files cited in this study are available only in the West Pakistan Record Office (hereinafter WPRO), most files are found both in the WPRO and the India Office, London.

government relied upon a low revenue demand and land acts favouring the zamindars to ensure the economic prosperity and hence the loyalty of the rural classes, but gradually debt and loss of land to moneylenders undermined their economic and social position. Fearing agrarian disaffection over the mounting debt, the government committed itself to a programme of paternal legislation to protect the zamindars and to shore up the traditional prop of British rule in the Punjab.

The first paternal legislation in 1900 was the greatest single piece of social engineering ever attempted in India, the Punjab Alienation of Land Act. The act took away the zamindar's right to sell or mortgage his land except with the prior approval of the district officer. These officers generally gave permission only when the recipient of the land could prove that he belonged to a tribe designated in the government gazette as 'agricultural'. According to the Imperial Revenue and Agriculture Department, the restrictions on land transfer would immediately halt the flow of land outside the agricultural community. The act would also prevent further indebtedness by curtailing cultivators' credit. The Hindu commercial castes which monopolized moneylending in the Punjab fought the bill with memorials and public meetings, but they failed to mobilize the cultivators against legislation designed to protect the cultivator's own interests. The alienation bill became law in November of 1900.[4] Encouraged by the limitation of agitation to the urban trading class, the Punjab government enacted additional bills designed to improve the economic condition of the zamindars and the small aristocratic class. The Punjab Pre-Emption Act provided that agnates and agriculturists residing in a village had first claim on land sold by a villager, the Court of Wards Act gave the local government the right to place insolvent aristocratic families under an official court of wards without the prior consent of the family, and the Agricultural Debt Limitation Act extended the period during which debt suits could be instituted from three to six years. These bills stressing paternal protection of the rural classes were enacted within four years after the Alienation Act without arousing agitation or even adverse comment from the Punjab native press.[5] The only opposition, ironically, came from the normally pro-government *Pioneer* of Allahabad, which warned in an editorial entitled 'Bureaucracy and Ballast' that the local government had

---

[4] The bill and subsequent agitation are discussed in N.G. Barrier, *The Punjab Alienation of Land Bill of 1900*, Durham, N.C., 1966, Chapters 1-3.

[5] J. Wilson minute, 22 February 1904, Financial Commissioner File (WP Board of Revenue, Lahore; hereinafter cited as FCF) 441/108A.

thrown away its intellectual and economic ballast in misguided zeal and devotion to paternalism. Class legislation, the *Pioneer* predicted, would ultimately destroy the peace of the Punjab.[6]

Only one piece of paternal legislation was yet to be enacted, an amendment to the 1893 Punjab Colonization of Land Act which would extend the role of the government in the daily life of the Chenab canal colony. The Punjab Government began the impressive colony in 1887. The diversion of the Chenab river into a system of perennial canals rapidly turned the barren wasteland of the central Punjab into fertile farmland. The British had committed themselves to this large capital expenditure of over Rs. 30,000,000 for at least three reasons. First, the authors of the project hoped that the additional land revenue from the once desolate desert would swell the provincial budget. At the same time colonization of the canal area would relieve the acute population pressure in the central districts bordering the colony. Finally, the Chenab colony was to be a social and economic experiment, a model farm for the rest of the Punjab. Healthy agricultural communities 'of the best Punjab type' would be established and kept under constant supervision. These in turn would demonstrate to other Punjabis how proper sanitation, careful economic planning, and cooperation with the government could result in a higher standard of living.[7]

The Chenab colony quickly fulfilled the first two official aspirations. By 1907 the colony had proven a financial success. The capital outlay had been repaid, while the government received over Rs. 700,000 annually as net profit from water charges and land revenue.[8] Besides this profit, the government had distributed over 2,000,000 acres of cultivable land to Muslim, Sikh, and Hindu zamindars from the crowded districts of Amritsar, Gurdaspur, Jullundur, Sialkot, and Hoshiarpur. District officers chose the new settlers on the basis of their ability as agriculturists and less tangible qualifications such as family and 'proven loyalty' to the British.[9] The

[6] *Pioneer Mail*, 27 January 1905. Also noting in Judicial and Public File (hereinafter J & P) 1380, 1905.

[7] Popham Young minute, 15 November 1895, Government of India (hereinafter GI) Revenue File, December 1896, 22-47A; Popham Young minute, 16 April 1907, FCF 441/108A. All GI files are from the National Archives of India, New Delhi.

[8] *Revenue Report of the Punjab Irrigation Department, 1902-1905*, Lahore, 1905, p. 5. For the background on the colonies and irrigation, P.W. Paustian, *Canal Irrigation in the Punjab*, New York, 1930.

[9] J. Beazley and F. Puckle, *The Punjab Colony Manual*, Lahore, 1922, I, pp. 5-22.

Colonization Officer was in sole charge of the colony. He divided the settlers on the basis of district, caste, and religion into small administrative areas (*chaks*). Within each village in the chak, the officer had made previous arrangements for sanitation and orderly bazaars. Similarly, each chak had been carefully demarcated into agricultural zones, townships, and village greens. Each villager received at least a square of land (approximately 27 acres) and a designated plot for a house and stable. He was required to build a house, to dispose of manure, and to build a wall around his animal yard. In addition to the 'peasant grant' of one square, provisions were made for three other types of colonists. Retired soldiers from the Punjab or the neighbouring United Provinces received 'military grants' of one to three squares. Grants of similar size were allotted to 'yeomen', agriculturists with enough capital to maintain more than one square. The British felt that these yeomen were necessary to the success of the colony because they would supposedly supply the initiative and leadership required to build up community spirit. 'Capitalist grants' were large plots ranging from 5 to 100 squares which were distributed by auction to urban investors. Because the Punjab Government wanted to preserve the province as a 'country of peasant farmers', however, it distributed the bulk of the Chenab land to peasant grantees.[10] In the Rakh and Mian Ali branches, for example, 341,998 acres were held by peasants as compared to 30,473 acres by capitalists, 36,630 acres by yeomen, and 6,313 acres by military grantees.[11]

Beginning with the Chenab colonization project, the Punjab Government altered its policy of waste land distribution. Colonists in earlier projects had been permitted to purchase proprietary rights for a nominal sum. The subsequent loss of land and the government's fear that this loss would increase in the future resulted in the decision to end the sale of land to colonists. In the Chenab and subsequent colonies, settlers were generally granted only occupancy rights, not full title to their property.[12] With the exception of the capitalists and yeomen who continued to purchase proprietary rights, all colonists were crown tenants holding their grants as long as they paid revenue and fulfilled the conditions laid down at the

[10] PG to GI, 337S, 22 July 1891, GI Revenue, November 1891, 35-37A; *Chenab Colony Gazetteer*, Lahore, 1905, pp. 29-33.

[11] Statistics from 'Report on the Rakh and Mian Ali Branches of the Chenab Colony', GI Revenue, September 1897, 59-62A.

[12] PG to GI, 129, 12 March 1889, and Buck minute (26 April 1889), GI Revenue, May 1891, 15-18A.

initiation of the grant. Conditions varied with each type óf grant. Peasant grantees, for example, had to live on their land, cut wood from specified areas, maintain a clean compound, and make arrangements for sanitary disposal of night soil.[13] The Colonization Officer and his subordinate Indian staff supervised all the details of colony life in order to ensure that each colonist fulfilled his conditions and contributed to the well-being of the entire community. The officer was virtually a dictator. His word was final in all disputes over revenue or conditions, because civil courts were considered barred from intervening in government-tenant relations.[14] The officer also interfered in the social affairs of the colonists. In a period of communal tension, for example, Popham Young, the beloved Canal Officer of the Chenab, called together all the religious leaders and used his influence to prevent a riot.[15]

For several years the colony remained an idyllic example of 'good administration', a project to which Punjab officers proudly pointed as an indication of the beneficial and humanitarian effects of British rule. The first colonists did not resent the paternal attitude of the Colonization Officer—in their eyes he was an economic saviour, their *ma-bap* (parent) who loved them and gave protection against all evil.[16] Nevertheless, trouble subsequently began to appear in the colony which threatened the tranquil and prosperous atmosphere of this Punjab show-place. First, the Irrigation Department ran out of good land and distributed plots not readily accessible to canal branches. From 1902 onwards officials were faced with irate colonists in possession of inferior land.[17] Moreover, landholdings tended to become fragmented. When the original colonists died, the occupancy right of their land was sub-divided among sons. The British became alarmed at a process which if unchecked might eventually leave the colonists with only a few acres. 'A purely peasant colony with holdings getting smaller in each generation would not... be a good thing. The people might be prosperous up to a certain point, but their lives would be narrow and sordid, and Government would look in vain to them for any active help in time of trouble'.[18] Finally, the Canal Officer was finding it increasingly

---

[13] Conditions and contracts in Puckle, *Colony Manual*, vol. II.

[14] FC to PG, 19 January 1903, and noting, FCF 441/108A.

[15] C.L. Tupper minute, 1 March 1901, FCF 441/108A and Keep-With (KW).

[16] Popham Young minute, 15 November 1895, GI Revenue, December 1896, 22–47A. Young was immortalized in a sixty-stanza poem frequently read in the colony.

[17] FC to PG, 19 January 1903, FCF 441/108A and KW.

[18] Douie letter, *Reports for the Chenab, Jhalum, Chunian Sohaq Para Canals,*

difficult to enforce discipline. Many colonists evaded the residence requirements, illegally cut trees, built houses in the farming plots, and led (by British standards) unsanitary lives. This tendency was particularly evident in the recently settled Gugera branch, where yeomen grantees were for the first time required to reside on their property.[19] After unsuccessfully attempting to frighten the colonists by confiscating the grants of perennial offenders, the colonization staff developed an informal system of fines which they charged in lieu of confiscation. The fine system went uncontested until 1903, but when a recording of occupancy rights generated a sharp rise in penalties, several yeomen who were lawyers or relatives of lawyers took the matter to the civil courts. To the astonishment of the Punjab Government, several legal decisions went against the Canal Officer. The 1893 act made conditions binding on the tenant and the government, but it apparently did not rule out an appeal to the courts. As the Financial Commissioner summarized:

> Heretofore the colony has resembled a new country. The prestige and authority of the Colonization Officer has been very great and the colonists have generally acquiesced in the somewhat patriarchal methods of administration which have been extremely well suited to the requirements of the case. But in no long time, as the country becomes more settled, the colonists will begin, indeed, some of them already have begun, to use the weapon of the law to resist the orders of the Colonization Officer.[20]

The Punjab Government's initial response to the erosion of paternal rule in the colony was to ensure that future colonists would be governed by more stringent conditions. The fine system, compulsory tree planting, and a higher occupancy fee were included in the contracts signed after 1902. If a new settler died without gaining occupancy rights (generally before five years), the land lapsed to the government, which reserved the right to re-allot the property to lineal heirs or to unrelated tenants. Yeomen were prohibited from transferring property rights without previous sanction of the government, a troublesome restriction apparently calculated to keep out 'undesirables' (moneylenders or urban dwellers unpopular with the British).[21] Most of the fertile land had already been distributed,

---

1903, Lahore, 1904, pp. 8-9. Also, Douie minute, FCF 441/108A.

[19] B.H. Dobson, *Report on the Chenab Colony Settlement*, Lahore, 1915, pp. 9, 13-14; C.L. Tupper minute, 18 March 1907, KW to FCF 441/108A.

[20] C.L. Tupper minute, 1 March 1901, FCF 441/108A.

[21] Dobson, *Chenab*, p. 13.

234 / PEASANT RESISTANCE IN INDIA 1858-1914

however, and the Punjab Government decided to prepare a bill legalizing
the fine system and remedying the numerous administrative problems
which had appeared since 1893. Normally agrarian legislation was prepared in lengthy consultation with district officers and revenue experts, but
this procedure was not followed in the case of the colonization bill. The
colonies had always had an anomalous place in the Punjab administrative
structure. The Irrigation Department controlled the building of the canals
and supervised allocation of water, while the Revenue Department collected taxes. Because neither department had full jurisdiction over the
colonies, they frequently fought over their administration.[22] The Financial
Commissioner's office therefore drew up the bill without consulting either
department in order to avoid renewing the quarrels. It only consulted the
Colonization Officer, who naturally insisted upon a defence of his personal powers. In addition to avoiding a bureaucratic struggle over jurisdiction, the Financial Commissioner hoped that the truncated legislative
process would permit the government to enact a measure before its
relations with the colonists disintegrated. 'Benevolent despotism' had to
be prolonged, and therefore the normal consultative process was abandoned.[23]

Many of the bill's provisions were retroactive, openly abrogating the
conditions originally covenanted with the colonists. Fresh conditions were
added, including the planting of trees and sanitary rules. The bill forbade
transfer of property by will. Only strict primogeniture as interpreted by the
Canal Officer would be permitted in the future. Fines were legalized, while
the courts were explicitly barred from interfering with executive orders.
Although the Government of India requested clarification of several
sections, it approved the bill and sent it to the Secretary of State. Two
Punjab officers, a council member, Denzil Ibbetson, and the Revenue
Secretary, James Wilson, defended the bill and speeded its passage
through the Indian Government Secretariat.[24] In London a former Punjab
Lieutenant Governor, Dennis Fitzpatrick, warned his colleagues that the
bill would create a disturbance because it gave individuals too much
discretionary power and interfered with 'vested interests'. The India

[22] J.F. Connolly minute, 5 September 1903, FCF 441/108A; FC to PG, 858, 17
October 1901, GI Revenue, May 1902, 32-34A.
[23] C.L. Tupper minute, 1 March 1901, and noting, GI Revenue, December
1904, 8-9A.
[24] Wilson minute, 13 September 1904, and council noting, GI Revenue,
December 1904, 8-9A.

Council and the Secretary of State agreed and ordered the Punjab by telegraph to suspend discussion of the bill.[25] Denzil Ibbetson, who had temporarily replaced Charles Rivaz as Lieutenant Governor, redrafted the objectionable portions of the bill but defended the principles of setting new conditions and interfering in succession. By the time the Government of India reviewed the modified proposals, Ibbetson had returned to the Council and supported his own bill. Curzon and his Council strongly backed the Punjab, and the India Council in London grudgingly accepted the slightly modified version.[26]

The Punjab Government introduced the bill in the local Legislative Council on 25 October 1906. Section 8 stated that tenants held their land under old conditions and the supplementary conditions to be published in the gazette.[27] Section 18 said that in future grants, a District Commissioner would resume land upon death of a grantee if the latter had no legal heirs. In the future, land could only be transferred to a son, and if no sons survived, a widow. Sections 19-27 established retroactive conditions concerning sanitation, tree planting, and construction. In case of breach of the conditions, fines would be collected 'in the same fashion as revenue'. Section 31 ruled that the courts would have no jurisdiction in the colony. The Council appointed a select committee to discuss the bill and to make necessary amendments.

The year 1906 was the wrong time to present a measure extending official powers. Discontent over maladministration in the colony, which had been spreading since 1900, had gradually come to a head. The colonists hated the extra-legal fine system that had cost them over Rs.300,000 in penalties. More offensive than fines, however, was the rampant corruption among lower officials. The cumbersome system of assessment and records in the colony necessitated the maintenance of a large staff of Punjabi officers, whose 'pay, position, training and traditions', in the words of an experienced British official, were such that 'they are practically certain to make the greatest possible use of any opportuni-

[25] Fitzpatrick minute, 1 January 1905; Sec. of State Leg. Despatch 5, 3 March 1905, J & P 3075, 1905.
[26] PG to GI, 18, 22 January 1906; Ibbetson minute, 10 February 1906; GI Revenue Despatch 9, 24 May 1906, GI Revenue, May 1906, 43-44A; Sec. of State Leg. Despatch 150, Revenue and Statistics File (hereinafter R & S), 1439, 1906.
[27] This paragraph drawn from *Return of the Punjab Land Colonization Bill*, London, 1907, pp. 3-14.

ties they may have for extorting bribes'.[28] The fine system offered such an opportunity. Indian officials forced innumerable colonists to pay bribes for petty offences. The situation degenerated to the point that if a zamindar wished to visit his home village for a week, he frequently paid off the canal official so that his residency requirement would not be jeopardized. Insistence upon cleanliness and sanitation opened up additional paths for corruption. The graft became so oppressive that in 1903 a newspaper, the *Zamindar*, was started to publicize the colonists' plight.[29] The economic pressure had almost reached a breaking point by 1906. Following a fresh investigation of occupancy rights, many colonists had lost their land because they could not prove their residency status. Crop failure was the final blow. The boll worm destroyed the 1905 and the 1906 cotton crop, the chief crop in the irrigated areas.[30]

The provincial government did not take these factors into consideration primarily because of the administrative assumption that it knew what was best for the rural Punjabi. The men responsible for the measure and much of the earlier legislation—Denzil Ibbetson, James Wilson, and Charles Rivaz—conceived of the British as patriarchal defenders of the cultivating class. As Financial Commissioner, Wilson prepared the first version of the bill, Lieutenant Governor Rivaz supported it, and Ibbetson defended its principles in the Executive Council.[31] Wary of interdepartmental quarrelling and confident that the bill was the only means of preserving governmental authority, the Secretariat did not circulate the measure. The Lieutenant Governor and the Financial Commissioner realized that fines and corruption occasionally upset the colonists, but they attributed the memorials and newspaper attacks to 'unreasonableness' and incurable

[28] Popham Young note, in Minto to Morley, 3 July 1907, *Minto Collection*, National Library of Scotland, Edinburgh. All Minto correspondence hereinafter is from the Minto Collection.

[29] *Zamindar*, 18 August 1905, *Selections from the Punjab Vernacular Press* (Hereinafter *SPVP*), 1905, p. 215. Siraj-ud-Din Ahmad, a retired postal officer who ran the paper, had no previous connexion with agitation or public criticism of the British.

[30] Dobson, *Chenab*, pp. 13-14; confidential circular, 7 May 1907, and noting, PG Revenue file (Printed), 108A.

[31] The attitudes of these men are reflected in the following: Ibbetson letter to PG, 7 March 1889, GI Judicial, December 1891, 234—300A; Rivaz minute, 24 May 1899, GI Revenue, July 1899, 44-45A; minute, 1 September 1904, attached to PG Revenue (Printed), 108A; Wilson minute, 4 May 1907, GI Leg., June 1907, 4-8A.

defects in the administrative system.[32] Even if there were limited unrest, they noted, the colonists would soon realize that the British passed the seemingly harsh legislation only to protect the agriculturists. Punjabis had meekly accepted earlier legislation, and they would similarly accept the new colonization bill.[33]

By coincidence the Punjab Government also announced in November 1906 a drastic increase in the occupier rate (the charge on canal water) on the Bari Doab canal running through the districts of Amritsar, Gurdaspur, and Lahore. Water rates had previously been kept lower in these districts than in the western Punjab because the Punjab Government hoped a policy of leniency would ensure the loyalty of the Sikh Jats who supplied many recruits for the Indian army. The districts watered by the Bari Doab canal were the heartland of the *manjha*, the chief recruiting ground for Sikh soldiers.[34] The 1903 Irrigation Commission had questioned the wisdom of this sacrifice of revenue and convinced the efficiency-minded Curzon that the Punjab must raise its rates. The Punjab Government accordingly deputed an irrigation official to investigate. He suggested a sharp increase, which, after slight modification, was announced in the winter of 1906. Although the enhancements averaged 25 per cent, increases on cash crops such as sugar cane and on vegetable gardens bordering urban areas ran as high as 50 per cent.[35]

As in the case of the canal colonies, the local government was not fully acquainted with the peculiar circumstances in the areas affected by its measure. The canal had been over-extended, with the result that the zamindars were increasingly bitter about the irregular supply of water. The Chief Engineer had inaugurated in 1905 a redistribution programme to correct the problem, but the resulting confusion over which branches would be altered and bribery demands by subordinate officials only

[32] 'Report on the Punjab Press, 1903', GI Public, September 1904, 218B. A voluminous file, FCF 441/108A, is filled with clippings and comments on editorials and attacks on the government.

[33] Rivaz minute, 18 November 1906, KW, FCF 441/108A.

[34] PG to GI, 5377, 3 October 1890, GI Irrigation, March 1891, 25-30A. Also discussion in GI to PG, 511, 23 March 1891, same file; A.H. Bingley, *Sikhs*, Simla, 1899, pp. 29, 106.

[35] Sec. noting, PG Revenue, December 1902, 7-9A; Wilson minute, 8 August 1905, GI Revenue, June 1905, 11B; Sec. noting in GI Irrigation, August 1906, 6-8A.

intensified the discontent.[36] In addition, sharp rises in labour costs due to widespread deaths from plague and a series of poor harvests threatened the prosperity of the zamindars dependent upon canal irrigation.[37] The Chief Secretary, Edward Maclagan, tried to warn the Lieutenant Governor that the enhancements were inopportune. He noted that the irrigation experts had not taken into consideration human factors in recommending the increase in water rates, and that district officers in the Delhi division had questioned the wisdom of enhancements in their territory a year earlier. If district officers in the regions affected by the increases were given a similar opportunity to express their views, Maclagan argued, perhaps the government would find that the enhancements should be postponed until further study could be completed. Rivaz rejected his opinion. The Irrigation Department was in charge of the canal, the Lieutenant Governor concluded, and had taken a fair survey of local conditions. When the Irrigation Department defended the enhancements and maintained that they would be accepted without protest, Rivaz published the notification of the increases in water rates. He categorically refused to reopen the issue and denied that district officers had insight into the working of the irrigation project.[38]

In 1906 a local government which prided itself on good relations with the cultivating class was therefore embarking upon controversial legislation and an increase in taxation without adequate study of the problems involved and without consideration of rural reaction. The Punjab authorities self-confidently assumed that they knew what was best for the cultivator. Nothing could shake the latter's trust in British goodwill. Nevertheless, within four months a virulent and unexpected response to the two measures indicated that the miscalculations of the government had alienated the political support of the very class upon which it predicated continued rule. The Punjab countryside was about to burst into flames.

Rural Agitation and the Punjab Disturbances of 1907

The Chenab colonists saw the new bill as an unjust intrusion of govern-

[36] Notes by Lahore DC (22 July 1907), FC (10 October 1907), GI Revenue, June 1908, 15B.

[37] Bari Doab Committee Report, in PG to GI, 20RI, 21 January 1908, GI Revenue, June 1908, 15B.

[38] Maclagan minutes, 22 April, 2 May 1906; Rivaz minute, 3 May 1906, printed notes, PG Irrigation, December 1902, 7-9A. Also Walker minute, 10 October 1907, GI Revenue, June 1908, 15B.

ment into their daily lives. The bill also legalized the hated fine system and thereby widened the area within which officers could extract bribes. The editor of the *Zamindar*, Siraj-ud-Din Ahmad, and several prominent yeomen belonging to a farmer's association, the Bar Zamindar Association, initiated a systematic protest against the bill. Colonists held mass meetings and signed memorials denouncing the government proposals. Many of the meetings centred in the newly developed Gugera branch, where harsh residency and sanitation regulations as well as water scarcity had cut most deeply into the colonists' faith in British intentions.[39] On 29 January, for example, a meeting was held at Sangla with an estimated attendance of 3,000 highly indignant colonists. They passed resolutions attacking alteration of previous contracts, protested against 'illegalities committed by local officers in imposing and recovering coercive fines and penalties amounting to lakhs of rupees' and suggested that 'the same should be refunded and the repetition stopped'.[40]

The agitation was unlike any other agitation previously witnessed in the Punjab. The local government had for the first time aggrieved a large portion of the rural population, and in response the Chenab zamindars and their sympathizers organized mass demonstrations. The unrest was not limited to the colony. The leading agricultural families of the central districts who had relatives in the colony or aspired to a future grant added their voice to the protests. Disaffection also penetrated the Indian army. Ex-government servants or educated Punjabis living in the colony headed the agitation, not urban politicians. Ram Chand, a Lyallpur lawyer and landowner, Shahab-ud-Din, a Muslim lawyer with a yeomen grant, and Siraj-ud-Din Ahmad, a retired postal officer, led the Zamindar Association meetings. They sent out printed invitations before each meeting, released news to the local press, and got several rich yeomen to donate a 'war chest' to support colonists who refused to pay fines.[41] The *Zamindar* was in charge of publicity. The paper's editor was particularly adroit at printing and distributing cartoons and literature appealing to the generally illiterate colonists. An example of the effective propaganda was a picture of an

[39] Reports of meetings in *Civil and Military Gazette*, 13 and 31 January 1907; Popham Young note, Minto to Morley, 3 July 1907; G. De Montmorency minute, 28 March 1907, GI Revenue, October 1907, 13-25A.

[40] *Tribune*, 1 February 1907.

[41] De Montmorency minute, 28 March 1907; minutes by A. Diack (2 April 1907), A. Kettlewell (17 April 1907), notes attached to FCF 441/108A. Also comments in *Panjabee*, 12 December 1906; *Tribune*, 4 November 1906.

open-mouthed alligator labelled 'Colonization Act' preparing to swallow terrified and befuddled colonists.[42] Besides being a rural movement, the agitation was also differentiated from past political activity because of its non-communal nature. Hindus and Muslims were equally affected by the bill and mixed freely in protest meetings. Even the *Observer*, perennial critic of Muslim participation in politics, censured the measure as 'shaking public confidence' and called for demonstrations.[43] The culmination of this show of communal unity was a resolution passed on 3 February by 10,000 colonists at Lyallpur, the capital of the colony. The resolution called on Hindus and Muslims to refrain from acts offensive to the other community and to unite against British *zaloom* (oppression).[44]

Representatives from the Lahore Indian Association, a branch of the Indian National Congress, attended the Lyallpur meeting. Headed by Lajpat Rai, a leader of a Hindu revivalist sect called the Arya Samaj, the urban politicians came as spectators and had no role in planning the agitation.[45] The presence of Lajpat Rai and his associates at the rural demonstration, however, convinced the Punjab Government that they had masterminded the entire agitation. It therefore resolved that the rural areas were not seriously upset over the measure and passed the bill through the legislative mill. Rivaz was determined to pass the bill before his retirement in March, and forced the Select Committee to present a report without giving it time to study the accumulating number of protests and memorials. Several Punjabi members of the Committee did not even have a chance to meet the amending body before it presented the report. Rushed by Rivaz and dominated by official members, the Select Committee made no significant alterations in the legislation.[46] In the heated debate the government claimed that the bill was being 'misrepresented' by 'some who claim to reflect and direct public opinion'. The three Punjabi members of the Council disagreed and demanded a postponement of the debate, but Rivaz called for an immediate vote, and the bill passed with the aid of the official majority.

[42] *Zamindar*, 8 April 1907, *SPVP 1907*, pp. 102-103.

[43] *Observer*, 9 March 1907, *SPVP 1907*, p. 65.

[44] Reports in *Tribune*, 7 February 1907; *Panjabee*, 6 February 1907.

[45] Lajpat Rai, *Autobiographical Writings of Lajpat Rai*, ed. by V.C. Joshi, Delhi, 1965, pp. 119-20; Lajpat Rai to G.K. Gokhale, 29 January 1907, *Gokhale Collection* (Friends of India Society, on permanent loan to the National Archives of India, New Delhi).

[46] Rivaz minute, 18 January 1906, printed notes, FCF 441/108A. Discussion of bill from proceedings of legislative council, Vol. I, pp. 178-82.

Having convinced itself that urban agitators engineered the rural demonstrations, the Punjab Secretariat assumed that the unrest would immediately subside. The Financial Commissioner confidently prepared draft notifications to be distributed as soon as the Indian Government formally approved the Punjab's final venture into paternal legislation. Approval was considered a matter of form because the two governments had corresponded over the principles of the bill since 1902.[47] To the dismay of the Punjab Government, however, the agitation spread and became more virulent. Urban politicians helped the rural leaders prepare fresh demonstrations to influence the Government of India in its final decision on the bill. They called a monster meeting on the occasion of a livestock show at Lyallpur on 22 and 23 of March and publicized the affair with printed invitations and eight-page pamphlets in Punjabi accusing the British of treachery.[48] The Lieutenant Governor realized a week before the meeting that he had at least partially misjudged the seriousness of the agitation and attempted to conciliate the colonists with a notification defining the aims and the limitations of the act. Supposedly the circular would remove the half-truths being spread 'by outside agitators'. The circular had the opposite effect, for the organizers of the meeting painted the action as a weak attempt to gloss over the real intentions of the bill.[49] Approximately 9,000 colonists gathered at Lyallpur to protest against 'government tyranny'. A local editor, Prabh Dyal, opened the programme with an inflammatory poem entitled *Pagri Sambhal O Jatto* (Oh Jat, take care of thy turban). The turban was a symbol of self-respect for the warlike Jat zamindars, and the poem appealed to them to strengthen their resistance to the bill.[50] After a number of patriotic songs, Lajpat Rai and Ajit Singh, a bitter revolutionist from Lahore, made the major speeches. Lajpat Rai attempted to be moderate, but as frequently happened with his speech-making, the crowd's frenzy drove him to use phrases and ideas verging on what the British termed 'sedition'. Lajpat Rai claimed that the colony belonged to the King, not his servants the civil service. The colonists should not quietly accept the ICS's dictates. If Punjabis abandoned the law

[47] Douie minute, 5 April 1907, and noting, FCF 441/108A. Under the Indian Councils Act the Viceroy had to assent to a local act before it came into operation. *Practice and Procedure of the Indian Government*, Calcutta, 1913, pp. 177-80.

[48] Reports in *Zamindar*, 24 March 1907, *SPVP 1907*, p. 97; *Panjabee*, 20 April 1907, *Civil and Military Gazette*, 19 March 1907.

[49] Notification of 13 March 1907, and sec. noting, FCF 441/108A.

[50] Report in *Panjabee*, 27 March 1907.

courts, developed native industries, and took vows of unity, the British would have to back down.[51] Ajit Singh was less temperate. He spoke of the need for bloody sacrifice and demanded a boycott of all government posts. Colonists should also band together and not pay revenue.[52] Immediately after the meeting Lajpat Rai went on a lecture tour in the United Provinces, while Ajit Singh continued to roam the central districts whipping up enthusiasm for the anti-bill agitation and giving rousing speeches. He and his revolutionary friends hoped that the canal bill would be a tool with which they could spark an uprising against the British.[53]

There are no indications that Ajit Singh played a significant role in the canal bill agitation either before or after the March meeting, but he and his organization, the Anjuman-i-Mohibban-i-Watan (Society of Lovers of the Homeland) did engineer the demonstrations against the Bari Doab enhancements. A local zamindar association headed by Lahore aristocrats had already discussed the proposals, but serious organization of the Amritsar and Lahore farmers began only after Ajit Singh made their cause his cause. On 4 April he held a mass meeting of 12,000 zamindars.[54] Two days later Ajit Singh was cordially provided with a larger audience by the annual convocation of the Singh Sabha (a Sikh religious organization with strong rural backing) in Lahore. At a meeting on 6 April he called on the zamindars to disobey all British commands. The colonists passed resolutions supporting a boycott of British goods and started a committee to ensure that no zamindar paid the new water rate. The price of disobedience was social ostracism by the offender's caste or a fine of Rs.500.[55]

[51] Lajpat Rai speech, appended to history sheet, GI Home-Political File, August 1907, 148-235A (hereinafter cited as 'Disturbances File'). Also, *Autobiographical Writings*, pp. 112-22. Lajpat Rai and Ajit Singh were enemies and had not planned the meetings or the agitation.

[52] Ajit Singh's speech, appended to history sheet, 'Disturbances File'. Also Ajit Singh, 'Autobiography' (unpublished ms., V.C. Joshi, National Archives of India), pp. 10-11.

[53] Lajpat Rai history sheet, Ajit Singh history sheet, 'Disturbances File'. Also Ajit Singh autobiography, pp. 13-14.

[54] Telegram in R & S 1036, 1907. It is unclear why Ajit Singh waited until April to agitate. Other papers and organizations had taken up the cry a month earlier.

[55] Ajit Singh history sheet and proceedings of the meeting, 'Disturbances File'. Meeting reports in *Tribune*, 11 April 1907; *Panjabee*, 1 May 1907.

The British Response to Unrest in the Punjab

The new Lieutenant Governor of the Punjab, Denzil Ibbetson, was convinced that the agitation had been engineered by 'Lahore pleaders' and after taking office in early March he began to examine measures for stemming the protests.[56] His first step was the publication of a resolution postponing for a year the rate enhancements on the ground that crop conditions and plague made the increases unwise. This did not stop the agitation.[57] In fact, the unrest became more serious with riots in Amritsar, Lahore and Rawalpindi. Ibbetson then re-examined the situation and decided that the rural demonstrations, talk of sedition in the army, and rioting were the result of a secret plot to overthrow the British government. He consequently sent the Viceroy, Lord Minto, a lengthy minute on the deteriorating political condition of the Punjab and requested extensive executive powers to quell a rebellion.[58] Sedition, according to Ibbetson, took two directions in the Punjab. First, Ajit Singh was trying to spread disaffection among the troops and the students and secondly the 'fomenters of unrest' were corrupting the 'yeomanry'. Ibbetson maintained that the Colonization Bill and Bari Doab agitation had been entirely worked up. The Arya Samaj, under the leadership of Ajit Singh and Lajpat Rai, had succeeded in getting the rural population to suspend revenue payments. The next step would be open revolt. Ibbetson therefore requested the power to ban public meetings, to seize presses, and to arrest individuals inciting zamindars not to pay revenue. He also asked that Ajit Singh and Lajpat Rai be deported 'to strike terror into the minds of those concerned'.[59]

Poor channels of communication contributed to Ibbetson's mistaken impression concerning an overall plot to overthrow the British. He had only just returned to the Punjab after an extended absence of fifteen years and, without personal knowledge, he was dependent largely upon his men and the CID (Criminal Intelligence Department) for assessment of the unrest. The local CID drew its news chiefly from informants who exagger-

[56] Ibbetson to Minto, 23 March 1907.

[57] Resolution in *Civil and Military Gazette*, 28 April 1907; comments in *Panjabee*, 1 May 1907.

[58] Ibbetson minute, in PG to GI, 695, 3 May 1907, 'Disturbances File'.

[59] Ibbetson minute. The deportations were to be carried out under Regulation III of the 1818 Bengal Regulations, which gave the Governor-General in Council the right to deport Indians to a place in India or outside India on the grounds that they endangered the peace.

ated and hopelessly confused what was actually happening. Ram Bhaj Datta, the conservative barrister who headed one of the Arya Samaj branches, was reported to have called on students to kill Englishmen. Lajpat Rai and Ajit Singh, in fact antagonistic to one another, if not active enemies, were pictured as co-conspirators who organized the outbreaks through secret societies.[60] The district officer reports on which Ibbetson heavily relied for the minute were equally misinformed. The favourite method of assessing what had happened in the districts was to call in prominent Muslim or Hindu aristocrats and to ask them about the agitation. The Muslims not surprisingly accused the Hindus of spreading sedition and lies, while the Hindu aristocrats (who generally hated the liberal Arya Samaj and Lajpat Rai) said that the Aryas were behind the disturbances.[61] In one key district, Rawalpindi, an officer who hated lawyers and Arya Samajists, P.D. Agnew, wrongly charged five pleaders with planning the local riot under the direction of Lajpat Rai.[62]

Ibbetson also misjudged the situation because of his attitude towards political agitation. He believed that since the British had the 'contentment of the masses' as a basic goal, the zamindars would never turn against the rulers unless they had been misled. Rural agitation could never be genuine—it had to be engineered by urban politicians for unscrupulous reasons.[63] Ibbetson would not admit that the Colonization Bill was a real grievance. He felt that intervention in the canal colonies was necessary for the good of the people, just as necessary as the Alienation Bill of which he had personally been the author.[64] To question the government once it had carefully weighed an administrative decision was pure and simple sedition. A patriarchal ruler schooled in the Punjab tradition of the 'iron fist', Ibbetson was singularly incapable of differentiating between 'constitutional agitation' such as public meetings and methods such as rioting and inciting

[60] Drawn from the first three 'Information' telegrams sent to Morley. The telegrams were based on CID reports, principally those of 4,5 and 6 May (in appendices to notes, 'Disturbances File').

[61] Enclosures to PG to GI, 695, 3 May 1907, 'Disturbances File'. Also Ibbetson to Maclagan, printed notes, FCF 441/108A.

[62] Agnew letter to PG, 8 May 1907, GIPOL, (Home-Political), July 1907, 8A. Agnew was an intimate friend of Ibbetson.

[63] Ibbetson to PG, 7 March 1889, GI Judical, December 1891, 234-300A; Ibbetson to Minto, 23 March 1907.

[64] Ibbetson minute, 26 April 1907, printed notes FCF 441/108 AKW; PG to GI, 22 January 1906, and Ibbetson note, 10 February 1906, GI Revenue, May 1906, 43-44A.

troops to mutiny. Any challenge to authority and British *izzat* (honour and standing) was treason and must be met with absolute repression.[65]

Ibbetson's minute was totally unexpected. Minto and Lord Kitchener, the Commander in Chief of the Indian Army, had seen reports of rumoured sedition and tampering with the loyalty of the native troops, but they had not imagined that the Punjab situation was as critical as Ibbetson portrayed.[66] Minto could not legally give the Lieutenant Governor the powers he requested without the consent of his Executive Council, and so he circulated the Punjab letter and called a meeting for 6 May.

Personal relations and the time-honoured principle of supporting the man on the spot eventually resulted in the Council's approval of the deportation of Ajit Singh and Lajpat Rai. The Council had little information except for the Punjab letter and its enclosures. The local government had been able to support its case only with a few reports of speeches, history sheets of Lajpat Rai and Ajit Singh, and excerpts from district officer correspondence. Because the Imperial CID was prevented by statute from collecting information within the areas falling under the jurisdiction of the provincial governments, the CID and the Council were virtually dependent upon the Punjab police agency for news and analysis.[67] The problem was whether to trust Ibbetson's interpetation and back the Punjab Government on the basis of skimpy material, or not to heed the plea for urgency and call for a detailed investigation. The stakes were high. The Punjab was the recruiting ground for more than a third of the Indian army. If disaffection had spread throughout the army or among families of soldiers, a revolt might be imminent.[68] Rumours of a pending military uprising were particularly rampant, for 10 May was the fiftieth anniversary of the 1857 Mutiny. On the other hand, deportation and extension of executive power would, as one Council member noted, create 'great commotion' in Parliament and among Indian politicians whom Minto was

[65] As Minto later observed, 'Ibbetson appears to me to entirely misunderstand the position. He appears to assume that we can stamp out the unrest. This we can never do. It has come to stay, in the shape of new ideas and aspirations of which everyone who has thought seriously over the subject ought to be aware. He confused this with sedition...', Minto to Morley, 5 November 1907.

[66] Minto to Morley, 16 May 1907; Kitchener to Minto, 5 May 1907; Minto to Lord Roberts, 6 June 1907.

[67] *Punjab CID Manual*, Lahore, 1915, pp. 7-8, 43.

[68] Approximately 30,000 Sikhs (23 per cent of the army) and 18,000 Punjab Muslims (13 per cent of the army) were in the army in 1907. Statistics from *Indian Army List, 1907*.

trying to woo with constitutional reform.[69] Edward Baker, the future Lieutenant Governor of Bengal and a liberal friend of Surendranath Banerjea, distrusted Ibbetson and urged his fellow Council members to oppose the Punjab request. Harvey Adamson also challenged the Punjab case for deportation and suggested instead a stronger sedition law. Erle Richards, a close friend of Ibbetson, maintained that he alone must be the judge of the situation. Minto's chief advisors, Kitchener and Herbert Risley, the Home Secretary, urged full support for the local government.[70] Minto had two personal reasons for initially defending Ibbetson's demand. Ibbetson had been a close associate of the Viceroy, and Minto believed that he knew more about the Punjab, and possibly India, than any other civil servant.[71] Moreover, his judgement was influenced by the earlier decision not to back Fuller, the Lieutenant Governor of East Bengal and Assam. Fuller had bullied Bengali nationalists and aggravated political tension by banning schools connected with agitation from participation in the Calcutta University examinations. When John Morley, the Liberal Secretary of State, attacked Fuller, Minto first defended his 'man on the spot', but subsequently decided that 'imperial interests' were threatened by this 'dangerous man' whose retention would produce a 'conflagration'.[72] Ibbetson had been an intimate friend of Fuller and defended him with the argument that if the local government were not supported, the prestige of every Bengal civil servant would be impaired. Ibbetson noted that as future Lieutenant Governor of the Punjab he might be 'confronted by a state of things similar to that which Sir B. Fuller has to face'. If Fuller were not backed, Ibbetson said he feared the Government of India would not stand behind his government.[73] After lengthy correspondence with Ibbetson, who tried to convince his chief that a resignation would be a 'political blunder', Minto finally forced Fuller to resign.[74] While Minto's decision shocked Ibbetson, he had been touched by the Viceroy's profuse assurances that his opinion was highly valued. Minto promised Ibbetson full

[69] Erle Richards minute, 5 May 1907, 'Disturbances File',

[70] Adamson minute (5 May), Baker minute (6 May), Finlay minute (6 May), Kitchener minute (6 May), 'Disturbances File'. Also Minto to Lady Minto, 9 May 1907, Minto to Morley, 8 May 1907.

[71] Minto to Morley, 8 May 1907.

[72] Minto to Valentine Chirol, 18 May 1910.

[73] Ibbetson to Minto, 25 June 1906.

[74] Richards Diary, 6 August 1906. *Richards Collection* (Eur. Mss. F. 122, India Office Library).

support in whatever action he might have to take in the Punjab.[75] Mindful of his promise and the dangers inherent in frequent failure to back the subordinate governments, Minto convinced Adamson and Baker that the Punjab evidence justified deportation. Risley prepared a carefully worded telegram drawn heavily from the Ibbetson minute, the Council approved it, and the telegram was sent to Morley and the India Council in London.[76] The Executive Council notified Morley two days later that in addition to deportation it planned to extend an ordinance on public meetings to selected Punjab districts. The ordinance empowered a magistrate to cancel a meeting without announcing his reason. Plans for meetings had to be given to the government seven days in advance, and persons planning or attending illegal meetings were subject to six months' imprisonment.[77]

John Morley had been following events in the Punjab with great interest. The exaggerated Reuter report on riots in Rawalpindi aroused his fear that the Indian Government might do something hasty, and he telegraphed Minto on 6 May not to take any steps without consultation.[78] The telegram crossed with the news of the deportation of Lajpat Rai. Morley was 'suspicious' of the decision, but trusting Ibbetson and lacking first-hand information, he supported Minto.[79] The telegram on the meeting ban arrived on 10 May. Morley angrily threw the message down, screaming 'No, I can't stand that: *I will not have that*'. His private secretary and advisor, F.A. Hirtzel, convinced him that the home authorities were obligated to back Minto during the emergency or until they received a detailed report. Morley bitterly acquiesced.[80]

In the Punjab Ibbetson was frantically trying to calm the European population and to pacify the zamindars. Officers held darbars in each district and explained the Colonization Act. Because the colonists and agriculturists felt they had been wronged, however, they continued to hold meetings and draft petitions.[81] With meetings banned in five major

[75] Ibbetson to Minto, 21 and 24 July 1906; Minto to Ibbetson, 21 and 23 July 1906.

[76] Order in Council, 6 May 1907, and 8 May telegram, 'Disturbances File'.

[77] H. Risley minute, 11 May 1907, 'Disturbances File'.

[78] Telegram to Minto, 6 May 1907.

[79] Hirtzel Diary, 8 May 1907, Home Misc. Series 864, India Office Library; Morley to Minto, 9 May 1907.

[80] Hirtzel Diary, 10 May 1907.

[81] Printed circular, 7 May 1907, and noting by FC Secretariat, PG Revenue (Printed) 108A; CID report, 13 June 1907, West Pakistan CID Archives, Punjab CID Reports of 1907.

districts, the urban agitations quietened. While corresponding with Minto about further executive power and reassuring the zamindars of government intentions, Ibbetson discovered that his persistent lip ulcer was malignant. Ibbetson informed Minto that he must leave immediately for London to consult specialists; he hoped he would be able to return in three months.[82] Minto agreed and appointed Gordon Walker, the senior Punjab officer, as acting Lieutenant Governor. With Ibbetson gone and an inexperienced and untested subordinate holding the reigns, initiative for further action in pacifying the Punjab passed to Minto and his Council.[83] After the departure of Ibbetson, the Government of India considered a move from coercion to conciliation.

Minto continued to believe for several days following the deportation that the disturbances were part of a sinister plot. Police informants in the Punjab—whose continued employment in effect depended upon their ability to unearth or to fabricate sedition—sent in incredible tales of pending mutiny and an Afghan-Russian-Punjabi entente to seize the border districts. The Home Department unquestioningly sent the reports to Morley.[84] At approximately the time of Ibbetson's departure, however, the Home Department Secretariat and the Viceroy began to suspect that much of the Punjab information was inaccurate. Minto baulked at emphasis upon a Russian plot in a telegram of 17 May, for example, while the embarrassed Home Department Secretariat frequently had to admit that much of the supporting evidence sent home to London had been alarmist.[85] Kitchener decided to send out his own investigators to determine the depth and causes of disaffection among Punjabi soldiers. The officers talked with the sepoys and found that existing tension was due primarily to the Colonization Bill, which had affected either soldiers or their families. On the basis of these reports and scattered memorials from Punjabi soldiers, Kitchener urged Minto to veto the controversial measure. Obviously press laws and

[82] Ibbetson to Minto, 13 May 1907.

[83] Minto to Morley, 18 June 1907; Minto to Lady Minto, 15 May 1907.

[84] CID report, 15 May 1907, and resulting telegram, 16 May 1907, GIPOL, July 1907, 29-117B. Risley, the Home Secretary, also perpetuated the mistakes because he believed that Punjabis and Bengalis had joined in an all-India plot.

[85] Minto minute, 17 May 1907; telegram to Morley, 12 May, later corrected by 21 May telegram and Punjab demi-official letter, 19 May 1907, GIPOL, July 1907, 29-117B. The latter issue was over a Reuter telegram which the Punjab somehow used as a source of news for what was happening. The telegram claimed that thousands of 'rustics' had invaded Lahore.

deportation alone would not strike at the root of military unrest.[86]

Minto had begun to weigh the vetoing of the bill before receiving Kitchener's suggestion. Nevertheless, he postponed final judgement until his Council could note on the issue. The Revenue Secretary and a former Punjab man, James Wilson, vehemently defended the legislation, maintaining that 'most colonists' were not disturbed by the new restrictions. Even if they were, the 'colonists who have been brought from narrow poverty in their old homes and placed, by the beneficence of Government, in a position of prosperity, unimagined in their fondest dreams' had no 'legal or moral right' to the land.[87] The Revenue Member, Miller, claimed that Minto must support the men on the spot, but admitted that assent to the bill might accelerate disaffection.[88] Minto was confused by the issues and alternatives, and therefore he called for fresh noting by all members of his Council.[89]

Minto was in a difficult position. The final decision on assent was his alone, for under the India Councils Act the Governor-General and not the Governor-General in Council had the power and the responsibility to accept or reject local legislation.[90] There were growing indications that the Punjab situation did not resemble the image presented in the Punjab correspondence. On the other hand Minto still respected Ibbetson's opinion, and just before leaving India Ibbetson had warned that a veto would only open the path for further demands. Realizing that Minto was considering a veto, however, Ibbetson suggested a compromise. The Lieutenant Governor was empowered by section 35 of the Act to suspend the measure in notified areas. Ibbetson proposed that he suspend the Act immediately after assent, thus giving both governments further opportunity to investigate the charges against the legislation without undermining the izzat of the local authorities.[91] Weighing against Ibbetson's offer was Kitchener's demand that only a veto would stop disaffection among his soldiers.[92] Morley also favoured a veto. He had consulted with T.W. Holderness, the India Council revenue specialist, and Hirtzel, and then

---

[86] Kitchener to Minto, 12 May 1907. Also memorials to Kitchener, in appendix to GI Leg., June 1907, 4-8A.

[87] Wilson minute, 4 May 1907, GI Leg., June 1907, 4-8A.

[88] Miller minute, 6 May 1907, GI Leg., June 1907, 4-8A.

[89] Minto minute, 14 May 1907, GI Leg., June 1907, 4-8A.

[90] *Practice and Procedure of the Government of India*, pp. 29-30.

[91] Ibbetson to Minto, 14 May 1907.

[92] Kitchener to Minto, 12 May 1907.

telegraphed Minto that an assent would be 'impolitic'.[93] The additional material coming into the Secretariat summer office at Simla tended to support such action. For example, Minto received a Punjab revenue letter with enclosures suggesting that the colonists themselves had organized the demonstrations and were sincerely afraid of the bill.[94]

Minto wrote Morley on 16 May that a veto would be preferable to compromise. The Government of India had to act decisively to regain the trust of the Punjabi rural class. He privately informed his wife a day earlier that the bill was in his opinion a gross mistake. Although Ibbetson and perhaps other local governments would denounce the Indian Government's refusal to support subordinates, 'it is a question whether it may not be better for me to "play to the gallery" for now, and disallow the Act off my own bat on the grounds that the occasion is so urgent one must do something to impress upon the Native mind as to our fair play'.[95]

The Executive Council argued heatedly over the issue of veto. Wilson's defence of the Act was so impassioned that Minto completely discounted his opinion.[96] Miller called for a fresh investigation, while Baker denied that any legitimate reason existed for postponing settlement of the matter. Baker added that because the Punjab Government rushed the bill and made no effort to weigh public opinion, he was 'out of sympathy' with defending their action or their izzat. Adamson noted cynically that the Act was arbitrary, 'but this was the usual nature of Punjab Government agricultural legislation'. Finlay urged backing the man on the spot. Kitchener claimed that the image of British justice required the veto of 'a poor piece of legislation'. Ibbetson's friend, Richards, concluded the noting by supporting assent, suspension of the Act, and a thorough inquiry.[97]

Minto weighed the opinions for five days and then vetoed the Act on 26 May. Despite his personal attachment to and sympathy with Ibbetson, he decided to risk weakening the power of the provincial government in order to regain the loyalty of the Punjab rural population and the Punjab contingent in the army. Minto based his minute of veto upon the earlier minutes of Baker and Kitchener, whom he considered to be the best men

[93] Morley telegram to Minto, 14 May 1907; Hirtzel Diary, 14 May 1907.

[94] Enclosures to PG to GI, 60 RS, 29 April 1907, GI Revenue, October 1907, 13-28A.

[95] Minto to Lady Minto, 15 May 1907.

[96] Minto to Morley, 16 May 1907; Wilson minute, 14 May 1907; Minto minute, 7 July 1907; GI Revenue, October 1907, 13-28A.

[97] Minuting, 14-20 May 1907, GI Leg., June 1907, 4-8A.

on the Council. The Governor-General said the bill was so poorly worded and the principles so questionable that for the first time the politically inert Punjab had been torn by agitation.[98] Acting decisively 'to impress the public imagination with the absolute justice of our intentions', Minto denied the accusation that he was giving way to agitators. The bill was a bad one, and the Government of India dare not back the mistaken policies of subordinate officials:

My grounds for refusing assent are based on the belief that we have found ourselves about to be committed to the approval of a very faulty piece of legislation—legislation which would be inadvisable at any time, but which at the present moment, if it became law, would add fuel to the justifiable discontent which has already been caused—whilst the appearance of surrender to agitation, should any portions of the public entertain such reasoning, would in my opinion be far less dangerous than to insist upon enforcing the unfortunate legislation proposed upon a warlike and loyal section of the Indian community.[99]

Minto's minute was clearly written for publication. The protests of the Punjab Government and Wilson made the Viceroy change his mind, however, and all the public saw of the intense ten-day debate was the simple fact that the Colonization Bill would not become law.[100]

The Punjabi agriculturists' response to the veto was instantaneous. The Punjab vernacular press and resolutions from public meetings praised the action as a vindication of British justice. The zamindars did not understand how or why the bill had been repealed—they only knew that the threat had been removed. Many attributed the veto to a change of heart by Denzil Ibbetson. Ibbetson returned in August to find himself a hero. As the *Zamindar* observed, contributions were even collected to build a statue to the Lieutenant Governor in Lyallpur.[101] When the former Colonization Officer, now Settlement Commissioner, Popham Young, visited the Chenab colony ten days after the veto, he was greeted with garlands and pledges of undying loyalty. The Bar Zamindar Association and the pleaders who led the agitation apologized for their action and praised the

[98] Minto to Lady Minto, 20 June, 5 July 1907. Official minute, 26 May 1907, GI Leg., June 1907, 4-8A.

[99] Minto minute, 26 May 1907, GI Leg., June 1907, 4-8A.

[100] Punjab unofficial letter to GI, 5 June 1907; Wilson minute, 10 June 1907; Minto minute, 11 June 1907, GI Leg., June 1907, 4-8A.

[101] *Zamindar*, 1 June 1907, *SPVP 1907*, p. 222. Also comments in *Tribune*, 15 June 1907; H. Stuart minute, 2 July 1907, GIPOL, August 1907, 113A.

integrity of the British government.[102] In his subsequent report on the colony, Popham Young confirmed that the burst of loyalty toward the British raj was sincere. Nevertheless, he warned that the rulers must now study conditions in the colony and remove perennial grievances. The veto had prevented an explosion, but the government had the further task of determining long-range policy toward the colonists and the Bari Doab rates.[103] Minto accordingly called for a thorough investigation. A commission of inquiry would be called, he informed the Punjab Government, because 'there has been much, verging almost on the tyrannical, in the Government of the Punjab that calls for careful inquiry'.[104] The Bari Doab question was also to receive attention. Ibbetson suspended the enhancements, but the matter had to be finally settled.

Aftermath of the Disturbances

The Government of India informed the local government in late May that it would call a committee consisting of a Punjab representative, a military officer, and a revenue officer from another province to study several issues connected with the Colonization Act and canal administration in general: the retrospective aspects of future legislation, the sufficiency of existing laws, penalties, and whether the rule by a colonization officer and his staff had been arbitrary and overbearing.[105] The officiating Lieutenant Governor wrote to Ibbetson and complained that Minto planned to judge the Punjab Government before the public. He also informed the Indian Government that his administration did not require an investigation because the colony was a local problem. If the Government of India interfered, the prestige of the Punjab civil service would be further damaged. In no circumstances should a military man be on the committee.[106] The Punjab reply sparked another Council debate. The Viceroy and his associates almost unanimously agreed to overrule the local government in view of the mounting evidence of maladministration. Baker, Adamson, and Miller vehemently demanded that the Punjab view of the matter be contested—as Adamson phrased the Council position, 'the other

[102] Popham Young minute, enclosed in Minto to Morley, 3 July 1907.

[103] *Ibid.*

[104] Minto to Morley, 5 June 1907; GI to PG, 1027, 29 May 1907, GI Leg., June 1907, 4-8A.

[105] GI to PG, 995/205-3, 24 June 1907, GI Revenue, October 1907, 13-28A.

[106] PG to GI, 882, 2 July 1907, *ibid.* Also Ibbetson to Minto, 27 June 1907.

side of the shield' had to be seen.[107] Only Wilson persistently fought the decision with the argument that, but for outside agitators, there would have been no trouble. Minto had lost all patience with the Punjab point of view and sarcastically replied that the Indian Government was therefore indebted to the agitators. Since the Punjab Government did not know what was happening, he welcomed authentic information and advice from any source.[108] Despite its evident displeasure with the Punjab officials, however, the Council still hesitated at pushing the local government before the return of Ibbetson. Owing to Council pressure and his own mixed feelings towards the ailing head of the government, Minto agreed to suspend discussion about a committee for a month.[109] Ibbetson returned in August, and he and Minto reached a compromise. Minto sympathized with his plea that the Punjab Government should not be 'publicly arraigned and placed in the dock, on its trial, in its own province'.[110] The committee would therefore not judge past administration but only outline defects and indicate the direction for new legislation. Minto also agreed to keep the final report confidential.[111]

While the two governments spent months preparing a directive for the committee and choosing members acceptable to each other, Ibbetson appointed a small committee to study the water rate problem. The committee reported in January 1908 that much of the agitation had been justified by poor water supply, unwise judgements on increases, and crop failure. Future enhancements should be undertaken concurrently with revenue settlements and not separately. The government should fearlessly face the water distribution problem by stopping extension of the canal and ending reduction of water supply in existing branches. Even at the cost of revenue the British should honour the pledge to furnish water to zamindars currently relying upon irrigation.[112] The Punjab Government prepared a draft resolution which incorporated most of the committee recommendations. The resolution stated that because of bad crops and water shortage,

[107] Minutes by Miller (13 June), Baker (18 June), Adamson (19 June), *ibid.*
[108] Minutes by Wilson (3 July), Minto (7 and 11 July), *ibid.* Also Minto telegram to Morley, 15 July 1907.
[109] GI to PG, 1121, 15 July 1907, *ibid.*
[110] Ibbetson to Minto, 27 June 1907; PG to GI, 1764S, 19 August 1907, *ibid.*
[111] PG to GI, 1764S, 19 August 1907; GI to PG, 1380, 8 September 1907; Minto minute, 31 August 1907, *ibid.*
[112] Report of committee, in PG to PG, 20 RI, 21 January 1908, GI Revenue, June 1908, 15B.

the enhancements would be dropped until the next revenue settlement.[113]

The Government of India accepted the committee's findings but refused to permit the local government to print their resolution. Miller and Minto felt that the real reason for cancelling the enhancements had been that they were defective. The local authorities were not prepared to admit this publicly, while the excuse about bad weather was an obvious falsehood: 'We have made a mistake; that is admitted in the Punjab letter; our attention has been called to it by agitation, and we mean to retrace our steps. The explanation about bad seasons will deceive nobody'.[114] The Council agreed and ordered the Punjab Government to announce its decision without an explanation.[115]

A more significant reversal of policy soon followed. The Punjab Canal Colony Committee presented its report in the winter of 1908. The Committee had toured the central Punjab for several months, circulated questionnaires, interviewed over two hundred zamindars, and listened to complaints in large open air meetings. It found that agitation had disappeared, and the attitude of the colonists was now 'moderate'.[116] The Committee recommended that full proprietary rights be given to the colonists and that from 1909 onwards the chaks should be integrated into the regular administrative system. The colonists who were interviewed claimed that they previously considered themselves proprietors and wanted official title to their land. The Committee agreed. The Punjab Government could no longer expect to treat the colonies as a special preserve for experimentation and paternalism. When the colonists showed that they were good tenants, they should be given an opportunity to purchase the land. Afterwards the regular laws should cover them, not the patriarchal decisions of an officer unfettered by the courts.[117] Fines were preferable to expulsion, but the committee suggested relaxation of sanitary requirements. The ordinary laws of succession should also become operative. If a tenant died before acquiring property rights, the government should permit his heirs

[113] Draft resolution in PG 20 RI, 21 January 1908, *ibid*.

[114] Minutes by Miller (19 April), Minto (18 May), GI to PG, 1162-1, 25 May 1908, *ibid*.

[115] Order in Council, 20 May 1908, *ibid*.

[116] Report of committee, GI Revenue, April 1909, 1-4A. The remainder of the paragraph is drawn from this report.

[117] This innovation had been anticipated and favoured by the Indian Revenue Department. For example, Miller minute, 13 June 1907; Minto minute, 7 July 1907; GI Revenue, October 1907, 12-28A.

to take over the land. Finally, future administrators must protect the colonists from graft.

The Punjab Government understandably denounced the recommendations that colonists should own the land and that paternalism as a system of administration must be abandoned in the colonies. It argued that the British should remain the landlords and guardians of the people.[118] Minto's Council, however, had finished listening to the local government and unanimously demanded a bill along the lines suggested by the Committee.[119] Miller's comment on the Punjab attitude summarized the viewpoint of the entire Council:

> Are we to keep these colonies for ever as a sort of enclave in the province governed by special rules of its own? Perhaps this might have been possible if the Punjab had not alarmed the colonists to such an extent by its system as to lead to a very serious agitation. That point having been reached there seems to me to be nothing for it, but to place the colonists on the same footing as far as may be as their brethren in the districts from which they have come.[120]

The Council's disgust with the Punjab revenue and administrative ideas was also reflected in the noticeable change within the personnel of the Imperial Revenue Department. From the appointment of Denzil Ibbetson to the Revenue Secretariat in 1894 until the retirement of James Wilson in late 1907, the Department had been 'Punjabized'. The series was broken sharply and self-consciously in 1907, and Punjab officers never again regained their once prominent position in Indian revenue affairs.[121]

The Punjab pushed its ideas on administration to the breaking point. The resulting agitation caused the Government of India to denounce patriarchal rule, and it also cost the Punjab a hold on revenue policy. Without a single spokesman on the Executive Council, the local government had to accept the dicates of the Viceroy and his associates. A new committee prepared a bill closely resembling the recommendations of the previous committee. Colonists could buy land rights after fifteen years residence, penal codes were lightened, and residency requirements were abolished for all yeomen tenants. Revenue decisions could be appealed to

[118] PG to GI, 2141S, 21 August 1908, GI Revenue, April 1909, 1-4A.
[119] Order in Council, 15 January 1909, and prior noting, *ibid.* The death of Ibbetson and the elimination of his influence on the Council undoubtedly strengthened its attitude toward the Punjab authorities.
[120] Miller minute, 4 January 1910, GI Revenue, May 1910, 10-12A.
[121] Adamson minute, 19 June 1907, GI Revenue, October 1907, 13-28A. Also Minto to Morley, 3 July 1907.

the civil courts.[122] The Punjab vainly fought the decision, but eventually the bill reached the local legislative council in the revised form. Following a brief and non-controversial introduction of the bill, it was sent to a select committee and circulated for almost two years. As the new Lieutenant Governor, Louis Dane, commented upon its enactment in 1912, 'all possible suggestions' had been incorporated into the final act.[123] A measure formulated in 1902, the cause of rural unrest in the Punjab, and the source of endless controversy and intra-government manoeuvring, finally became law.

Concluding Remarks

A serious failure of communication between ruler and ruled and between the Indian and Punjab Governments lay at the bottom of the 1907 disturbances. The local authorities made a number of administrative blunders without attempting to gauge the effect of the measures and without explaining official intentions. The Government of India was normally a check against such mistakes, but in the canal colony affair it defended the local government because of Punjab influence on the Executive Council. Rivaz and Ibbetson saw no reason to consult Punjabis about administrative matters or legislation. Living in an administrative ethos inherited from the days of John Lawrence, they assumed that since the British knew what was right for the zamindar, he would, as in the past, accept the dictate unquestioningly. This assumption blinded the Punjab Government to the spreading unrest. When the unrest pressed itself on Ibbetson's attention in the form of demonstrations, rioting, and disaffection in the Indian army, the Lieutenant Governor could not differentiate between real grievances and the efforts of Ajit Singh. He unintentionally misled Minto and his Council. Minto initially supported Ibbetson because he did not have first-hand information on the agitation. His Council's first reaction was to honour the principle of backing the man on the spot. Minto's relationship with Ibbetson and Ibbetson's friendship with several Council members significantly reinforced this tendency. Upon receipt of better information, however, it became evident that the Punjab authorities had muddled the affair. The resulting debate over possible remedies brought into focus the conflicting assumptions and goals of the two

---

[122] Committee report, PG to GI, 633S-RA-1, 23 June 1909, GI Revenue, December 1909, 10-11A.

[123] *Proceedings of the Punjab Legislative Council*, 1912, p. 103.

governments. The Punjab believed an admission of failure would irrevocably weaken its authority and intensify the unrest. Minto and his Council, on the other hand, realized that the unrest was genuine and threatened the army. Minto placed more importance upon reviving the zamindar's faith in British integrity and goodwill than upon a possible weakening of local influence. Still without full information on the agitation or the causes for the demonstrations, Minto agonizingly decided on a veto. His judgement proved correct, and the rural unrest subsided.

The handling of the disturbances was but one in a series of incidents in the daily task of governing the British Empire in India. It illustrates that when the Governor-General and his Council possessed information suggesting that a grievance was real and threatened the peace of a region, they were ready to remove the source of discontent at any cost. No price was too large to pay for the political stability of the subcontinent, nor the price of forcing a subordinate government to lose face nor the price of vetoing plans for rural reconstruction. Whether the Punjab programme of paternalism and agrarian reform would have helped significantly towards preventing the growing indebtedness in the province was irrelevant to Minto and the Executive Council. The Punjab had pushed too hard, and the Government of India backed down in the face of the resulting agitation.

Three lessons *emerged* from the 1907 experience. First, Morley never again fully trusted the Indian Government. Its mistakes over the disturbances exposed him to a year of parliamentary ridicule. Thereafter the vain Secretary of State with pronounced liberal leanings scrutinized every act of repression suggested by Minto and frequently vetoed his plans. The Government of India also became wary of reports from subordinate agencies and governments. It particularly watched the Punjab Government for several years and prevented local authorities from putting patriarchal principles into practice. Secretariats, Councils and Viceroys frequently changed, however, and the mistakes of 1907 were gradually forgotten. That the supreme government eventually relaxed the grip on the Punjab during the First World War helps to explain how Michael O'Dwyer—the man so closely resembling Ibbetson that the dying Lieutenant Governor suggested *him* as a successor—was able to establish a reign of terror in 1919. The third and most vital lesson went unheeded. The government continued to ignore the problem of trying to rule India with faulty channels of communication. On the one hand this meant that the British were unable to explain fully their goals and intentions, with the result, as in the 1907 disturbances, that rumour and the spread of suspicion threatened to create unrest. The fears of the colonists effectively prevented further attempts to improve their lot, just as breaks in communication and misunderstanding contributed to the failure of land reform and the limited success of the agricultural banks programme. The British were also frustrated in their efforts to look behind the native curtain, to assess the views of the Indian people and their probable response to official action. This lack of basic information was particularly important as the British tried to counter the growing force of Indian nationalism. Although the CID became more efficient, the Government of India continued to rely upon subordinate agencies and governments for news and political reports. Because at each level of administration there was opportunity for personal bias and misinterpretation, the type of mistakes made in 1907 were to be repeated again and again. More often than not, however, the Viceroy and the local authorities corrected the mistakes before they *sparked* a conflagration. Frequently cut off from actual events, the *men* who ruled India *made* command decisions based upon inadequate information—the Indian empire in the last analysis rested upon their ability to make an educated guess.

Chapter Eight

State Forestry and Social Conflict in British India*

MADHAV GADGIL and RAMACHANDRA GUHA

Introduction: The Ecological Perspective

Geographically speaking, India is a land of tremendous diversity—from
bare and snowy mountains in the north to tropical rain forests in the south,
from arid desert in the west to alluvial flood plain in the east. Although the
United States has, arguably, a comparable range of ecological regimes,
what is especially striking about India is its diversity of human cultures,
corresponding to different agro-climatic and vegetative zones. These
cultures exhibit diverse technologies of resource use and also of social
modes of resource control, spanning the entire range of productive
activities known to humans. These range from stone-age hunter-gatherers
at one end of the spectrum, through shifting cultivators, nomadic pastoralists, subsistence and cash-crop agriculturalists, and planters, to every
form of industrial enterprise—at the other. There is, too, a great variety of
property relations which match different techniques—private, communal,
corporate or state management of resources, as the case may be.

An awareness of this diversity is heightened by the acute natural-
resource crisis faced by the country in recent years: shortages of prey for
hunters and fishermen, of land for shifting cultivators, of grazing for
pastoralists, of fuel, fodder and manure for subsistence plough agriculturalists, of power and water for cash-crop agriculturalists, and of power,
water and raw materials for industry. These shortages have generated a
variety of conflicts—and collusions—as different segments of Indian
society exercise competing claims over scarce resources. Inevitably such

* The authors would like to thank Michael Adas, Arjun Appadurai, Bill Burch,
Jim Scott and Timothy Weiskel for their helpful comments on an earlier draft of
this article.

conflicts, which show no signs of abating, strongly affect the quality both of human life and of the natural environment.[1]

These contemporary concerns have led several scholars, including ourselves, to try to reconstruct Indian history using insights derived from recent debates in human ecology. It has been suggested that British colonial rule marks an important watershed in the ecological history of India. The encounter with a technologically advanced and dynamic culture gave rise to profound dislocations at various levels of Indian society. However, the essential interdependence of the ecological and social changes that came in the wake of colonial rule has not been accorded due recognition. The agrarian history of British India has focused almost exclusively on social relations around land and conflicts over the distribution of its produce, to the neglect of the ecological context of agriculture—for example, fishing, forests, grazing land and irrigation—and of state intervention in these spheres.[2] Thus the second volume of the *Cambridge Economic History of India*, an impressive and in many ways valuable survey of colonial agrarian history, has no section devoted to the management and utilization of the forest; it thus leaves out of its purview over one-fifth of India's land area, controlled and monitored by the state in ways that crucially affected agrarian social structure. It shows little awareness of the existence of this vast wooded estate of the government—let alone of the elaborate bureaucratic and technical apparatus that governed it—and mentions only in passing the bitter and intense conflicts around forest resources between the state and its subjects. However, as a synthetic review of colonial economic history, the Cambridge volume is here only reflecting a more general deficiency in the literature. One indication of this gap is the fact that, to the best of our knowledge, not one of the many reviews of the volume has mentioned what to us is its most obvious flaw.[3]

It is beyond the scope of this study to investigate the reasons for the

---

[1] For an overview, see Centre for Science and Environment, *India: The State of the Environment, 1984-85: A Citizens' Report* (New Delhi, 1985).

[2] A partial exception is irrigation, for which some good studies exist. See especially Elizabeth Whitcombe, *Agrarian Conditions in Northern India*, i, *The United Provinces under British Rule, 1860-1900* (New Delhi, 1971); Nirmal Sengupta, 'The Indigenous Irrigation Organization of South Bihar', *Indian Econ. and Social Hist. Rev.*, xvii (1980).

[3] Dharma Kumar (ed.), *The Cambridge Economic History of India*, ii (Cambridge, 1983). Major review symposia appeared in the two premier journals in the field: *Modern Asian Studies*, xix (1985); *Indian Econ. and Social Hist. Rev.*, xxi (1984).

almost universal neglect of Indian ecological history, though it is quite clear that it stems from both methodological and theoretical limitations. Suffice it to say that as far as this article is concerned, what are ostensibly 'social' changes need to be viewed against the backdrop of concomitant changes in patterns of the utilization of natural resources. Here the significance of the British intervention lies in the novel modes of resource extraction made possible by the political dominance of the raj and the availability of technologies previously foreign to India.

The increasing intensity of natural-resource use fostered by colonialism was accompanied too, by equally dramatic changes in forms of management and control. By far the most significant of these was the takeover of woodland by the state. While state management had not been unknown in the pre-colonial period, it was usually restricted in its application and oriented towards highly specific ends: the reservation of elephant forests in the Mauryan period, for example, or later edicts affirming a state monopoly over commercial species such as teak and sandalwood.[4] Now state control, notably over forests, was extended over large tracts and throughout the subcontinent. Moreover, while asserting formal rights of ownership over various natural resources, the colonial government brought to bear on their management a highly developed legal and administrative infrastructure.

It is by now well established that the imperatives of colonial forestry were essentially commercial. Its operations were dictated more by the commercial and strategic utility of different species than by broader social or environmental considerations. For what follows, it is important to understand the mechanisms of intervention—the institutional framework which governed the workings of state forestry in British India.[5]

In the early decades of its rule, the colonial state was markedly indifferent to forest conservancy. Until well into the nineteenth century, forests were viewed by administrators as an impediment to the expansion

[4] For Mauryan elephant forests, see Thomas R. Trautmann, 'Elephants and the Mauryas', in S.N. Mukherjee (ed.), *India: History and Thought: Essays in Honour of A.L. Basham* (Calcutta, 1982).

[5] For a detailed analysis of colonial forestry science, legislation and management, see Ramachandra Guha, 'Forestry in British and Post British India: A Historical Analysis', 2 pts., *Econ. and Polit. Weekly*, 29 Oct. 5-12 Nov. 1983; Ramachandra Guha, 'Scientific Forestry and Social Change in Uttarakhand', *Econ. and Polit. Weekly*, special no. (Nov. 1985); Madhav Gadgil, 'Forestry with a Social Purpose', in W. Fernandes and S. Kulkarni (eds.), *Towards a New Forest Policy* (New Delhi, 1983).

of cultivation. With the state committed to agricultural expansion as its major source of revenue, the early decades of colonial rule witnessed a 'fierce onslaught' on India's forests.[6] The first show of interest in forestry—the reservation of teak forests in Malabar in 1806—was dictated by strategic imperial needs. With the depletion of oak forests in England and Ireland, the teak forests of the Western Ghats were utilized for shipbuilding. Indian teak, the most durable of shipbuilding timbers, was used extensively for the royal navy in the Anglo-French wars of the early nineteenth century and by merchant ships in the later period of maritime expansion.[7]

These isolated and halting attempts at the systematic and sustained production of roundwood, however, did not constitute a general policy of forest management; that had to await the building of the railway network in the last decades of the nineteenth century. It was the pace of railway expansion (from 7,678 kilometres of line in 1870 to 51,658 kilometres in 1910) which brought home forcefully the fact that India's forests were not inexhaustible. The writings of forest officials of the time are dominated by the urgent demand for sleepers. Dubbing early attempts at forest working a 'melancholy failure', the governor-general, Lord Dalhousie, had in 1862 called for the establishment of a department that could meet the enormous requirements of the railway companies (nearly a million sleepers annually). Impending shortages, Dalhousie observed, had made the 'subject of forest conservancy an important administrative question'.[8]

As Britain itself had no tradition of managing forests for sustained timber production, the Forest Department was started with the help of German foresters in 1864. However, the task of reversing the deforestation of the past decades required the forging of legal mechanisms to curtail the exercise of use rights by village communities. After an earlier act had been found wanting, state monopoly over forests was safeguarded by the stringent provisions of the Indian Forest Act of 1878. This was a comprehensive piece of legislation—later to serve as a model for other British colonies—which by one stroke of the executive pen attempted to obliterate centuries of customary use of the forest by rural populations all over India. Several officials within the colonial administration were sharply critical of

[6] E.A. Smythies, *India's Forest Wealth* (London, 1925), p. 6.

[7] R.G. Albion, *Forests and Sea Power* (Cambridge, Mass., 1926), pp. 35-6, 363-8.

[8] Dispatch, government of India to secretary of state, Nov. 1861, quoted in C.G. Trevor and E.A. Smythies, *Practical Forest Management* (Allahabad, 1923), p. 5. The railways were built to facilitate both troop movements and trade.

the new legislation, calling it an act of confiscation and predicting (accurately, as we shall see) widespread discontent at its application. Their objections, however, were swiftly overruled.[9] Essentially designed to maintain strict control over forest utilization from the perspective of strategic imperial needs, the Act also enabled the sustained working of compact blocks of forest for commercial timber production. It provided, too, the underpinnings for the scientific management of the forests. But the logical corollary of the combined operations of law and 'scientific' management was sharp restrictions on customary use. For rationalized timber production could only be ensured through the strict regulation of traditionally exercised rights. Under the provisions of the 1878 Act, each family of 'rightholders' was allotted a specific quantum of timber and fuel, while sale or barter of forest produce was strictly prohibited. This exclusion from forest management was, therefore, both physical—denying or restricting access to forests and pasture—and social—allowing 'rightholders' only a marginal and inflexible claim on the produce of the forests.[10]

In so far as the main aims of the new department were the production of large commercial timber and the generation of revenue, it worked willingly or unwillingly to enforce a separation between agriculture and forests. This exclusion of the agrarian population from the benefits of forest management had drawn sharp criticism from within the ranks of the colonial intelligentsia. In the words of an agricultural chemist writing in 1893, the Forest Department's objects 'were in no sense agricultural, and its success was gauged mainly by fiscal considerations; the Department was to be a revenue paying one. Indeed, we may go so far as to say that its interests were opposed to agriculture, and its intent was rather to exclude agriculture than to admit it to participate in its benefits'.[11]

In order that forests should more directly serve the interests of the rural population, Dr. Voelcker advocated the creation of fuel and fodder reserves, using the characteristic justification that the consequent increased revenue from land tax would more than compensate for any loss

[9] See D. Brandis, *Memorandum on the Demarcation of the Public Forests in the Madras Presidency* (Simla, 1878), pp. 41-2, minute by W. Robinson, 3 Feb. 1878, minute by governor of Madras, 9 Feb. 1878.

[10] 'Rightholders' denote those villagers who were conceded to have legal right of use.

[11] J.A. Voelcker, *Report on Indian Agriculture*, 2nd edn. (Calcutta, 1897), pp. 135-6.

of revenue from a decline in commercial timber operations. As the writings of other contemporary critics also suggest, by bringing about an escalation in the intensity of resource exploitation and control, state forestry sharply undermined the ecological basis of subsistence cultivation, hunting and gathering.[12] It must be stressed that the ecological and social changes that came in the wake of commercial forestry were not simply an intensification of earlier processes of change and conflict. Clearly many of the forest communities, especially hunter-gatherers and shifting cultivators described in this article, had for several centuries been subject to the pressures of the agrarian civilizations of the plains. Yet while these pressures themselves ebbed and flowed with the rise and fall of the grain-based kingdoms of peninsular India, they scarcely matched in their range or scope the magnitude of the changes that were a consequence of the state takeover of the forests in the late nineteenth century. Prior to that the commercial exploitation of forest produce was largely restricted to commodities such as pepper, cardamom and ivory, whose extraction did not seriously affect either the ecology of the forest or customary use. It was the emergence of timber as the major commodity that led to a qualitative change in the patterns of harvesting and utilization of the forest.

Thus when the colonial state asserted control over woodland which had earlier been in the hands of local communities, and proceeded to work these forests for commercial timber production, it intervened in the day to day life of the Indian villager to an unprecedented degree. First, since by 1900 over 20 per cent of India's land area had been taken over by the Forest Department, the working of state forestry could not fail to affect almost every village and hamlet in the subcontinent. Secondly, the colonial state radically redefined property rights, imposing on the forest a system of management whose priorities sharply conflicted with earlier systems of local use and control. Lastly, one must not underestimate the changes in forest ecology that resulted from this shift in methods of management. For a primary task of colonial forestry was to change the species composition of the largely mixed forests of India in favour of component species that had an established market value. Silvicultural techniques, for example, attempted with success to transform the mixed coniferous broad-leaved forests of the Himalaya into pure coniferous stands, and to convert the rich evergreen vegetation of the Western Ghats into single-species teak forests.

---

[12] See, for example, Jotirau Phule, *Shetkarya Asud: The Whipcord of the Farmer* (1882-3), reprinted in Marathi in *The Collected Works of Mahatma Phule*, ed. D. Keer and S.G. Malshe (Pune, 1969).

While these induced changes in forest ecology have in the long term had a slow but imperceptible effect on soil and water systems, they immediately ran counter to the interests of surrounding villages, since the existence of several species rather than one could better meet the varied demands of subsistence agriculture. Significantly, the species promoted by colonial foresters—pine, cedar and teak in different ecological zones— were invariably of very little use to rural populations, while the species they replaced (such as oak) were intensively used for fuel, fodder and small timber.

In these various ways, colonial forestry marked an ecological, economic and political watershed in Indian forest history. The intensification of conflict over forest produce was a major consequence of the changes in patterns of resource use it initiated. The present article analyses some of the evidence for conflict over forests and pasture in colonial India. While it does not pretend to be comprehensive in its coverage, it attempts to outline the major dimensions of such conflict, by focusing on its genesis, its geographical spread and the different forms in which protest manifested itself. As a contribution to the sociology of peasant protest under colonialism, it is intended to provide a set of preliminary findings and to encourage more detailed research on the ecological history of different parts of the subcontinent.

I. *Hunter-Gatherers: The Decline to Extinction*

Until the early decades of this century almost a dozen communities in the Indian subcontinent depended on the original mode of sustenance of human populations, hunting and gathering. They were distributed over almost the entire length of India, from the Rajis of Kumaun in the north to the Kadars of Cochin in the south. The abundant rainfall and rich vegetation of their tropical habitats facilitated the reproduction of subsistence almost exclusively through the collection of roots and fruit and the hunting of small game. While cultivation was largely foreign to these communities, they did engage in some trade with the surrounding agricultural population, exchanging forest produce such as herbs and honey for metal implements, salt, clothes and very occasionally grain. With minimal social differentiation and restraints on over-exploitation of resources through the partitioning of territories between endogamous bands, these hunter-gatherers, if not quite the 'original affluent society', were able to subsist quite easily on the bounties of nature, as long as there existed sufficient areas under their control.[13]

Predictably, state reservation of forests sharply affected the subsistence activities of these communities, each of them numbering a few hundred and with population densities calculated at square miles per person rather than persons per square mile. The forest and game laws affected the Chenchus of Hyderabad, for example, by making their hunting activities illegal and by questioning or even denying their existing monopoly over forest produce other than timber. The cumulative impact of commercial forestry and the more frequent contacts with outsiders that the opening-out of such areas brought about virtually crippled the Chenchus. As suspicious of mobile populations as most modern states, in some places the colonial government forcibly gathered tribal peoples into large settlements. Rapidly losing their autonomy, most Chenchus were forced into a relationship of agrestic serfdom with the more powerful cultivating castes. Further south, the Chenchus of Kurnool, almost in desperation, turned to banditry, frequently holding up pilgrims to the major Hindu temple of Srisailam.[14]

Like the Chenchus, other hunter-gatherer communities were not numerous enough actively to resist the social and economic changes that followed state forest management. Forced sedentarization and the loss of their habitat induced a feeling of helplessness as outsiders made greater and greater inroads into what was once an undisputed domain. Thus the Kadars succumbed to what one writer called a 'proletarian dependence' on the forest administration, whose commercial transactions and territorial control now determined their daily routine and mode of existence. In this way, the intimate knowledge of their surroundings that the Kadars possessed was now utilized for the collection of forest produce marketed by the state. In the thickly wooded plateau of Chotanagpur the commercialization of the forest and restrictions on local use had meanwhile led to a

[13] The phrase is that of Marshall Sahlins. See his *Stone Age Economics* (Chicago, 1971).

[14] C. Von Furer Haimendorf, *The Chenchus: Jungle Folk of the Deccan* (London, 1943), pp. 57, 295, 311-12, 321, *et al.*; A. Aiyappan, *Report on the Socio-Economic Conditions of the Aboriginal Tribes of the Province of Madras* (Madras, 1948), p. 32.

precipitous fall in the population of the Birhor tribe—from 2,340 in 1911 to 1,610 in 1921.[15]

While the new laws restricted small-scale hunting by tribal peoples, they facilitated more organized *shikar* expeditions by the British. From the mid-nineteenth century there began a large-scale slaughter of animals, in which white *shikaris* at all levels, from the viceroy down to the lower echelons of the British Indian army, participated. Much of this shooting was motivated by the desire for large bags. While one British planter in the Nilgiris killed four hundred elephants in the 1860s, successive viceroys were invited to shoots in which several thousand birds were shot in a single day in a bid to claim the 'world record' Many Indian princes sought to emulate the predatory instincts of the British. The maharaja of Gwalior, for example, shot over seven hundred tigers in the early 1900s.[16]

Although it is difficult to estimate the impact of such unregulated hunting on faunal ecology, the consequences of shikar were apparent by the time India gained independence, reflected in the steadily declining populations of game species such as the tiger and elephant. More relevant to this study is the disjunction between the favours shown to the white shikari and the clamp-down on subsistence hunting. While there were few formal restrictions on the British hunter until well into the twentieth century, hunter-gatherers as well as cultivators for whom wild game was a valuable source of protein found their hunting activities threatened by the new forest laws.

The Baigas of central India, for example, were famed for their hunting skills—'expert in all appliances of the chase'. Early British shikaris relied heavily on their 'marvellous skill and knowledge of the wild creatures'. Yet the stricter forest administration, dating from the turn of the century, induced a dramatic decline. Writing in the 1930s, Verrier Elwin noted that while their love for hunting and meat persisted, old skills had largely perished. There remained, however, a defiant streak: as one Baiga said, 'even if Government passes a hundred laws we will do it. One of us will keep the official talking; the rest will go out and shoot the deer'.[17] In the

---

[15] U.R. Ehrenfels, *The Kadar of Cochin* (Madras, 1952), pp. 8, 13-24, 47-8, *et al.*; S.C. Roy, *The Birhors: A Little-Known Jungle Tribe of Chota Nagpur* (Ranchi, 1925). p. 549.

[16] Scott Bennet, 'Shikar and the Rai', *South Asia*, new ser., vii (1984), pp. 72-88; J.G. Elliott, *Field Sports in India, 1800-1947,* (London, 1973); R. Sukumar, *Ecology and Management of the Asian Elephant* (Cambridge, forthcoming), ch.1.

[17] H.C. Ward, *Report on the Land Revenue Settlement of the Mundlah District*

Himalayan foothills, too, where there was an abundance of game, villagers continued to hunt despite government restrictions, taking care to be one step ahead of the forest staff—a task not difficult to accomplish, given their familiarity with the terrain.[18]

Among shifting cultivators, there was often a ritual association of hunting with the agricultural cycle. Despite game laws, the Hill Reddis of Hyderabad clung to their ritual hunt—called Bhumi Devata Panduga or the hunt of the earth god—which involved the entire male population and preceded the monsoon sowing. The reservation of forests also interfered with the movement of hunting parties across state boundaries. In 1929 a police contingent had to be called in to prevent a party of Bison Marias from the state of Bastar, armed with bows and spears, from crossing into the British-administered Central Provinces. This, of course, constituted an unnatural intervention as the ritual hunt was no respecter of political boundaries. Nevertheless in later years the authorities were successful in confining the Maria ritual hunt to Bastar, the amount of game killed steadily declining in consequence.[19]

II. The 'Problem' of Shifting Cultivation

Shifting or *jhum* cultivation was the characteristic form of agriculture over large parts of north-eastern India, especially the hilly and forested tracts where plough agriculture was not always feasible. Jhum typically involves the clearing and cultivation of patches of forest in rotation. The individual plots are burned and cultivated for a few years and then left fallow for an extended period (ideally a dozen years or longer), allowing the soil to recoup and recover lost nutrients. Cultivators then move on to the next plot, abandoning it in turn when its productivity starts declining.[20] It was usually

of the Central Provinces, 1868-9. (Bombay, 1870), p. 37; J.W. Best, *Forest Life in India* (London, 1935), pp. 123-4; Verrier Elwin, *The Baiga* (London, 1939), p. 84.

[18] See 'Gamekeeper', 'Destruction of Game in Government Reserves during the Rains', *Indian Forester*, xiii (1887), pp. 188-90. Cf. also Jim Corbett, *My India* (Bombay, 1952).

[19] C. Von Furer Haimendorf, *The Reddis of the Bison Hills: A Study in Acculturation* (London, 1943), pp 191-3. W.V. Grigson, *The Maria Gonds of Bastar* (London. 1938), pp. 158-9.

[20] See Michael Eden, 'Traditional Shifting Cultivation and the Tropical Forest System', *Trends in Ecol. Evolution*, ii (1987), Shifting cultivation is known by various names—*jhum, podu, dhyal, bewar,* etc. We shall use jhum here.

practised by 'tribal'[21] groups for whom jhum was a way of life encompassing, beyond the narrowly economic, the social and cultural spheres as well. The corporate character of these communities was evident in the pattern of cultivation, where communal labour predominated and where different families adhered to boundaries established and respected by tradition. The overwhelming importance of jhum in structuring social life was strikingly manifest too in the many myths and legends constructed around it in tribal cosmology.[22]

As in many areas of social life, major changes accompanied the advent of British rule. For almost without exception, colonial administrators viewed jhum with disfavour as a primitive and unremunerative form of agriculture in comparison with plough cultivation. Influenced both by the agricultural revolution in Europe and the revenue-generating possibilities of intensive as opposed to extensive forms of cultivation, official hostility to jhum gained an added impetus with the commercialization of the forest. Like their counterparts in other parts of the globe, British foresters held jhum to be 'the most destructive of all practices for the forest'.[23] There was good reason for this animosity: 'axe cultivation was the despair of every forest officer',[24] largely because timber operations competed with jhum for territorial control of the forest. This negative attitude was nevertheless tempered by the realization that any abrupt attempt to curtail the practice would provoke a sharp response from jhum cultivators. Yet the areas cultivated under jhum often contained the most-valued timber species.[25] Here was an intractable problem for which the colonial state had no easy solution.

[21] In India 'tribal' is a legal rather than a social category, encapsulating those ethnic groups believed to be autochthonous and which are economically and socially distinct (to a lesser or greater extent) from the 'caste' society of settled agriculture.

[22] For a fine ethnographic study of one of the last communities of shifting cultivators in peninsular India, see Savyasachi, *Agriculture and Social Structure*: *The Hill Maria of Bastar* (mimeo), World Inst. Development Economics Research (Helsinki, Jan. 1987).

[23] C.F. Muhafiz-i-Jangal (pseud), 'Jhooming in Russia', *Indian Forester*, in (1877), pp. 418-19.

[24] Verrier Elwin, *The Aboriginals* (Oxford Pamphlet on Indian Affairs, No. 14, Bombay, 1943), p.8.

[25] As the chief commissioner of the Central Provinces put it, 'the best ground for this peculiar cultivation is precisely that where the finest timber trees like to grow': Sir Richard Temple, quoted in J.F. Dyer, 'Forestry in the Central Provinces and Berar', *Indian Forester*, 1 (1925), p. 349.

A vivid account of the various attempts to combat jhum can be found in Verrier Elwin's classic monograph on the Baiga,[26] a small tribe of the Mandla, Balaghat and Bilaspur districts of the present-day Madhya Pradesh. The first serious attempt to stop shifting cultivation in the 1860s had as its impetus the civilizing zeal of the chief commissioner of the province, Sir Richard Temple. In later years, though, it was the fact that the market value of forest produce 'rose in something like geometrical proportions' which accounted for the 'shifting of emphasis from Sir Richard Temple's policy of benevolent improvement, for their own sake to a frank and simple desire to better the Provincial budget'. A vigorous campaign to induce the Baiga to take to the plough culminated in the destruction of standing jhum crops by an over-enthusiastic deputy commissioner. When many tribal people fled to neighbouring princely states, the government advised a policy of slow weaning from axe cultivation.

Such difficulties had in fact been anticipated by the settlement officer in 1870 who observed that 'it has been found quite impracticable as well as hard and impolitic to force the Baigas to give up their dhya (jhum) cultivation and take to the plough'. He advised a limiting of jhum rather than a total ban. A more cautious policy was dictated, too, by the dependence of the Forest Department on Baiga labour for wood-cutting and the collection of forest produce. As a consequence, the government established the Baiga *chak* (reserve) in 1890 covering 23,920 acres of forest where it planned to confine all jhum cultivators. The area chosen was described as 'perfectly inaccessible [and] therefore useless as a timber producing area'. While permitting jhum within the reserve, the administration stressed an overall policy of discouraging it elsewhere. In this it was partially successful, as Baiga villagers outside the chak, faced with the prospect of leaving their homes, accepted the terms of plough cultivation. As many Baigas continued to migrate into neighbouring princely states, within the chak itself the population of jhum cultivators steadily dwindled.

The Baigas' opposition took the form of 'voting with their feet' and other means of resistance that stopped short of open confrontation, such as the non-payment of taxes and the continuance of jhum in forbidden areas. The new restrictions inculcated an acute sense of cultural loss, captured in a petition submitted to the British government in 1892. After jhum had been stopped, it said: 'We daily starve, having had no food grain in our

---

[26] Elwin, *Baiga*. The following account is drawn from this source (as are all quotations), ch. 2, esp. pp. 111-30. See also Ward, *Report on the Land Revenue Settlement*, pp. 35. 38-9, 160, *et al.*

possession. The only wealth we possess is our axe. We have no clothes to cover our body with, but we pass cold nights by the fireside. We are now dying for want of food. We cannot go elsewhere as the British government is everywhere. What fault have we done that the government does not take care of us? Prisoners are supplied with ample food in jail. A cultivator of the grass is not deprived of his holding, but the Government does not give us our right who have lived here for generations past'.

In some areas tribal resistance to the state's attempt to curb jhum often took a violent and confrontationist form. This was especially so where commercialization of the forest was accompanied by the penetration of non-tribal landlords and moneylenders who came to exercise a dominant influence on the indigenous population. Elwin himself, talking of the periodic disturbances among the Saora tribal people of the Ganjam Agency, identified them as emanating from two sources: the exactions of plainsmen and the state's attempts to check axe cultivation. Thus Saoras were prone to invade reserved forests and clear land for cultivation. In the late 1930s several villages endeavoured to fell large areas of reserved forests in preparation for sowing. The Saoras were ready for any penalty: when the men were arrested and gaoled, the women continued the cultivation. After returning from gaol, the men cleared the jungle again for the next year's crop. As repeated arrests were unsuccessful in stopping Saoras from trying to establish their right, the Forest Department forcibly uprooted crops on land formally vested in the state.[27]

Perhaps the most sustained resistance, extending over nearly a century, occurred in the Gudem and Rampa hill tracts of present-day Andhra Pradesh. Inhabited by Koya and Konda Dora tribes, predominantly jhum cultivators, the hills were subjected under British rule to a steady penetration by the market economy and the influx of plainsmen eager to exploit its natural wealth. Road-construction led to the rapid development of trade in tamarinds, fruit, honey and other forest produce exported to urban centres and even to Europe. Traders from the powerful Telugu caste of Komatis took over from local chiefs the leases of tracts of forest as well as the trade in palm liquor. As in other parts of India, they were actively helped by the colonial government which had banned domestic brewing of liquor (an important source of nutrition in the lean season) and farmed out liquor contracts in a bid to raise revenue. At the same time, commercial

[27] Verrier Elwin, 'Saora Fituris', *Man in India*, xxv (1945), pp. 154-7; A.L. Bannerjee, 'A Note on the Parlakamadi Forest Division', *Indian Forester*, ixviii (1942), pp. 71-2.

forest operations were begun on a fairly large scale and, as elsewhere, the creation of forest reserves conflicted with the practice of jhum. Slowly losing control over their lands and means of subsistence, many tribes-people were forced into relations of dependence on the more powerful plainsmen, either working as tenants and share-croppers in the new system of market agriculture or as forest labourers in the felling and hauling of timber.[28]

Several of the many small risings or *fituris* documented by David Arnold were directly or indirectly related to forest grievances. The Rampa rebellion of 1879-80, for example, arose in response to the new restrictions concerning liquor and the forest. Complaining bitterly against the various exactions, the tribespeople said that 'as they could not live they might as well kill the constables and die'. Led by a minor tribal chieftain, Tammam Dora, the rebels attacked and burned several police stations, executing a constable in an act of ritual sacrifice. Although Tammam Dora was shot by the police in June 1880, the revolt spread to the Golconda Hills of Vishakapatnam and the Rekepalle country in Bhadrachalam. The transfer of the latter territory from the General Provinces to Madras had led to greater restrictions on the practice of jhum there. Protest emanated directly from forest grievances and, as in other fituris, police stations—highly visible symbols of state authority—were frequent targets. It took several hundred policemen and ten army companies to suppress the revolt, a task not finally accomplished until November 1880.

The last recorded fituri in 1922-3 was, like its predecessors, closely linked to restrictions on tribal access to the forest. Its leader, a high-caste Hindu from the plains called Alluri Sita Rama Raju, was able to transform a local rising into a minor guerrilla war, recruiting dispossessed landhold-ers and offenders against the forest laws and gaining help from villagers who gave them food and shelter. After raids on police outposts had netted a haul of arms and ammunition, Raju's men evaded capture thanks to their superior knowledge of the hilly and wooded terrain. Unsuccessful in his attempts to spread the rebellion into the plains, Rama Raju was finally taken and shot in May 1924.

When the Indian princes sought to emulate their British counterparts in realizing the commercial value of their forests, they too came in conflict

[28] This account is largely based on David Arnold, 'Rebellious Hillmen: The Gudem Rampa Rebellions, 1829-1914', in Ranajit Guha (ed.), *Subaltern Studies I* (Delhi, 1982); supplemented by C. Von Furer Haimendorf, 'Aboriginal Rebellions in the Deccan', *Man in India*, xxv (1945).

with shifting cultivators. Regarding the state takeover as a forfeiture of their hereditary rights, tribespeople in several chiefdoms rose in revolt against attempts to curb jhum. A major rebellion took place in Bastar in 1910, directed against the new prohibition of the practice, restrictions on access to forest and its produce, and the *begar* (unpaid labour) exacted by state officials. The formation of reserved forests had resulted in the destruction of many villages and the eviction of their inhabitants. In order to draw attention to their grievances, some tribespeople went on hunger strike outside the king's palace at Jagdalpur. Affirming that it was an internal affair between them and their ruler, the rebels—mostly Marias and Murias—cut telegraph wires and blocked the roads. At the same time, police stations and forest outposts were burned, stacked wood looted and a campaign mounted against *pardeshis* (outsiders), most of them low-caste Hindu cultivators settled in Bastar. Led by their headmen, the rebels looted several markets and attacked and killed both state officials and merchants. In a matter of days, the rebellion engulfed nearly half the state, an area exceeding 6,000 square miles. Unnerved, the king called in a battalion of the 22nd Punjabis (led by a British officer) and detachments of the Madras and Central Province police. In a decisive encounter near Jagdalpur, over nine hundred tribesmen armed only with bows, arrows and spears, and of all ages from sixteen upwards were captured.[29]

In 1940 a similar revolt broke out in the Adilabad district of Hyderabad. Here Gonds and Kolams, the principal cultivating tribes were subjected to an invasion of Telugu and Maratha cultivators who flooded the district following the improvement of communications. Whole Gond villages fell to immigrant castes. In the uplands, meanwhile, forest conservancy restricted jhum, with cultivated land lying fallow under rotation being taken into forest reserves. Following the forcible disbandment of Gond and Kolam settlements in the Dhanora forest, the tribespeople, led by Kumra Bhimu, made repeated but unsuccessful attempts to contact state officials. After petitions for authorized resettlement were ignored, the tribespeople established their own settlement and began to clear forests for cultivation. An armed party sent to burn the new village was resisted by Bhimu's Gonds, who then took refuge in the mountains. When the police

---

[29] National Archives of India, New Delhi, Foreign Department, Secret-I Progs., Nos. 16-17 for Sept. 1910, Nos. 34-40 for Aug. 1911; Grigson, *Maria Gonds*, pp. 16-17; Clement Smith, 'The Bastar Rebellion, 1910', *Man in India*, xxv (1945). For a fuller treatment of the 1910 Bastar revolt, see Ramachandra Guha, 'Raja Praja as Pita/Putra: Forms of Customary Rebellion in Princely India', forthcoming.

asked them to surrender, they were met with the counter-demand that Gonds and Kolams should be given possession of the land they had begun to cultivate. The police thereupon opened fire, killing Bhimu and several of his associates.[30]

Elsewhere in Hyderabad, the Hill Reddis of the Godavari Valley were also at the receiving end of the new forest laws. The restriction of jhum to small demarcated areas forced the Reddis to shorten fallow cycles or to prolong cultivation on a designated patch until deterioration set in. They made their feelings plain by moving across the Godavari to British territory, where the forest laws were not quite so stringent, returning to Hyderabad when the ban on jhum was lifted.[31] An ingenious method of protest, similarly questioning forest policy without quite attempting to combat the state, is reported from several coastal districts in Madras Presidency where cultivators, supported by several officials, insisted that the ban on jhum had resulted in a greater incidence of fever.[32]

These repeated protests had a significant impact on government policy. In some parts of Madras Presidency, certain patches of land were set aside for tribespeople to continue jhum. For although 'the Forest Department would welcome the complete stoppage of podu [jhum] it is not done for fear of fituris [tribal uprisings]'.[33] Elsewhere the state found a novel way of pursuing commercial forestry without further alienating tribal cultivators. This was the *taungya* method of agro-silviculture—developed in Burma in the nineteenth century—in which jhum cultivators were allowed to grow food crops in the forest provided they grew timber trees alongside. When, after a few years, the cultivator moved on to clear the next patch, a forest crop had been established on the vacated ground. Taungya thus made possible the establishment of the labour force necessary for forest works at a 'comparatively low cost' and it is still widely in operation. It helped to forestall the very real possibility of revolt if tribespeople were displaced by a prohibition of their characteristic form of cultivation although even taungya cultivators sometimes thwarted the state by planting up only those areas likely to be inspected by touring officials. Ironically enough, its success has even led to the reintroduction of jhum in

[30] Haimendorf, 'Aboriginal Rebellions', pp. 213-16.

[31] Haimendorf, *Reddis of the Bison Hills*, pp. 307-8, 318-19; C. Von Furer Haimendorf, *Tribal Hyderabad: Four Reports* (Hyderabad, 1945), pp. 3. 11. *et al.*

[32] Anon., *A. Selection of Despatches... on Forest Conservancy in India. pt. 2: Madras* (London, 1871), pp. 148-50.

[33] Aiyappan, *Report on the Socio-Economic Conditions of the Aboriginal Tribes*, pp. 16-17.

tracts where it had died out or been put down at an earlier stage.[34]

More commonly, however, the cumulative impact of market forces and state intervention forced the abandonment of jhum in favour of the plough or wage labour. Even where the practice continued, the disruption of the delicate balance between humans and forests initially through the usurpation of forests by the state and later through a secular rise of population led to a sharp fall in the jhum cycle. A form of agriculture practised for several millenniums had become unsustainable in the face of external forces.

III. *Settled Cultivators and the State*

Notwithstanding the spatial separation between field and forest, over the most part of India plough agriculturalists (mostly caste Hindus) were scarcely less affected by forest reservation than jhum cultivators. For they too depended on their natural habitat in a variety of ways. An adequate forest cover was ecologically necessary to sustain cultivation, especially in mountainous tracts where terrace farming predominated; and since animal husbandry was a valuable appendage to cultivation, the forest was a prime source of fodder in the form of grass and leaves. The forests also provided such necessities as fuel, leaf manure and timber for construction and agricultural implements.

Here, too, state reservation enforced changes in the traditional pattern of resource utilization even if these changes were not quite as radical as in the case of shifting cultivators. Under the provisions of the 1878 Act, the takeover of a tract of forest involved settling the claims of surrounding villages. Under the new 'legal' (that is, codified) arrangements, the previously unlimited use rights were severely circumscribed. These restrictions affected two distinct classes of agriculturalists, and in somewhat different ways. In areas dominated by cultivating proprietors, and where social differentiation was not strongly marked, those affected by state forestry consisted primarily of middle to rich peasants, many of whom were graziers rather than agriculturalists. On the other hand, in tracts exhibiting more advanced forms of class differentiation, a different social stratum was at the receiving end. These were *adivasi* (tribal) and low-caste communities who supplemented their meagre earnings as tenants and

[34] See H.R. Blanford, 'Regeneration with the Assistance of Taungya in Burma', *Indian Forest Records*, xi. pt. 3 (1925); B.H. Baden-Powell, *The Forest System of British Burma* (Calcutta, 1874), p. 36; H.G. Champion and S.K. Seth, *General Silviculture for India* (Delhi, 1968), esp. pp. 315-16.

share-croppers with the extraction and sale of fuel, grass and other minor forest produce.

An example of the first form of deprivation comes from the Madras Presidency. There, several decades after forest reservation, villagers had vivid memories of their traditional rights over the forest, continuing to adhere to informal boundaries demarcating tracts of woodland claimed and controlled by neighbouring villagers.[35] The tenacity with which they clung to their rights was visibly manifest, too, in the escalation in forest offences (averaging 30,000 per annum) with the killing of forest personnel a not infrequent occurrence. A committee formed to investigate forest grievances was puzzled to find that villagers interpreted the term 'free grazing' quite differently from the committee itself. While quite prepared to pay a small fee, peasants understood 'free grazing' to mean 'the right to graze all over the forests': that is, the continuance of the territorial control that they formerly enjoyed.[36] Thus the demand for grazing was accompanied by the demand for free fuel, timber and small timber, in effect 'for the abolition of all control and for the right to use or destroy the forest property of the state without any restriction whatever.' Commenting on the widespread hostility towards state forest management, the committee observed that 'the one department which appears at one time to have rivalled the Forest Department in unpopularity is the the Salt Department which, like the Forest Department, is concerned with a commodity of comparatively small value in itself but an article for daily use and consumption'.[37]

In the state of Travancore, bordering Madras on the Malabar coast, restrictions on village use of the forest stemmed from two sources: the desire to commercialize the forest and the sale, at extremely low prices, of vast expanses of woodland to European planters. These processes were interrelated. The development of a road and railway network to facilitate the export of tea, coffee and rubber also served to hasten the pace of timber exploitation. As a consequence, agriculturists faced acute distress through the loss of green manure (extensively used in paddy cultivation) and other

[35] On traditional systems of communal resource management in Madras, see Brandis, *Memorandum on the Demarcation of the Public Forests.*

[36] Under commercial forest management, areas with young saplings are completely closed to grazing, thus restricting grazing to specific blocks of forest where it cannot harm the reproduction of commercial timber species; see, for details, Guha, 'Scientific Forestry and Social Change'. See also J. McKee, 'On Grazing', *Indian Forester*, i. 1875.

[37] Anon., *Report of the Forest Committee*, i (Madras, 1913), pp. 2-3, 8, *et al.* See also C.J. Baker, *An Indian Rural Economy, 1880-1955* (Delhi, 1984), pp. 157-61.

forest produce. Denied access to pasture, the population of sheep and goats declined precipitously in the years following forest reservation. While there were no incidents of large-scale protest, the peasantry refused to co-operate with the Forest Department or to submit to the new regulations.[38]

Not surprisingly, opposition to state forestry was far more intense among lower castes and tribespeople. An important source of income for tribal households in the Thane district of coastal Maharashtra was the sale of firewood to Koli fishermen. This trade was severely affected by the stricter control exercised over the forest from the later decades of the nineteenth century. Typically, the early manifestations of discontent were peaceful: petitioning the local administration, for example. When this had no impact, however, collective protest turned violent. Surrounding the camp of a deputy collector, a group of villagers demanded that 'the forests be thrown open, palm tax be abolished, country liquor [be sold] at one anna a seer, salt at one anna a paiti, rice at R[upe]e 1 1/4 per maund and that the Government should redeem their mortgaged land and restore it to them'. In another incident, a large number of tribespeople carrying firewood to market were intercepted by the police. In protest, the adivasis stacked wood on a nearby railway line and refused to allow a train to pass. Sensing the prevailing mood of defiance, the officer in charge of the force allowed them to proceed.[39]

A similar turn of events is reported from the Midnapur district of Bengal Presidency. In an area called the Jungle Mahals, land owned by the Midnapur Zamindari Company (M.Z.C.)—an associate of the important British managing agency firm of Andrew Yule—was cultivated by Santhal tribal tenants. While early leases clearly specified that all land was to be handed over to the lessee, the coming of the railway and consequently of a thriving timber trade led the zamindars to impose sharp restrictions on the Santhals. Again, the tribespeople first tried the courts and other means of legal redress. However, the conditions of economic distress prevailing

---

[38] See M.S.S. Pandian, 'Political Economy of Agrarian Change in Nan-chilnadu: The Late Nineteenth Century to 1939' (Univ. of Madras Ph.D. thesis, 1985). The impact on local ecology of the massive expansion of tea plantations in north-east India has yet to be studied. Apart from the widespread deforestation they entailed, these plantations also displaced communities of hunter-gatherers and shifting cultivators.

[39] Raajen Singh, 'Dawn of Political Consciousness: Riots of Kalve-Mahim, 1896', in *Background Papers in Forestry* (mimeo, BUILD Documentation Centre, Bombay, n.d.). For attempts to enforce state monopoly over firewood trade in the south-western coastal districts, see D. Brandis, *Suggestions Regarding Forest*

in the aftermath of the First World War provoked a more militant response. In 1918 the forest-dwelling Santhals proceeded on a campaign of *haat* (market) looting, their principal targets being up-country cloth-traders who were moneylenders as well.

Some years later, and after the intervention of Congress nationalists, the Jungle Mahals witnessed a movement more sharply focused on the question of forest rights. Early in 1922 Santhals working as forest labour went on strike. Following a scuffle between employees of the M.Z.C. and the strikers, the Congress directed the Santhals to plunder the forests. Further incidents of haat looting (including the burning of foreign cloth) and attempts to restrict the export of paddy were also reported. In one subdivision of the area, Silda, Santhals began to plunder jungles leased to timber merchants and a police party trying to confiscate the newly cut wood was beaten up.[40]

Another form of assertion of traditional rights was the looting of fish from ponds controlled by individual zamindars. In April 1923 there was a wave of such activity and breaches of the forest law over an area of 200 square miles, from Jhargram in Midnapur to Ghatshila in the Singhbhum district of Bihar. While recognizing this to be 'illegal', the tribespeople argued that tank-raiding would force the zamindars to concede their customary rights over forests. The Santhal, commented the district magistrate, 'will tell you how in his father's time all jungles were free and *bandhs* (ponds) open to the public. Sometimes he is right...' When the protests were supported by a dispossessed local chieftain, even the recognition that they were illegal was abandoned. Indeed, as alarmed officials reported, 90 per cent of the crowd believed that they were merely restoring a golden age when all jungles were free.[41]

Defiance of forest regulations also formed part of the country-wide campaigns led by the Indian National Congress in 1920-2 and 1930-2. Gandhi's visit to Cudappah in south-eastern India in September 1921 was widely hailed as an opportunity to get the forest laws abolished. In nearby Guntur peasants invaded the forests in the belief that 'Gandhi-Raj' had

---

Administration in the Madras Presidency (Madras, 1883), pp. 313-15.

[40] Swapan Dasgupta, 'Local Politics in Bengal: Midnapur District, 1907-1923' (School of Oriental and African Studies, Univ. of London, Ph.D. thesis, 1980), pp. 127-44.

[41] Sumit Sarkar, 'The Conditions and Nature of Subaltern Militancy: Bengal from Swadeshi to Non-Cooperation', in Ranajit Guha (ed.), *Subaltern Studies III* (New Delhi, 1984), pp. 302-7.

been established and that the forests were now open. Ten years later, during the Civil Disobedience movement the violation of forest laws was far more widespread. In Maharashtra where women played a significant part, nearly 60,000 villagers in Akola district marched into government forests with their cattle. In Satara district peasants resolved not to pay the grazing fee, arguing that grazing restrictions deprived the sacred cow of its daily food. Encroachment on reserved forests was followed by the felling of teak trees and the hoisting of the national tricolour on a teak pole in front of a temple dedicated to the Hindu god Shiva. Women also played a key role in a similar campaign in the coastal district of North Kanara (in present-day Karnataka), garlanding and smearing ritual paste on men who went off to the forest to cut the valued sandal tree. There, too, the timber was loaded on to carts and stacked in front of a local temple. When the men were arrested the women symbolically breached the rules themselves, invoking the god Sri Krishna who had gone into the forest. In the Central Provinces, meanwhile, tribal peoples had come forth in great numbers to participate in the organized violation of forest laws. While formally conducted under the rubric of the Congress, these movements actually enjoyed a considerable degree of autonomy from that organization: the many violent incidents were clearly in defiance of nationalist leaders, wedded as the latter were to an ideology of non-violence.[42]

Perhaps the most sustained opposition to state forest management was to be found in the Himalayan districts of present-day Uttar Pradesh.[43] Dominated by magnificent stands of coniferous species, the hill forests have been the only source of softwoods and hence the most valuable forest property in the subcontinent. At the same time, they have also played a

[42] Sumit Sarkar, 'Primitive Rebellion and Modern Nationalism: A Note on Forest Satyagraha in the Non-Cooperation and Civil Disobedience Movements', in K.N. Pannikar (ed.), *National and Left Movements in India* (New Delhi, 1980); G.S. Halappa, *History of Freedom Movement in Karnataka*, ii (Bangalore, 1964), pp. 110-12; Sulabha Brahme and Ashok Upadhya, *A Critical Analysis of the Social Formation and Peasant Resistance in Maharashtra* (mimeo, Gokhale Inst. Politics and Economics, Pune, 1979), n.pp. 153-4, iii. pp. 679-80; D.E.U. Baker, 'A "Serious Time": Forest Satyagraha in Madhya Pradesh, 1930', *Indian Econ. and Hist. Rev.*, xxi (1984).

[43] For a detailed treatment of forest management and social protest in these districts, see Ramachandra Guha, 'Forestry and Social Protest in British Kumaun, c. 1893-1921', in Ranajit Guha (ed.), *Subaltern Studies IV* (New Delhi, 1985); Ramachandra Guha, *The Unquiet Woods: A Century of Protest in the Indian Himalaya* (New Delhi and Berkeley, forthcoming).

crucial role in sustaining agriculture in the mountainous terrain, a role strikingly reflected in the traditional systems of resource conservation evolved to inhibit over-exploitation of village forests.

In the period of colonial rule this region was divided into two distinct socio-political structures—the princely state of Tehri Garhwal and the British-administered Kumaun Division. Since the forests of Tehri Garhwal came under commercial management even earlier, *circa* 1865, however, peasant resistance to encroachment on customary rights was remarkably sustained and uniform in both areas. In Tehri, important, if localized movements occurred in 1904, 1906, 1930 and 1944-8 and forest grievances played an important and sometimes determining role in all of them. Through the collective violation of the new laws and attacks on forest officials, the peasantry underscored their claim to a full and exclusive control over forests and pasture. As in other pre-capitalist societies where the ruler relied on a traditional idiom of legitimacy, protest was aimed at forest management and its back-up officials and not at the monarch himself. In Kumaun Division, on the other hand, social protest was aimed directly at the colonial state and at the most visible signs of its rule: the pine forests under intensive commercial management and government buildings and offices. It reached its zenith in the summer of 1921, when a wide-ranging campaign to burn forests controlled by the Forest Department virtually paralysed the administration, forcing it to abolish the much-disliked system of forced labour and to abandon effective control over areas of woodland. Largely autonomous of organized nationalist activity as represented by the Congress, the movements of 1916, 1921, 1930 and 1942 in Kumaun Division brought to the fore the central importance of forests in peasant economy and society. Notwithstanding inevitable differences in the social idiom of protest in both Tehri Garhwal and Kumaun Division, forest restrictions were the source of bitter conflicts, unprecedented in their intensity and range, between the peasantry and the state.[44]

IV. *Everyday Forms of Resistance: The Case of Jaunsar Bawar*

In a penetrating study of rural Malaysia, the political scientist James Scott has observed that, while most students of rural politics have focused on agrarian revolt and revolution, these are by no means the characteristic

[44] Since 1973 these hill districts have been the epicentre of the Chipko (tree-hugging) movement, possibly the best-known environmental movement in contemporary India.

forms of peasant resistance. Far more frequently peasants resort to methods of resisting the demands of non-cultivating elites that minimize the element of open confrontation: non-cooperation with imposed rules and regulations, for example, giving false or misleading information to tax collectors and other officials, or migration.[45] In colonial India, too, the peasantry often resorted to violent protest only after quasi-legal channels, such as petitions and peaceful strikes, had been tried and found wanting. Although the historical record is heavily biased towards episodes of violent revolt in which peasants impose themselves rather more emphatically on the processes of state, it is important not to neglect other forms of protest that were not overtly confrontational in form.

These other forms of resistance often preceded, or ran concurrently with open conflict. Thus in many areas breaches of forest laws were the most tangible evidence of the unpopularity of state management: the available evidence shows that, typically, the incidence of forest 'crime' followed a steadily escalating trend. While this would be true of regions where sustained protest did occur (such as those described above), the absence of an organized movement plainly did not signify approval of state forestry.[46]

That the conflict between villagers and colonial forest management did not always manifest itself in open revolt is clearly shown by the experience of Jaunsar Bawar, the hilly segment of Dehra Dun district which bordered Tehri Garhwal on the west. From the early 1860s the forests of Jaunsar Bawar had attracted the attention of the state. They were important for three reasons: as a source of wood for the railway, as 'inspection' forests for training students at the Forest School in the nearby town of Dehra Dun, and as a source of fuel and timber for the military cantonment of Chakrata.[47] In a settlement made in 1868 the state divided the forests into three classes. While Class I forests were wholly closed for their protection, villagers had certain rights of pasturage and timber collection in the second

[45] J.C. Scott, *Weapons of the Weak: Everyday Forms of Peasant Resistance* (New Haven, 1986).

[46] Work on forest 'crime', so far unpublished, by three Indian historians should help clarify some of these issues: Neeladri Bhattacharya on Kulu and Kangra, Prabhu Mahapatra on Chotanagpur and Gopal Mukherjee on Chathisgarh.

[47] For the compulsions behind the state takeover of forests in this area, see N. Hearle, *Working Plan of the Deoban Range, Jaunsar Forest Division, Northwestern Provinces* (Allahabad, 1889); and D. Brandis, *Suggestions Regarding the Management of the Forests Included in the Forest School Circle, Northwestern Provinces* (Simla, 1879).

class. The third class was to be kept for the exclusive use of the peasants with the caveat that they were not allowed to barter or sell any of the produce.

Early protests were directed at this apparent government monopoly. The confused legal status of the Class III forests in which (village leaders argued) it was not clear who held actual proprietary right, the state or the village, was compounded by the refusal to allow rightholders to dispose of their timber as they pleased. While peasants believed that they could not dispose of the produce of the Class III forests as they liked and that their control was only a formal one, the government for its part was loath to give up its monopoly over the timber trade. Extending over three decades and conducted through a series of petitions and representations, this was in essence a dispute over the proprietary claims of the two parties. As the superintendent of the district observed, villagers were concerned more with the legal status of the Class III forests than with their extent: indeed 'they would be contented to take much less than they have now, if they felt it was their own'.[48]

The unsettled state of the forest boundaries had made the peasantry suspicious that the government would slowly take over the Class III forests and put them under commercial management. On a tour of the district, the lieutenant-governor of the province encountered repeated complaints concerning the 'severity of the forest rules,' dwelling chiefly on the fact that no forest or wasteland was made over to them in absolute proprietary right, and so they were afraid that at some future period 'government might resume the whole of it and leave them destitute'. As one hill man succinctly put it, 'the forests have belonged to us from time immemorial: our ancestors planted them and have protected them: now that they have become of value, government steps in and robs us of them'. The superintendent urged a revision of the forest boundaries and the confirmation of village proprietorship in Class III forests since 'nothing would tend to allay the irritation and discontent in the breasts of the people so much as giving them a full proprietary title to all lands not required by government'.[49]

[48] Uttar Pradesh Regional Archives, Dehra Dun (hereafter U.P.R.A.), Post Mutiny Records (hereafter P.M.R.), List No. 2 (hereafter L2), Dept. XI, file no. 71, H.G. Ross, superintendent of Dehra Dun, memorandum on verbal complaints made to the lieutenant-governor by *syanas* (headmen) of Jaunsar and Bawar (n.d., but probably 1871 or 1872).

[49] U.P.R.A., P.M.R., L2, Dept. I, file no. 2, no. 340, superintendent, Dehra Dun, to commissioner, Meerut Division, 15 September 1873. See also C. Bagshawe,

At the level of everyday existence, the restrictions on customary use under the Forest Act were regarded as unnecessarily irksome. Thus the government tried, not always with success, to restrict the use of deodar (Himalayan cedar, the chief commercial species) by villagers, arguing that, while the peasants were 'clearly entitled to wood according to their wants, nothing is said about its being *deodar*'. This legal sleight of hand did not always succeed, as villagers insisted on claiming deodar as part of their allotted grant, the wood being extensively used in the construction of houses.[50] Again, the takeover of village grazing lands and oak forests to supply the fuel and grass requirements of Chakrata cantonment was a grievance acknowledged by district officials to be legitimate, even if they could do little about it within the overall structure of colonial administration. Particularly contentious were proposals to regulate or ban the traditional practice of burning the forest floor before the monsoons for a fresh crop of grass. While this was regarded by the Forest Department as essential for the reproduction of timber trees, it led to the drying-up of grass and, consequently, a shortage of green fodder, as well as a proliferation of ticks.[51] Pointing to deodar forests where numerous young seedlings had sprung up despite the constant grazing and even occasional fires, villagers were openly sceptical of the department's claim that closure was 'scientific'.[52] An additional reason for the persistent hostility towards grazing restrictions was the liberal allowance extended to nomadic cattle herders from the plains. Important as suppliers of milk to the cantonment and to lumbermen working in the forest, these herdsmen from the Muslim community of Gujars were allowed access to forest pasture even in areas where sheep and cattle belonging to the local peasantry were banned.[53]

'Forest Rights in Jaunsar', in D. Brandis and A. Smythies (eds.), *Report on the Proceedings of the Forest Conference Held at Simla, October 1875* (Calcutta, 1876), p. 33.

[50] G.F. Pearson, 'Deodar Forests of Jaunsar Bawar', in *Selections from the Records of the Government of the Northwestern Provinces*, 2nd ser., ii (Allahabad, 1870).

[51] U.P.R.A., P.M.R. L.2. file no. 244, note by C. Streadfield, superintendent, Dehra Dun. 1 Nov. 1898. See also Guha, 'Scientific Forestry and Social Change', for attempts to resolve this conflict.

[52] See E.C. McMoir, 'Cattle Grazing in Deodar Forests', *Indian Forester*, viii (1882), pp. 276-7.

[53] U.P.R.A., P.M.R. L.2. file no 244, no. 483, B.B. Osmaston, deputy conservator of forests, Jaunsar Division, to assistant superintendent, Jaunsar Bawar, 19

The Forest Department also prohibited the use of the axe by peasants claiming their allotment of timber. Villagers demurred, arguing that the saw was too expensive, that they were not familiar with its use, that split wood lasted longer than sawn and, finally, that since their forefathers had always used the axe, so would they. As a consequence, attempts to insert a clause prohibiting the use of the axe in the land settlement of 1873 came to nothing. Although the settlement had considerably raised the land revenue, the main grievance expressed continued to be the infringement of village rights over forest. Village headmen first asked for a postponement of the settlement, and then drove a hard bargain, agreeing to the new revenue rates and the continuance of forest restrictions only on condition they were allowed to use axes in obtaining their grants of timber from forest land.[54]

If such petitions represented an appeal to the 'traditional' obligations of the state,[55] the peasants of Jaunsar Bawar also resorted to extra-legal forms of protest which defied the government's control over forest extraction and utilization. Before an era of motorized transport, commercial forestry depended on the fast-flowing hill rivers to carry felled logs to the plants, where they were collected by timber merchants and sold as railway sleepers. Nearly two million sleepers were floated annually down the Yamuna and its chief tributary, the Tons, and they were considered to be the property of the Forest Department. Although villagers dwelling on the river banks had been 'repeatedly warned that Government property is sacred', thefts were endemic. As 'every Jaunsari knows well all about the working of the Government forests and the floating of timber', officials tried to stop pilfering by levying heavier sentences than those sanctioned by the Forest Act. Thus, while each sleeper was worth only 6 rupees, it was

---

Mar. 1899. This clash between the pesantry and Gujars, with the Forest Department trapped in between, has persisted to this day. See Bharat Dogra, *Forests and People* (Rishikesh, 1980).

[54] U.P.R.A. P.M.R., L.2. file no. 244, no. 520, report on forest administration in Jaunsar Bawar, submitted by superintendent, Dehra Dun to commissioner, Meerut Division, 10 Dec. 1900: L.2. Dept. XXI, file no. 244, E.C. Buck, officiating secretary to Board of Revenue, N.W.P, to C.A. Elliot, secretary to government of N.W.P. L.2. file no. 2, no. 47, settlement officer, Jaunsar Bawar, to commissioner, Meerut Division, 17 Feb. 1872.

[55] That is, what James Scott has called the key reciprocal duty of non-cultivating elites in peasant societies, the guaranteeing of subsistence. See J.C. Scott, *The Moral Economy of the Peasant* (New Haven, 1976).

not unknown for villagers caught in possession of one to be sentenced to two months' rigorous imprisonment or a fine of 30 rupees. Stiff sentences needed to be enforced, magistrates argued, as 'river thieves are pests and a deterrent fine is necessary'. Such measures failed to have the anticipated effect and as late as 1930—a full sixty years after the state takeover of woodland—the superintendent of the district was constrained to admit that 'pilfering, misappropriating and stealing Government and State timber' was 'a chronic form of crime in Jaunsar Bawar'.[56]

As in eighteenth-century England, the infringement of forest laws, which was viewed as 'crime' by the state, was an assertion of customary rights and as such it represented an incipient form of social protest.[57] In Jaunsar Bawar the theft of floating timber and the defacement of government marks were accompanied by other forms of forest 'crime', notably the infringement of the laws preventing forest fires. In a fascinating incident, the head priest of the major temple of the area, dedicated to the god Mahashu Devta at Hanol,[58] organized a firing of the pine forest to get rid of the dry grass and the insects it harboured and of the deer who were a hazard to the adjoining croplands. Under the direction of the priest, Ram Singh, several villagers set fire to the forest on the night of 13 July 1915. Under Section 78 of the Forest Act villagers were liable to inform the forest staff of any fire in their vicinity. This they proceeded to do but only after several hours had elapsed. Ram Singh then advised a low-caste labourer, Dumon Kolta, to call the forest guard but to go slowly.

While early enquiries clearly revealed that the fire was not accidental, its occurrence near the Mahashu Devta temple and the involvement of its priest made it difficult for the state to convict those accused.[59] Indeed several prosecution witnesses, after a meeting with village headmen at the temple suddenly retracted their confessions in court. Expected by the state

[56] See Dehra Dun Collectorate, Criminal Record Room, Basta (Box) for 1927-30 for Chakrata Tehsil, trial nos. 98 of 1925, 36 of 1927, 53 of 1930, and unnumbered trials dated 1 May, 15 June 1922, 7 April 1923.

[57] See the fine studies by Douglas Hay and E.P. Thompson in Douglas Hay *et al.*, *Albion's Fatal Tree* (Harmondsworth, 1976); and Thompson's *Whigs and Hunters* (Harmondsworth, 1976).

[58] For the importance of the deity in the social and cultural life of the area, see the sensitive study by Jean Claude Galey, 'Creditors, Kings and Death', in Charles Malamud (ed.), *Debts and Debtors* (New Delhi, 1983).

[59] The oath in the court of Jaunsar Bawar was taken in the name of Mahashu Devta.

to act as a bulwark of the administration, the headmen underlined their partisan stance by appearing *en masse* for the defence. One elder, Ranjit Singh, (whose fields were closest to the forest fire) disavowed the *wajib-ul-arz* (record of rights) which required headmen personally to put out fires and collect other villagers for the same purpose. As he defiantly told the divisional forest officer, 'such a wajib-ul-arz should be burnt and ... his ancestors were ill-advised to have agreed to such a wajib-ul-arz with the Government'.[60]

Such organized and collective violations were hardly as frequent, of course, as the numerous acts of individual 'crime'. In Jaunsar Bawar, centuries of unrestricted use had fostered the belief that the forests were open and accessible to all villagers. Not surprisingly, the demarcation of forest land as government property aroused a 'great cry'.[61] What differentiates Jaunsar from other forest areas where protest took a more open and militant form is the reliance on individual and largely 'hidden' forms of resistance. But this was an equally effective strategy in thwarting colonial forest administration. As an official reflecting on the history of state forestry in Jaunsar Bawar remarked, 'prosecutions for forest offences, meant as deterrents, only led to incendiarism, which was followed by more prosecutions and the vicious circle was complete'.[62] Clearly, these ostensibly individual acts of violation relied on a network, however informal, of consensus and support within the wider community. Since all strata of village society were uniformly affected by commercial forestry, every violation of the Forest Act could draw sustenance from a more general distrust of state control; and since individuals could quite easily be subject to the due processes of colonial justice, this resistance could hardly 'hope to achieve its purpose except through a generalized, often unspoken complicity'.[63]

[60] See Dehra Dun Collectorate, Criminal Record Room, criminal case no. 98 of 1915. Ram Singh and five others were sentenced to terms of imprisonment ranging from three months to a year.

[61] See E.T. Atkinson, *The Himalayan Districts of the Northwestern Provinces of India*, i (1882; repr. Delhi, 1981), p. 870.

[62] M.D. Chaturvedi, 'The Progress of Forestry in the United Provinces', *Indian Forester*, li (1925), p. 365.

[63] Cf. J.C. Scott, 'Resistance without Protest and without Organization: Peasant Opposition to the Islamic and the Christian Tithe', *Comp. Studies in Soc. and Hist.*, xxix (1987), pp. 417-52.

V. *The Decline of Artisanal Industry*

Apart from its all too visible impact on the cultivating classes, state forest management also contributed to the decline of various forms of artisanal industry by restricting access to traditional sources of raw material.[64] Chief among these was bamboo, a resource vital to many aspects of rural life. Extensively used in house construction, basket-weaving, for the manufacture of furniture and musical instruments, and even as food and fodder,[65] this plant was initially treated as a weed by colonial foresters; and early management plans advocated its removal from timber-producing areas. With the discovery in the early decades of this century that bamboo was a highly suitable raw material for paper-making, there was a radical shift: foresters now encouraged industrial exploitation while maintaining restrictions on village use. Many weavers were forced to buy bamboo from government-run depots or on the open market.[66] Limited availability also led to new forms of social conflict within the agrarian population. Thus the Baigas, who had earlier supplemented slash-and-burn agriculture with bamboo-weaving, lost this subsidiary source of income when the Basors, an artisanal caste specializing in basketwork, asserted their 'trades union' rights to a monopoly of bamboo supplied by the Forest Department.[67]

While bamboo, whether obtained surreptitiously from the forest or bought in the market, continues to play an important role in present-day village society, one form of indigenous industry which collapsed under colonial rule was the manufacture of charcoal-based iron. Again we are indebted to Verrier Elwin for a sensitive study of the industry in its declining years. In his book on the Agaria, an iron-smelting tribe of the Central Provinces, Elwin describes in chilling detail how the high taxes on furnaces and diminished supplies of charcoal led to a sharp fall in the number of operating furnaces—from 510 to 136 between 1909 and 1938. Although peasants preferred the soft malleable ores of village smelters, changing circumstances virtually forced the Agaria out of business, especially since improved communications made local iron uncompeti-

---

[64] There is an extensive literature on the decline of Indian handicrafts under British rule. An early statement of the 'deindustrialization' thesis is D.R. Gadgil, *The Industrial Evolution of India in Recent Times* (Oxford, 1922).

[65] See S. Kurz, 'Bamboo and its Use', *Indian Forester*, i (1876).

[66] See S.N. Prasad and Madhav Gadgil, *Conservation of Bamboo Resources in Karnataka* (mimeo, Karnataka State Council of Science and Technology, Bangalore, 1981).

[67] Elwin, *Baiga*, p. 80.

tive when compared to imported British metal. Deeply attached to their craft, the Agarias resisted as best they could by defying forest laws concerning charcoal-burning or, alternatively, migrating to nearby chiefdoms where they were accorded more liberal treatment.[68] In an extensive survey of Madras Presidency, the first inspector-general of forests, Dietrich Brandis, provided confirmatory evidence of this decay of an industry that was formerly very widespread.[69]

Proposals to set up ironworks controlled by European capital did briefly evoke an interest in the conservation of trees for charcoal. Pointing out that the metallic content of Indian ores was nearly twice that of European, several administrators urged the reservation of large tracts of forests for the benefit of European-owned-and-managed works using the latest technological processes. Here the expansion of charcoal-based iron production was predicated on the assumption that 'iron-making by hand in India will soon be counted among the things of the past'.[70] While acknowledging that the abundance of wood in presently inaccessible areas made the promotion of charcoal iron a potential source of forest income, Brandis advocated a different form of utilization. Articulating an early version of 'intermediate' or 'appropriate' technology, he believed that any such attempt must build upon, rather than supplant traditional forms of manufacture. In the event both proposals came to nothing and the industry died an inevitable if slow death.[71]

[68] Verrier Elwin, *The Agaria* (Calcutta, 1942), pp. xxiv-xxv, 31-2, 39, 121-2, 241-3, *et al.* Cf also S. Bhattacharya, 'Iron Smelters and the Indigenous Iron and Steel Industry of India. From Stagnation to Atrophy', in S. Sinha (ed), *Aspects of Indian Culture and Society* (Calcutta, 1972). As is evident, this article draws heavily on the contemporary writings of anthropologists. Elwin and Von Furer Haimendorf, in particular, have portrayed with great sensitivity and skill the processes of economic and cultural deprivation whereby different communities lost control over nature and over their means of subsistence. As detailed and firsthand accounts of socio-ecological changes under colonialism, their writings should qualify as authentically 'primary' sources.

[69] Brandis, *Suggestions Regarding the Management of Forests*, pp. 53, 157-8, 182.

[70] Anon., 'Iron-Making in India', *Indian Forester*, vi (1880), pp. 203, 208. See also H. Warth, *Notes on the Manufacture of Iron and the Future of the Charcoal Iron Industry in India* (Simla, 1881).

[71] Brandis, *Suggestions Regarding the Management of Forests*, pp. 53-9, 136, 153-5; D. Brandis, 'The Utilization of the Less Valuable Woods in the Fire-Protected Forests of the Central Provinces, by Iron-Making', *Indian Forester*, v.

Other forms of artisanal industry, too, declined under these twin pressures: the withdrawal of existing sources of raw material and competition from machine-made, largely foreign, goods. Thus *tassar* silk industry, depending on the collection of wild cocoons from the forest, experienced a uniform decline through most of India from the 1870s onwards. Here, too, decay could be attributed to the new forest laws: specifically, the increased duties levied on weavers collecting cocoons from the forest. Although the tassar industry experienced a later revival under official patronage (chiefly in response to a growing export market), the household industry was in no position to compete with the newly formed centres of production operating from towns. A parallel case concerns the decline of village tanners and dyers, likewise denied access to essential raw materials found in the forest.[72]

Conclusion: The Social Idiom of Protest and its Mechanisms

As we indicated at the outset, in the absence of detailed studies of the socio-ecological history of different regions, the present study can only provide a preliminary mapping of the various dimensions of forest-based conflict in British India. Through a synthesis of the available evidence from both primary and secondary sources, we have tried to indicate the quite astonishing range of conflicts over access to nature, a range entirely consistent with the wide variety of ecological regimes and, correspondingly, of social forms of resource use prevalent in the Indian subcontinent. Yet even this initial survey reveals some interesting regularities in the form in which protest characteristically expressed itself, notably against the state's attempts to abrogate traditional rights over the forest.

(1879). The vision of a modern charcoal-fired iron furnace finally came to fruition in Karnataka: see M. Visvesvaraya, *Memoirs of my Working Life* (Bangalore, 1951), pp. 92-5.

[72] This paragraph is based on information kindly supplied to the authors by Tirthankar Ray of the Centre for Development Studies, Trivandrum, who is researching handicraft production during the colonial period. Fishing communities were also affected by forest laws, being forced to use inferior wood for canoes owing to the heavy duties levied on teak by the Forest Department. See Grigson, *Maria Gonds*, pp. 163-4. Among other artisanal castes, evidence from Khandesh in western India suggests that bangle-makers were almost ruined by the fee imposed on wood for fuel: see Maharashtra State Archives, Revenue Department, file 73 of 1884 (personal communication from Sumit Guha, St. Stephen's College, Delhi).

In essence, state monopoly and its commercial exploitation of the forest ran contrary to the subsistence ethic of the peasant. To adapt a contrast first developed by E.P. Thompson in his study of the eighteenth–century food riot, if the customary use of the forest rested on a moral economy of provision, scientific forestry rested squarely on a political economy of profit.[73] These two sharply opposed notions of the forest were captured in a perceptive remark made by Percy Wyndham, commissioner of Kumaun, during the uprising of 1921: he observed that the recurrent conflicts were a consequence of the 'struggle for existence between the villagers and the Forest Department; the former to live, the latter to show a surplus and what the department looks on as efficient forest management'. The same duality was invoked by someone ranged on the opposite side of the fence: Badridutt Pande, the leader of the movement, said that, with state management, tins of pine resin had replaced tins of ghee (clarified butter) as the main produce of the forest—a transition with telling consequences for the village economy.[74]

If state monopoly severely undermined village autonomy, then, what is striking about social protest is that it was aimed precisely at this monopoly. In many areas peasants first tried petitioning the government to rescind the regulations. When this had no visible impact, they issued a direct challenge to state control, in the form of attacks on areas controlled by the Forest Department and worked for profit. Whether expressed covertly, through the medium of arson, or openly, through the collective violation of forest laws, protest focused on commercially valuable species—pine, sal, teak and deodar in different geographical regions. Quite often these species were being promoted at the expense of tree varieties less valuable commercially but of greater use to the village economy. While challenging the proprietary right of the state, peasant actions were remarkably discerning. Thus in the Kumaun movement cited above, the 'incendiary' fires of the summer of 1921 covered 320 square miles of *exclusively* pine forests. In

[73] E.P. Thompson, 'The Moral Economy of the English Crowd in the Eighteenth Century', *Past and Present*, no. 50 (Feb 1971).

[74] Uttar Pradesh State Archives, Forest Department, file 109 of 1921, D.O. no. 67 II/21, Percy Wyndham to H.S. Crosthwaite, 27 Feb. 1921; file 157 of 1921, court of W.C. Dible, district magistrate, Almora, criminal case no. 7, 7 July 1921. For a perceptive study of alternate notions of property and resource use among English colonists and native Indians in North America, see William Cronon, *Changes in the Land: Indians, Colonists and the Ecology of New England* (1983; repr. New York, 1985), esp. ch. 4.

other words, by design rather than accident, the equally vast areas of broad-leaved forests *also controlled* by the state were spared, being of greater use to hill agriculture. As in peasant movements in other parts of the world, arson as a technique of social protest had both a symbolic and a utilitarian significance: the latter by contesting the claim of the state over key resources, the former by selectively choosing targets where the state was most vulnerable.[75]

Hisorical parallels with other peasant movements far removed in time and space are evident, too, in the close association of protest with popular religion. The ideology of social protest was heavily overlaid with religious symbolism. In the imagery of the famous Hindu epic, the *Ramayana*, for example, the British government was portrayed as a demonic government (*Rakshas Raj*) and the king emperor equated with the very personification of evil, the demon king Ravan.[76]

A religious idiom also reflected the sense of cultural deprivation consequent on the loss of control over resources crucial to subsistence. In many areas the customary use of nature was governed by traditional systems of resource use and conservation which involved a mix of religion, folklore and tradition regulating both the quantum and the form of exploitation.[77] The suppression and occasionally even obliteration of these indigenous systems of resource management under colonial auspices was acutely felt by different communities albeit in somewhat different ways. The Baigas, for example, resisted attempts to convert them into plough

[75] Modern environmentalists concerned with the abuse of nature for profit have also considered using forms of directed arson or 'ecotage' (ecological sabotage). Thus a group in the western United States which had previously fixed spikes in trees to thwart logging has now threatened to burn forests marked for felling, with the justification that, while fires were 'natural', logging brought in roads and more felling. See Nicholas D. Kristof, 'Forest Sabotage Is Urged by Some', *New York Times*, 22 Jan. 1986, p. A-21.

[76] Guha, 'Forestry and Social Protest', p. 92. On the religious idiom of peasant protest, see the important work by Ranajit Guha, *Elementary Aspects of Peasant Insurgency in Colonial India* (New Delhi, 1983). Religion was sometimes invoked to stall attempts to take over forests. Thus the manager of a temple grove in Malabar refused to lease the forest to the government, on the grounds that the temple deity had threatened him with dire consequences if he entered into such an agreement: see Anon., *Selection of Despatches*, pp. 213-15.

[77] For a review of traditional conservation practices in India, see Madhav Gadgil, 'Social Restraints on Resource Utilization: The Indian Experience', in D. Pitt and J.A. Mcneely (eds.), *Culture and Conservation* (Dublin, 1985).

agriculturalists by invoking their myth of origin in which they had been told specifically not to lacerate the breasts of mother earth with the plough. As Elwin observes, 'every Baiga who has yielded to the plough knows himself to be standing on *papidharti* or sinful earth'. However reluctant this conversion, it was not without divine retribution: as one Baiga put it, 'when the *bewar* (slash and burn) was stopped and we first touched the plough . . . a man died in every village'.[78]

The Gonds, aboriginal plough cultivators, were similarly afflicted by a melancholia or what Elwin has elsewhere called a 'loss of nerve'.[79] They were convinced that the loss of their forests signalled the coming of *Kaliyug*, an age of darkness in which their extensive medical tradition would be rendered completely ineffective. So insidious and seductive was the power of modern civilization that even their deities had gone over to the camp of the powerful. Unable to resist the changes wrought by that ubiquitous feature of industrial society, the railway, 'all the gods took the train, and left the forest for the big cities'—where with their help the town-dweller prospered.[80]

The belief that traditional occupations were sanctioned by religion was evident, too, in the obvious reluctance of the Agaria to abandon iron-smelting. According to *their* myth of origin, both slash-and-burn and plough cultivation were sinful. The Agaria believed that in the old days, when they were faithful to iron, they had enjoyed better health. Now that government taxes and scarcity of charcoal had forced many ironworkers to take to cultivation, their gods no longer provided immunity from disease. The real point of conflict with authority concerned charcoal-burning, and this was vividly reflected in the numerous dreams that hinged on surreptitious visits to the jungle, and which often culminated in the

[78] Elwin, *Baiga*, pp. 106–7. See also R.N. Datta, 'Settlement of Tribes Practising Shifting Cultivation in Madhya Pradesh (India)', *Indian Forester*, ixxxi (1955), p. 371. Drawing a parallel with attempts to settle American Indians, Elwin quotes Smohalla, prophet of a 'messianic' cult of the Columbia River Basin, who told his followers in 1870: 'You ask me to plough the ground. Shall I take a knife and tear my mother's bosom? You ask me to dig for stone. Shall I dig under her skin for her bones? You ask me to cut grass. But how dare I cut off my mother's hair?'

[79] Elwin, *Aboriginals, passim*.

[80] Verrier Elwin, *Leaves from a Jungle* (1936; repr. London, 1958), p. 58; Verrier Elwin, *A Philosophy for NEFA* (Delhi, 1960), p. 80: Shamrao Hivale and Verrier Elwin, *Songs of the Forest: The Folk Poetry of the Gonds* (London, 1935), pp. 16–17.

Agaria being intercepted and beaten up by forest officials.[81]

Researches over the past two decades have demonstrated that while peasants operate in a world largely composed of 'illiterates' whose movements lack a written manifesto, their actions are imbued with a certain rationality and an internally consistent system of values. It is the task of the scholar to reconstruct this ideology—an ideology that informs the peasant's everyday life as much as episodes of revolt—even where it has not been formally articulated.[82] From a reconstruction of the different episodes of social protest surveyed in this article, we can discern a definite ideological content to peasant actions. Protest against enforced social and ecological changes clearly articulated a sophisticated theory of resource use that had both political and cultural overtones.

Of special significance is the wide variety of strategies used by different categories of resource users to oppose state intervention. Hunter-gatherers and artisans, small and dispersed communities lacking an institutional network of organization were unable directly to challenge state forest policies. They did, however, try to break the new regulations by resorting chiefly to what one writer has called 'avoidance protest': petty crime or migration, for example, which minimized the element of confrontation with the state.[83] In the long term, however, these groups were forced to abandon their traditional occupations and to eke out a precarious living by accepting a subordinate role in the dominant system of agricultural production. Both slash-and-burn and plough agriculturalists were able to mount a more sustained opposition. Their forms of resistance ranged from individual to collective defiance, from passive or 'hidden' protest to open and often violent confrontation with instituted authority. Tightly knit in cohesive 'tribal' communities, jhum cultivators characteristically responded to forest laws with a militant resistance which was almost wholly outside the stream of organized nationalism. The fate of this protracted resistance varied greatly across different regions. Occasionally the colonial state capitulated, allowing traditional forms of cultivation to continue. More frequently the state reached an accommodation with these commu-

[81] Elwin, *Agaria*, pp. 264, 267-8.

[82] One may cite in this connection the work of Rodney Hilton, Eric Hobsbawm, George Rudé, Jim Scott and E.P. Thompson and, in India, the writings of the 'Subaltern Studies' School.

[83] Michael Adas, 'From Avoidance to Confrontation: Peasant Protest in Precolonial and Colonial Southeast Asia', *Comp. Studies in Soc. and Hist.* xxiii (1981), pp. 217-47.

nities, restricting but not eliminating jhum cultivation. The consequent shrinkage of the forest area available for swidden plots, coupled with rising population, led gradually to a reduction in fallow cycles and to declining yields. A large proportion of jhum cultivators have therefore had no alternative to becoming landless labourers.

Settled cultivators have perhaps been more successful in retaining some degree of control over forest resources. While sharply limiting access, the new laws did not seriously threaten the livelihood of agriculturalists and graziers. Since subordinate forest officials commonly hailed from the same castes, the peasantry was often able to obtain forest produce by bribing rangers and guards. While the *cost of access* may have increased significantly in such cases, the deprivation of forest resources was very rarely total. Moreover, Hindu peasants protesting against forest restrictions were more successful in using the resources and strategies of modern nationalism, such as petitions and litigation, to advance their own interests. Whatever the specific modalities of protest at different time periods, and across different regions and forms of resource use, it was in its essence 'social': it reflected a general dissatisfaction with state management of the forest and it rested heavily on traditional networks of communication and co-operation. It is noteworthy that traditional leaders of agrarian society— clan and village headmen—almost always played a key role in social mobilization and action. Since the colonial state regarded them as local bulwarks of power and authority, such leaders were subjected to conflicting pressures; but they usually decided to throw in their lot with their kinspeople. The tenuous hold exercised by the premier nationalist organization, the Congress, over most of the movements described in this article is also instructive. Although individuals like Gandhi may have recognized the importance of natural resources such as salt and forest produce in the agrarian economy, even protests formally conducted under the rubric of Congress often enjoyed a considerable autonomy from its leadership. Social protest over forest and pasture pre-dated the involvement of the Congress; and even when the two streams ran together they were not always in tune with one another. Finally, these conflicts strikingly presaged similar conflicts in the post-colonial period. Contemporary movements asserting local claims over forest resources have replicated earlier movements in their geographical spread, in the nature of their participation and in their strategies and ideology of protest.

A study of colonial history may thus have more than a fleeting relevance to contemporary developments. Nowhere is this more true than in the highly contentious sphere of forest policy. Here a vigorous debate among intellectuals, policy-makers and grass-roots organizations has in recent years brought to the fore two opposed notions of property and resource use: on the one hand, communal control over forests is paired with subsistence use, and on the other, state control with commercial exploitation. Yet this duality merely mirrors, albeit in a more formal and institutionalized fashion, the popular opposition to state control over forests which was endemic during the period of colonial rule. The movements described in this article may have been short-lived and unsuccessful, but their legacy is very much with us today.[84]

[84] Madhav Gadgil and Ramachandra Guha, 'The Two Options in Forest Policy', *Times of India*, 12-13 Sept. 1984.

Annotated Bibliography

1. GENERAL

Desai, A. R., *Peasant Struggles in India* (New Delhi: 1979). A collection of previously published essays—of very uneven quality—covering the entire period of British rule, with introductory essays to each section by A. R. Desai. Pages 125–237 cover the 1855–1918 period.

Dhanagare, D. N., *Peasant Movements in India 1920–50* (New Delhi: 1983). Chapter 1 deals with theoretical approaches to the study of peasant resistance; chapter 2 provides a brief history of peasant revolts in India before 1920.

Gough, Kathleen, 'Indian Peasant Uprisings', *Economic and Political Weekly*, special number, August 1974. Reprinted in A. R. Desai, *Peasant Struggles in India*. Taking the Naxalite upsurge in rural areas as the starting point, Gough argues that there is a long history of peasant resistance in India.

Guha, Ranajit, *Elementary Aspects of Peasant Insurgency in Colonial India* (New Delhi: 1983). Path-breaking study of the structure of peasant resistance in nineteenth-century India which should be read alongside this collection of essays.

Moore, Barrington, *Social Origins of Dictatorship and Democracy: Lord and Peasant in the Making of the Modern World* (Harmondsworth: 1973; first published in the U.S.A. in 1966). Chapter 6 is on India. Moore argues that the Indian peasantry was politically docile and that there was no strong tradition of peasant revolt in India, as in China.

Sarkar, Sumit, *Modern India 1885–1947* (New Delhi: 1983). Pages 43–54 provide an overview of peasant resistance in the period 1860–1905 and pages 153–6 for the period 1905–14.

Stokes, Eric, *The Peasant and the Raj: Studies in Agrarian Society and Peasant Rebellion in Colonial India* (Cambridge: 1978). The essays specifically on peasant resistance in this collection of previously-published articles are all on the revolt of 1857. The last chapter, 'The Return of the Peasant to South Asian History' (originally published in *South Asia* 6, December 1976), has something to say on peasant resistance after 1917.

2. THE PEASANT COMMUNITY

Chatterjee, Partha. (a) 'Agrarian Relations and Communalism in Bengal, 1926–1935' in Ranajit Guha (ed.), *Subaltern Sudies I* (New Delhi: 1982). (b) 'More on Modes of Power and the Peasantry', in Ranajit Guha (ed.), *Subaltern Studies II* (New Delhi: 1983). (c) 'For an Indian History of Peasant Struggles', *Social Scientist* vol. 16, no. 11, November 1988. These three essays set out Chatterjee's important concept of 'the communal mode of power'.

3. RESISTANCE TO INDIGO PLANTERS

(i) Bengal

Chowdhury, B. B., 'Agrarian Economy and Agrarian Relations in Bengal, 1859–1885', in N. K. Sinha (ed.), *History of Bengal 1757–1905* (Calcutta: 1967). Concentrates on the agrarian economy, but describes the indigo revolts in passing.

Guha, Ranajit, 'Neel-Darpan: The Image of a Peasant Revolt in a Liberal Mirror', *The Journal of Peasant Studies,* vol. 2, no. 1, October 1974. Analyses the peasant struggle, though the focus is on the middle-class reaction to the revolt.

Kaviraj, Narhari, *Wahabi and Farazi Rebels of Bengal* (New Delhi: 1982). Chaper 3 provides a history of the Fara'idi movement from the type of Marxian perspective which, in viewing religion as fundamentally divisive and informed by 'false consciousness', makes it hard for a historian otherwise sympathetic to the peasants' struggle to enter into their mental world.

Khan, Muin-ud-Din Ahmad, *History of the Fara'idi Movement in Bengal (1818–1906)* (Karachi: 1965). Detailed and scholarly account of the Fara'idi movement in East Bengal, which brings out the social background to the movement.

Kling, Blair B., *The Blue Mutiny: The Indigo Disturbances in Bengal 1859–1862* (Philadelphia: 1966). The best full-length study of the indigo revolt of 1859–62, though it focuses on the leading officials of the Bengal government rather than on the peasants and their struggle.

(ii) Bihar

Fisher, Colin, 'Planters and Peasants: The Ecological Context of Agrarian Unrest on the Indigo Plantations of North Bihar, 1820–1920', in Clive Dewey and A. G. Hopkins (eds), *The Imperial Impact: Studies in the Economic History of Africa and India* (London: 1978). Concentrates on the economics of the planter–peasant relationship in Bihar, but has a brief reference to resistance in the the late nineteenth century.

Misra, B. B., 'Introduction' to B. B. Misra (ed.), *Select Documents on Mahatma Gandhi's Movement in Champaran 1917–18* (Patna: 1963). Provides an overview of peasant resistance to planters in late-nineteenth and early-twentieth-century Bihar

Pouchepadass, Jacques, *Planteurs et Paysans dans L'Inde Coloniale: L'Inde du Bihar et le Mouvement Gandhien du Champaran (1917–18)* (Paris: 1986). In French. The most comprehensive study of the indigo system in Bihar and peasant resistance to it.

4. RESISTANCE TO LANDLORDS

(i) Bengal

Chowdhury, B. B., 'Agrarian Economy and Agrarian Relations in Bengal, 1859–1885', in N. K. Sinha (ed.), *History of Bengal 1757–1905* (Calcutta: 1967). Provides a history of the agrarian economy of Bengal as a background to the anti-zamindar movement of the 1870s. Describes the movement in passing.

Chowdhury, B. B., 'Peasant Movements in Bengal, 1850–1900', *Nineteenth Century Studies*, July 1973. Takes issue with K. K. Sengupta's argument that the anti-zamindar movement in Bengal was concerned primarily with defending occupancy rights, arguing instead that it was caused by rent-enhancement.

Chowdhury, B. B., 'The Story of a Peasant Revolt in a Bengal District', *Bengal Past and Present*, July–December 1973. On the anti-landlord movement in Bogra in 1873.

Sen, Sunil, 'Peasant Struggle in Pabna 1872–73' *New Age*, vol. 3, no. 2, February 1954. Five-page article with some useful information on the Pabna struggle.

Sen Gupta, K. K., *Pabna Disturbances and the Politics of Rent*

1873–1885 (New Delhi: 1974). Best existing history of the anti-zamindar movement in Bengal of the 1870s. Some chapters appeared as articles before the book came out, e.g. 'Peasant Struggle in Pabna, 1873. Its Legalistic Character', *Nineteenth Century Studies,* vol. 1, no. 3, July 1973 and the essay reproduced in this collection, 'The Agrarian League of Pabna, 1873', *Indian Economic and Social History Review,* vol. 7, no. 2, June 1970. Both these articles are also published in A. R. Desai, *Peasant Struggles in India.*

Ray, Rajat, *Social Conflict and Political Unrest in Bengal 1975–1927* (New Delhi: 1984). Chapter 1, part II (pp. 45–81) provides an overview of agrarian conditions and resistance to zamindars in late-nineteenth-century Bengal.

(ii) Kerala

Arnold, David, 'Islam, the Mappilas and the Peasant Revolt in Malabar', *The Journal of Peasant Studies,* vol. 9, no. 4, July 1982. Reviews the literature on the Mappila revolts.

Dale, Stephen, *Islamic Society on the South Asian Frontier: The Mappilas of Malabar 1498–1922* (Oxford: 1980). Also, 'The Mappila Outbreaks: Ideology and Social Conflict in Nineteenth-Century Kerala', *Journal of Asian Studies,* vol. 35, no. 1, 1975. See the Mappila uprisings as being primarily religious. The argument begs many questions.

Dhanagare, D. N., 'Agrarian Conflict, Religion and Politics: the Moplah Rebellions in Malabar in the Nineteenth and Early Twentieth Century', *Past and Present,* no. 74, February 1977. Substantially the same as chapter 3 of his *Peasant Movements in India 1920–1950* (New Delhi: 1983). Provides an overview of the subject, arguing that the revolts were an expression of agrarian discontent.

Kurup, K. K. N., *Wiliam Logan: A Study in Agrarian Relations of Malabar* (Calicut: 1981). A useful compilation, which focuses on Logan's investigations into the economic background to the Mappila risings, with excerpts from his writings.

Miller, Roland E., *Mappila Muslims of Kerala: A Study in Islamic Trends* (Madras: 1976). A broad history of the Mappilas, from the sixteenth century to 1976, with 17 pages on the Mappila risings of the nineteenth century.

Panikkar, K. N., *Against Lord and State: Religion and Peasant*

Uprisings in Malabar, 1836–1921 (New Delhi: 1989). The best full-length study of the Mappila risings, which argues that the grievances of the Mappilas against their landlords were translated into action through their religion.

Wood, Conrad, 'Peasant Revolt: An Interpretation of Moplah Violence in the Nineteenth and Twentieth Centuries', in Clive Dewey and A. G. Hopkins (eds), *The Imperial Impact* (London: 1978). Argues convincingly that the Mappila uprisings were caused by landlord exploitation. Also, 'The Moplah Rebellions between 1800–02 and 1921–22', *Indian Economic and Social History Review*, vol. 13, no. 1, 1976.

(iii) Mewar

Surana, Pushpendra, *Social Movements and Social Structure: A Study in the Princely State of Mewar* (New Delhi: 1983). Describes the movement by the peasants of Bijolia (Mewar State of Rajasthan) against their landlords, which began in 1897 and continued after our period.

5. RESISTANCE TO MONEYLENDERS

(i) Maharashtra

Catanach, Ian, 'Agrarian Disturbances in Nineteenth-Century India', *Indian Economic and Social History Review*, vol. 3, no. 1, March 1966. Attempts an analysis of the structure of the 1875 rising which is suggestive but often inadequate.

Catanach, Ian, *Rural Credit in Western India* (California: 1970). Chapter 1 is on the rising of 1875.

Charlesworth, Neil, 'The Myth of the Deccan Riots of 1875', *Modern Asian Studies,* vol. 6, no. 4, 1972. A controversial attempt to show that the 1875 rising was minor and of no long-term significance. Makes some telling points, but is generally questionable.

Charlesworth, Neil, *Peasants and Imperial Rule: Agriculture and Agrarian Society in the Bombay Presidency 1850–1935* (Cambridge: 1985). Chapter 4 is on the 1875 rising. The argument is substantially the same as in the article of 1972, though it is stated less polemically and with a more careful use of evidence.

Kumar, Ravinder, 'The Rise of the Rich Peasant in Western India', in D. A. Low (ed.), *Soundings in Modern South Asian History* (London: 1968). Argues that British rule led, in nineteenth-century Maharashtra, to the emergence of a rich peasantry. This has been subsequently challenged.

Kumar, Ravinder, 'The Deccan Riots of 1875', *The Journal of Asian Studies*, vol. 24, no. 4, August 1965. Analysis of the underlying causes of the rising of 1875, rather than the rising itself. Argues that a variety of grievances came together in 1875.

Kumar, Ravinder, *Western India in the Nineteenth Centurey: A Study in the Social History of Maharashtra*, (London: 1968). Provides more details on the background to the 1875 rising, with chapter 5 concentrating on the rising.

Report of the Committee on the Riots in Poona and Ahmednagar 1875 (Bombay: 1876). Report with three separate volumes of appendices containing evidence submitted to the committee. The later are of great value. The report blames the legal system for the rising of 1875, rather than British rule in a more general sense. The appendices contain details of attacks on moneylenders elsewhere in Bombay Presidency in the decades before 1875.

(iii) Elsewhere

Hardiman, David, 'The Bhils and Shaukars of Eastern Gujarat', in Ranajit Guha (ed.), *Subaltern Studies V* (New Delhi: 1987). Examines the history of the relationship between the Bhils and moneylenders and describes an uprising against the moneylenders during the famine of 1899.

6. RESISTANCE TO STATE TAXES

Barrier, N. Gerald, 'The Punjab Disturbances of 1907: The Response of the British Government in India to Agrarian Unrest', *Modern Asian Studies*, vol. 1, no. 4, 1967. This essay is strongest on the government reaction to the movement; on the peasants themselves it is sketchy.

Cashman, Richard, *The Myth of the Lokamanya: Tilak and Mass Politics in Maharashtra* (Berkeley: 1975). Chapter 4 describes the no-tax campaign in Maharashtra during the famine of 1896–7.

Copland, Ian, 'The Cambay "Disturbances" of 1890: Popular Protest in Princely India', *Indian Economic and Social History Review*, vol. 20, no. 4, October–December 1983. Poor analysis of a no-tax campaign in Cambay State in central Gujarat.

Guha, Amalendu, *Planter Raj to Swaraj: Freedom Struggle and Electoral Politics in Assam 1826–1947* (New Delhi: 1977). Has a description of resistance to land tax increases in Assam in 1893–4.

Kumar, Ravinder, 'The Deccan Riots of 1875' *The Journal of Asian Studies*, vol XXIV, no. 4, August 1965. describes the support of Pune politicians for the peasant agitation against land-tax increases in Pune District in 1873–4.

7. RESISTANCE TO FOREST LAWS

Bhattacharya, Neeladri, 'Colonial State and Agrarian Society', in R. Thapar and S. Bhattacharya (eds), *Situating Indian History* (New Delhi: 1986). One section of the essay shows how resistance by forest dwellers in the Punjab hills (now Himachal Pradesh) led to modifications of government forest policy.

Guha, Ramachandra, 'Forestry in British and Post-British India: A Historical Analysis', *Economic and Political Weekly*, 29 October 1983. Concentrates on the impact of British forest policy on the forests of India.

Guha, Ramachandra, *The Unquiet Woods: Ecological Change and Peasant Resistance in the Himalaya* (New Delhi: 1989). A detailed history of resistance to forest laws in the Uttar Pradesh hills. Parts of this were published in 'Forestry and Social Protest in British Kumaun, *c.* 1893–1921', in Ranajit Guha (ed.), *Subaltern Studies IV* (New Delhi: 1985).

Guha, Ramachandra and Gadgil, Madhav, 'State Forestry and Social Conflict in British India', *Past and Present*, no. 123, May 1989. Overview of protests by forest dwellers against colonial forest laws.

Tucker, Richard, 'Forest Management and Imperial Politics: Thana District, Bombay 1823–1887', *The Indian Economic and Social History Review*, vol. 16, no. 3, July–September 1979. Deals with the impact of forest policy in Thane District. Fails to describe the resistance by the adivasis of the area to this policy, though it did occur.

8. ADIVASI REVOLTS

Arnold, David, 'Rebellious Hillmen: The Gudem-Rampa Risings,

1839–1924', in Ranajit Guha (ed.), *Subaltern Studies I* (New Delhi: 1982). Examines the long history of *fituris* (uprisings) by hill dwellers of the present Andhra Pradesh region. During our period there were major uprisings in 1879–80 and 1886.

Fuchs, Stephen, *Rebellious Prophets: A Study of Messianic Movements in Indian Religion* (Bombay: 1965). Brief descriptions of a large number of adivasi revolts in India which are often sketchy and inaccurate. A good starting-point for further investigation, however.

Gazetteer of the Bombay Presidency, vol. III, *Kaira and Panch Mahals* (Bombay: 1879). Pages 255–8 describe the Naikda revolt in eastern Gujarat of 1868 led by Joria.

Sarkar, Sumit, *Modern India 1885–1947* (New Delhi: 1983). Pages 44–5 provide an overview of adivasi movements and revolts in India during the late nineteenth century.

Singh, K. Suresh, *The Dust Storm and the Hanging Mist: A Study of Birsa Munda and his Movement in Chhotanagpur* (Calcutta: 1966). The classic study of Birsa Munda's uprising of 1899–1900.

Smith, E. E. Clement, 'The Bastar Rebellion, 1910', *Man in India*, Rebellion Number, December 1945.

9. OTHER FORMS OF RESISTANCE

Arnold, David, 'Touching the Body: Perspectives on the Indian Plague, 1896–1900', in Ranajit Guha (ed.), *Subaltern Studies V* (New Delhi: 1987). Mentions—in passing—peasant resistance to British anti-plague measures.

Catanach, I. J., 'Plague and the Indian Village, 1896–1914', in Peter Robb (ed.), *Rural India: Land, Power and Society under British Rule* (London: 1983). Unsatisfactory examination of resistance by peasants to British anti-plague measures.

Hardiman, David, 'From Custom to Crime: The Politics of Drinking in Colonial South Gujarat', in Ranajit Guha (ed.), *Subaltern Studies IV* (New Delhi: 1985). Describes the imposition of colonial liquor laws on the peasants of South Gujarat in the late nineteenth century and the subsequent resistance.

10. MIDDLE CLASS NATIONALISTS AND PEASANT RESISTANCE

Cashman, Richard, *The Myth of the Lokamanya: Tilak and Mass Politics in Maharashtra* (Berkeley: 1975). Chapter 4 deals with

Tilak's work amongst the peasants of Maharashtra in 1896–7.

Chandra, Bipan, *The Rise and Growth of Economic Nationalism in India: Economic Policies of Indian National Leadership, 1880–1905* (New Delhi: 1966). Contains a detailed survey of early nationalist attitudes towards the peasant question. A defence of the liberal-nationalist position.

Guha, Ranajit, 'Neel-Darpan: The Image of a Peasant Revolt in a Liberal Mirror', *The Journal of Peasant Studies*, vol. 2, no. 1, October 1974. Strong critique of the attitude of the Calcutta intelligentsia towards the revolt of the peasants against indigo planters in 1859–62.

Joshi, V. S., *Vasudeo Balvant Phadke: First Indian Rebel against British Rule* (Bombay: 1959). Biography of V. B. Phadke from a strongly nationalist perspective.

Masselos, J. C., *Towards Nationalism: Public Institutions and Urban Politics in the Nineteenth Century (Bombay: 1974)*. Pages 109–31 on Pune politicians and the agrarian question during the 1870s.

McLane, John, *Indian Nationalism and the Early Congress* (Princeton: 1977). Contains by far the best available account of the attitude of the early Congress leaders towards the agrarian question.

Sen, Sunil, 'Peasant Struggle in Pabna 1872–73' *New Age*, vol. 3, no. 2, February 1954. Short description of the Pabna movement with some interesting material on the play *Zamindar Darpan* and Bankim Chandra Chattopadhyay's condemnation of it.

Source Material for a History of the Freedom Movement in India Collected from Bombay Government Records, vol. I, *1818–85* (Bombay: 1957). Contains the autobiography of V. B. Phadke.